Solvency

Models, Assessment and Regulation

Solvency

Models, Assessment and Regulation

Arne Sandström

Swedish Insurance Federation
Stockholm

CRC Press
Taylor & Francis Group
Boca Raton London New York

CRC Press is an imprint of the
Taylor & Francis Group, an **informa** business

A CHAPMAN & HALL BOOK

CRC Press
Taylor & Francis Group
6000 Broken Sound Parkway NW, Suite 300
Boca Raton, FL 33487-2742

First issued in paperback 2019

© 2006 by Taylor & Francis Group, LLC
CRC Press is an imprint of Taylor & Francis Group, an Informa business

No claim to original U.S. Government works

ISBN 13: 978-1-58488-554-2 (hbk)
ISBN-13: 978-0-367-39214-7 (pbk)

Library of Congress Card Number 2005048465

Library of Congress Cataloging-in-Publication Data

Sandström, Arne.
  Solvency : models, assessment and regulation / by Arne Sandström.
      p. cm.
  Includes bibliographical references and index.
  ISBN 1-58488-554-8
    1. Finance--European Union countries. 2. Insurance companies--European Union countries--Finance. I. Title.

HG186.A2S26 2005
658.15'094--dc22
                                                                                          2005048465

Visit the Taylor & Francis Web site at
http://www.taylorandfrancis.com

and the CRC Press Web site at
http://www.crcpress.com

Preface

As a member of the Comité Européen des Assurances's (CEA) solvency working group, I received a number of working papers written by the former chairman of the working group, Jukka Rantala. At the same time, early 2003, the International Actuarial Association's (IAA) working group on solvency assessment produced a draft report. During the summer that year, I tried to write a common theory as a platform for my own further work. Later that summer and during the autumn, I tried to formalize some of the risk categories proposed in different papers (IAA and CEA). An actuary, who had read one of my early drafts, told me that he had never before read anything as concrete as this on solvency. His remarks encouraged me to continue writing.

Having read paper after paper on solvency assessment, I realized that no summaries or reviews had been published of the solvency work done in Europe. For this reason, and because the EU Solvency II project was one of the hottest topics for the European insurance industry, I decided to write this book.

Writing on this topic could easily have resulted in a 10-volume encyclopedia. My aim, however, was to fill the gap I mentioned above and to provide inspiration for further work.

It is my hope that the reviews of work in different countries and the smorgasbord of ideas will help to provide this inspiration.

Proposals and opinions stated in this book are mine and do not necessarily reflect the ideas of my employer (Swedish Insurance Federation) or of any committee or working group of which I am a member.

I am grateful to many friends and colleagues for their valuable comments on earlier drafts. Especially, I thank Jukka Rantala, who encouraged me to finish this book. I also thank Erik Alm, Nigel Boik, Ellen Bramness Arvidsson, Allan Brender, Malcolm Campbell, Boualem Djehiche, Jeremy Dunn, Gundula Grießmann, William Hewitson, Philipp Keller, Lasse Koskinen, Jens Pagter Kristensen, Arild Kristiansen, Jarl Kure, John Kah Kern Lim, Helen Martin, Peter Millington, Teus Mourik, Jörgen Olsén, Peter Skjødt, Rolf Stölting, and Robert Thomson. They have all contributed to making this a more accurate and readable book through their useful comments and feedback.

I also thank my colleagues Karin Chenon, Ellinor Forslund, Birgitta Holmin, and Birgitta Nordström for valuable discussions and support.

Arne Sandström

Contents

Part B   Present: Modeling a Standard Approach (Chapters 7–11)

Part C Present and Future: EU Solvency II — Phase 2: Groups and Internal Modeling in Brief (Chapters 12–14)

Part D   Appendices

1

Introduction

1.1 General Outline of the Book

The first chapter of this book briefly introduces some of the organizations mentioned in the following chapters and also includes a reading list. Chapter 2 discusses the concept of solvency, and the remaining chapters are divided into four main parts.

1.1.1 Part A: Past and Present: A Historical Review and Different Approaches to Solvency (Chapters 3–6)

In this part we discuss the models, assessment, and regulations from a historical perspective. Chapters 3 and 4 are devoted to solvency regulation and accounting within the EU. The background of the three non-life and life directives (Solvency 0) is discussed in Chapter 3, and the newly adopted Solvency I in Chapter 4. The work toward Solvency II looks at some international approaches in banking, accounting, supervision, and the actuarial field and is reviewed along with the first phase of the European Solvency II project in Chapter 5. The solvency systems of 12 different countries are summarized in Chapter 6.

1.1.2 Part B: Present: Modeling a Standard Approach (Chapters 7–11)

A basis for solvency modeling is discussed in Chapter 7, and the valuation of assets and technical provisions (liabilities) in Chapter 8. Dependency and different conservative approaches are discussed along with a baseline and a benchmark approach in Chapter 9.

To illustrate the theoretical discussion, an example of risk structure and the effects of diversification are included in Chapter 10. The benchmark approach is used to construct an example of a standard solvency model that is converted into a simple spreadsheet approach, presented in Chapter 11. Some parameter estimates and a simple example are also used as illustrations.

1.1.3   Part C: Present and Future: EU Solvency II — Phase 2: Groups and Internal Modeling in Brief (Chapters 12–14)

Insurance groups, financial conglomerates, and reinsurance are discussed in brief in Chapter 12. In Chapter 13 we discuss the current status of the second phase of the European Solvency II project. The importance of internal modeling and stress testing is highlighted in the final chapter.

1.1.4   Part D: Appendices

The fourth part starts with the basic model for the standard approach discussed in Chapters 10 and 11. The other eight appendices include excerpts from several EU directives.

1.2   Organizations

Many organizations have played an important part in discussions on solvency and its development. Some of the ones mentioned in this book are presented in brief below.

1.2.1   BIS and BCBS

The *Bank for International Settlements* (BIS) was established in 1930 and is the oldest international financial institution. As the center for international central bank cooperation, BIS hosts the regular meetings of the central banks in Basel, Switzerland.

The governors of the central banks of the Group of Ten (G10) established the *Basel Committee on Banking Supervision* (BCBS) in 1974 as a result of the disturbances in international currency and banking markets. One important objective of the BCBS's work was to close gaps in international supervision. An outcome of the collaboration was the introduction of the *1988 Basel Capital Accord*, a solvency system for banks. This Basel Accord was revised in what was called the New Basel Capital Accord, *Basel II* (2001–2006). A framework for a new capital adequacy system was published in 2004 and will be implemented at the end of 2006 (see Section 5.1).

For more information, see the BIS Web site: http://www.bis.org/index.htm, and the Basel II Web site, http://www.basel-ii.info/.

1.2.2   CEA

The *Comité Européen des Assurances* (CEA) was established on March 5, 1953. The idea was that European insurers should exchange information and be represented on the Insurance Committee of the Organization for European

Economic Cooperation (later the OECD Insurance Committee). Its work in the 1960s helped to bring about the first EU insurance directive. Today CEA consists of 32 national associations of insurance companies. Its mission is to resolve issues of strategic interest to all European insurers, focusing on the regulatory environment.

More information can be found on the CEA Web site: http://www.cea.assur.org/.

1.2.3 CEIOPS and EIOPC

The *Insurance Committee* (IC) of the European Union (EU) was a regulatory and legislative policy body created in December 1991 under Council Directive 91/675/EEC.

To simplify and improve decision making and implementation in the financial services sector, the European Commission launched a package of seven measures in 2003: a proposal for a directive (COM/2003/659) and six commission decisions. With this package, the approach that was already used in the securities sector was extended to the insurance sector through the establishment of two new committees:

1. The *European Insurance and Occupational Pensions Committee* (EIOPC[1]) (2004/9/EC), which was set up to replace the Insurance Committee and to assist the commission in adopting implementing measures for EU directives
2. The *Committee of European Insurance and Occupational Pensions Supervisors* (CEIOPS) (2004/6/EC; see below), which was established to act as an independent advisory group on insurance and occupational pensions (formerly the Insurance Conference)

CEIOPS was established in 2003 as the successor to the Insurance Conference, the Conference of Insurance Supervisory Authorities of the European Union, established in Paris in 1958. CEIOPS, located in Frankfurt am Main, Germany, acts as an independent advisory group on issues related to insurance and occupational pensions for the European Commission.

Similar organizations for the banking and securities sectors are the Committee of European Banking Supervisors (CEBS) and the Committee of European Securities Regulators (CESR).

More information can be found on the CEIOPS Web site: http://www.ceiops.org.

1.2.4 Groupe Consultatif

The *Groupe Consultatif* was established in 1978 to bring together the actuarial associations in the European Union to represent the actuarial profession in

---

[1] EIOPC was formally organized in April 2005.

discussions with the European Union institutions on existing and proposed EU legislation that has an impact on the profession. The name of the organization was originally Groupe Consultatif des Associations d'Actuaires des Etats Membres des Communautés Européennes. In 2002 the name was changed to Groupe Consultatif Actuariel Européen. The Groupe Consultatif was one of the leading players in formulating the third life directive.

For more information, see the Groupe Consultatif Web site: http://www.gcactuaries.org/.

1.2.5   IAA

The *International Actuarial Association* (IAA) was founded in 1895 as a worldwide association of individual actuaries. Since 1895, when IAA held its first International Congress of Actuaries (ICA) in Brussels, its congresses have provided a platform for actuaries from all over the world to meet and discuss actuarial research. The 29th ICA will be held in Paris in 2006.

In 1995 an International Forum of Actuarial Associations (IFAA) was set up within the IAA. Three years later, the IAA was reorganized as an association of associations and subsequently replaced the IFAA. The IAA issues international actuarial principles, guidelines, and standards.

For more information, see the IAA Web site: http://www.actuaries.org/.

1.2.6   IAIS

The *International Association of Insurance Supervisors* (IAIS) was created in 1994 and represents insurance supervisory authorities in about 100 jurisdictions. The aim of the IAIS is to promote cooperation among the authorities and to set international standards for insurance supervision and regulation. The IAIS also provides training to members and coordinates work with regulators in other financial sectors and institutions. IAIS issues international insurance principles, standards, and guidance papers.

Since 1999, IAIS has welcomed other insurance professionals (e.g., IAA) as observer members. The IAIS is a member of IAA.

For more information, see the IAIS Web site: http://www.iaisweb.org/.

1.2.7   IASC and IASB

In May 1973 the *International Accounting Standards Committee* (IASC) was founded as a result of an agreement by accounting bodies in 10 countries, 5 of which are European countries.

In 1974 IASC published its first exposure draft (ED) and the first International Accounting Standard (IAS 1, "Disclosure of Accounting Policies").

In 1977 the International Federation of Accountants (IFAC) was set up to expand international accounting activities. In 1981 it was decided that IASC would have full and complete autonomy in the setting of international accounting standards. The membership link between IASC and IFAC was discontinued from 2000, when the IASC constitution was changed.

In the same year, 2000, the European Commission announced its plan to require IASC standards for all EU listed companies by no later than 2005.

A new structure came into effect on April 1, 2001, with the establishment of the *International Accounting Standards Board* (IASB). The board is responsible for setting accounting standards, which are designated *International Financial Reporting Standards* (IFRS). In 2004 IASB published "IFRS 4 Insurance Contracts."

For more information, see the IASB Web site: http://www.iasb.org/.

1.3   A Selection of Solvency Readings

A large number of papers and books have been written on solvency. Some of the main sources of information are reviewed below.

1.3.1   The 1980s

At the beginning of 1980, a research group (RG) was established to review the Finnish rules concerning the equalization reserve and to conduct a general study of solvency. This pioneering work in the field of non-life insurance was summarized by the two books edited by Pentikäinen (1982) and Rantala (1982). At the end of 1982, a solvency working party (SWP) of the General Insurance Study Group was established under the chair of Chris Daykin to develop a solvency assessment approach for the U.K. similar to the one developed by the Finnish RG. In the 1980s, close cooperation between the two groups resulted in a considerable number of papers.

One paper, Daykin et al. (1984), reviewed the uncertainty affecting non-life insurance undertakings and the adequacy for technical provisions. After this, the SWP started to adapt the Finnish solvency approach to the British environment. In its first stage the group modeled a run-off situation where no further business was written. The working party then developed a simulation model of a non-life undertaking — a model that could be used not only for solvency purposes, but also as an analytical tool for assessing the financial strength of a company (see Daykin et al., 1987). It also became clear that the model could be used as a management tool for use in conjunction with a company viewed as a going concern. This is discussed in Daykin and Hey (1990).

Greater flexibility in modeling can be achieved by using simulation methods. This was studied by a Finnish working group that presented a new report in 1989 (see Pentikäinen et al., 1989).

The third edition of Beard et al. (1984) contained a number of new approaches, which were later summarized in Daykin et al. (1994).

Kastelijn and Remmerswaal (1986) have reviewed a large number of solvency studies. They present a comparison of 19 different approaches, of which 14 are based on a going-concern approach, 4 on a run-off approach, and 1 on both approaches.

During this period, Norwegian actuaries were also working on solvency modeling (see, for example, Norberg and Sundt, 1985; Norberg, 1986). Other studies from this period include Ramlau-Hansen (1988).

1.3.2   The 1990s

During this period many studies were carried out in many different countries both outside and within the EU; see Chapters 5 and 6 for references to further reading.

A large number of papers were presented at congresses and colloquiums in the 1990s; see, for example, Norberg (1993) on life insurance.

1.3.3   Since 2000

In early 2000, the EU Commission Services and EU member states jointly initiated a new solvency project (Solvency II). The first phase in this project was devoted to gathering information and facts about systems used worldwide. In May 2002, KPMG presented a report (KPMG, 2002) that discussed a range of issues, including risks and risk models, technical liabilities, asset valuation, reinsurance, and the impact of future accounting changes. Other documents of interest may be found on the Web site of the Insurance Unit of the Commission: http://europa.eu.int/comm/internal_market/en/finances/insur/index.htm.

In 2002 the IAIS asked IAA for support on solvency assessment. IAA's Insurance Regulation Committee formed the Insurer Solvency Assessment Working Party (WP) chaired by Stuart Wason. This WP published its final report in 2004 (IAA, 2004), when the IAA's Insurance Regulation Committee also formed a Solvency Subcommittee (SSC).

Other literature, books, and papers will be referred to in the following chapters. For the Basel Accord, i.e., the banking solvency rules, the reader should refer to the BIS and Basel II Web sites (see above).

2

Solvency: What Is It?

The insurance sector is moving from a system of direct supervisory control to a more deregulated environment. This step requires new systems of risk control and risk management. The supervisors also need new and improved techniques to control the insurance companies. As these institutions are also large and major investors, their soundness has a clear impact on the financial market. The key benchmark of an insurance business is its *solvency* or its *financial strength*. Other terms that have been used are *financial health* or *solidity*.

The main liabilities of an insurance undertaking are its anticipated insurance claims and associated costs. These are usually calculated using actuarial methods, guided by regulations. These calculations are, of course, only estimates, with some probability of error.

In order to *protect the policyholders* and to *ensure the stability of the financial markets*, it has been required that insurance undertakings should hold a certain amount of additional assets as a buffer. This buffer, the so-called solvency margin, is the main concern of this book.

The concept of solvency is old. According to *Webster's Ninth New Collegiate Dictionary*, it originates from ca. 1727 as "the quality or state of being solvent." But the latter concept goes back 100 years earlier (1630) and is defined as "able to pay all legal debts." We illustrate this by an example from the 18th century when a man from Germany wanted to start a company in Sweden. In the second part of this chapter we will discuss the concept of solvency a bit closer. As just a buffer, it does not say anything about its nature. It is when we say that this buffer should be in place to protect the policyholders that we give it any substance. This gives rise to several questions, such as:

- How large should it be?
- For what time horizon should it be calculated?
- What kind of assets could be included in the buffer?

2.1   In the 18th Century

Sweden became a seafaring nation during the 17th century (Hägg, 1998). Risks such as storms, robbery, captivity, and diseases always threatened the sea expeditions and their cargo. As there was no way to transfer the risks to a third party, the sea companies were organized as shipping corporations. Out in Europe, e.g., in London, Amsterdam, and Hamburg, there were underwriters offering international marine insurance. Underwriters were the winners, as the establishment of insurance companies usually failed.

In 1724 a merchant from Hamburg proposed to establish a Swedish marine insurance company in Stockholm (Hägg, 1998, p. 116). The German presented his proposal to the government, and following his business idea, the company should be well established with large capital stock, and thus it would be *solvent*. The company should have better security and superior service than others, and hence it could attract both Swedish and foreign merchants to buy policies at even higher premiums than elsewhere. According to the German, the company would, for example, outdo Dutch insurers.

The proposal was never realized. But 9 years later, Swedish merchants brought up the proposal in a newspaper that highlighted the large amount of money that flew out of Sweden only because of marine insurance. They knew how to organize a marine insurance company, but needed help from the central government to realize the project. Four reasons for getting help from the government were presented (Hägg, 1998, p. 117):

1. The Bank of Sweden had refused to allow investors to deposit a planned fund of 250,000 daler silvermynts (the currency at the time) in the bank.

2. The merchants demanded a new marine act that replaced the former one of 1667.

3. The merchants wanted a monopoly.

4. The merchants wanted the right to be excused from taxes.

As a matter of fact, in 1739–1740 the marine insurance company Assecurance-Compagniet was established as a sort of stock company (with limited liability for investors). The business, according to Hägg (1998, p. 116), was regulated by a royal company code. One explanation for the approving of the royal privilege was the concern about the balance of trade.[1]

The concept of solvency, although perhaps in other terms, is not new, as could be seen from this story. When the Swedish regulatory system was set up in 1903, it was established on a *solvency principle*; i.e., it "should safeguard the performance of all entered insurance agreements" (Hägg, 1998, p. 264).[2]

---

[1] For example, in 1743 the total liability of marine insurance was 35% of the sum of Swedish export and import (Hägg, 1998, p. 119).

2.2   What Does Solvency Mean?

The solvency margin is a buffer in a company's assets covering its liabilities. For the supervisor, it is important that the policyholders are protected, but it is also important for him to ensure the stability on the financial market. In view of this, the definition of the *solvency margin* (SM) given by Pentikäinen (1952) is our benchmark (see also Figure 2.1):

> The solvency margin, *SM*, is the difference between assets, *A*, and liabilities, *L*: $SM = A - L$.

If we put some restrictions on the assets, e.g., that they should be of good quality, we have by this definition what could be called the *available solvency margin* (ASM). Note that in this definition there is no discussion on either the time horizon or the relative size of the buffer. The definition of solvency is also discussed in Campagne (1961, chap. 1).

Benjamin (1977) refers to the *Oxford Dictionary*, where the definition of solvency is "having money enough to meet all pecuniary liabilities." In an insurance context, this definition gives rise to two concepts of solvency (see Benjamin, 1977, p. 267). They are the two extremes of a range of possibilities; i.e., at one end the liabilities are those paid on an immediate liquidation of the company (*break-up* or *run-off* approach). At the other end, cf. also the

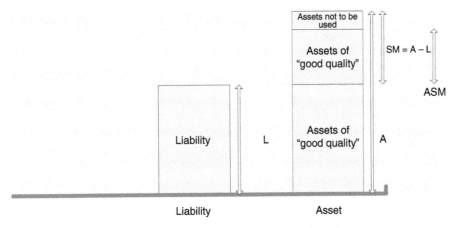

FIGURE 2.1
A short depiction of the solvency margin (SM) and the available solvency margin (ASM).

---

[2] The Swedish concept that was used was *soliditet*, i.e., "solidity." The principle stated that insurance should be organized and managed so that all insurance agreements entered could be correctly performed. In brief, the solvency principle was used as a generic term embracing all measures motivated to safeguard that insurance businesses under all circumstances were solvent (Hägg, 1998, p. 89).

definition by the *Investor Dictionary* below, a company could be regarded as solvent if it pays all its debts as they mature (*going-concern* approach). This means that a company is solvent when its solvency margin is positive. The so-called *ruin problem* would thus be the probability that the solvency margin of a company at any time in the future would become negative (cf. Pentikäinen, 1952).

According to the *Investor Dictionary*[3] on the Internet, solvency is defined as "the financial ability to pay debts when they become due. The solvency of a company tells an investor whether a company can pay its debts."

There are various other ways of looking at solvency. Pentikäinen (1967) presents two different ways of looking at the concept:

1. *From the point of view of the management of the company.* The continuation of the function and existence of the company must be secured.
2. *From the point of view of the supervising authorities.* The benefits of the claimants and policyholders must be secured.

Definition 2 is narrow, as it does not demand continuity of the company but allows it to be wound up. Definition 2 can be approved as a basis of the legal system: "the supervising authorities and the legal security measures shall be restricted to the minimum, i.e., to secure the insured benefits only, but otherwise each company shall have freedom to develop its function as it itself desires." As stated in Pentikäinen (1984), the latter case indicates the maintenance of the insurer's ability to meet his obligations for a short period, say, 1 year. In the former case, the objective is to guarantee the continued existence of the insurer. This is a more complex situation than the latter, and it includes this latter case as well.

If we take definition 2 as the basis of the legal system, then the company's existence can be left to management. This could be done by means of adequate reserves, loadings of premiums, and reinsurance. In the new environment, as proposed by EU (see Section 5.5), the second pillar with supervisory qualitative measures will build a bridge between the two approaches.

In the new three-pillar system of the EU (see Section 5.5), these two ways of looking at solvency are combined through the internal models, but also through the pillar system per se.

2.2.1   International Association of Insurance Supervisors

The International Association of Insurance Supervisors (IAIS) defined solvency as follows: "An insurance company is solvent if it is able to fulfil its obligations under all contracts under all reasonably foreseeable circumstances" (IAIS, 2002). The definition was later slightly changed to the "ability

---

[3] http://www.investordictionary.com/.

of an insurer to meet its obligations (liabilities) under all contracts at any time" (IAIS, 2003a).

In its definition it is also stated:

> Due to the very nature of insurance business, it is impossible to guarantee solvency with certainty. In order to come to a practicable definition, it is necessary to make clear under which circumstances the appropriateness of the assets to cover claims is to be considered, e.g., is only written business (run-off basis, break-up basis) to be considered, or is future new business (going-concern basis) also to be considered. In addition, questions regarding the volume and the nature of an insurance company's business, which time horizon is to be adopted, and what is an acceptable degree of probability of becoming insolvent should be considered.

2.2.2 The EU Directives

The concept of solvency margin has changed with the development of the EU directives. From the beginning it was seen as a *supplementary reserve*. In the Solvency I non-life directive it is defined "to act as a buffer."

In the first non-life directive of 1973 (EEC, 1973) it is said that "it is necessary that insurance undertakings should possess, over and above technical reserves of sufficient amount to meet their underwriting liabilities, a supplementary reserve, to be known as the solvency margin, and represented by free assets, in order to provide against business fluctuations." A similar statement was also made in the first life directive of 1979 (see EEC, 1979).

The third non-life directive (EEC, 1992a) introduced a change in the introduction of its Article 16 (see Appendix C) and is now focused on the solvency margin for the entire business, which was not expressed in the first directive. The introduction now states: "The home Member State shall require every insurance undertaking to establish an adequate solvency margin in respect of its entire business. The solvency margin shall correspond to the assets of the undertaking free of any foreseeable liabilities less any intangible items."

In the Solvency I non-life directive (COM, 2002b), the definition of the solvency margin is described in terms to *act as a buffer against adverse business fluctuations*: "The requirement that insurance undertakings establish, over and above the technical provisions to meet their underwriting liabilities, a solvency margin *to act as a buffer* against adverse business fluctuations is an important element in the system of prudential supervision for the protection of insured persons and policyholders." In this directive the *time horizon* is set to "all times," i.e., for a going concern.

A similar approach is given in the Solvency I life directive (COM, 2002c):

> It is necessary that, over and above technical provisions, including mathematical provisions, of sufficient amount to meet their underwriting liabilities, assurance undertakings should possess a supplementary reserve, known as the solvency margin, represented by free assets and,

with the agreement of the competent authority, by other implicit assets, which shall act as a buffer against adverse business fluctuations.

The time horizon is the same here as in the non-life directive.

The time horizon in the Solvency I directives is for "all times," i.e., for a going concern. There are other ways to define solvency in terms of the time horizon; e.g., the Dutch supervisory authority (see Section 6.5) proposed in its first outline of a new solvency system a three-part assessment:

1. On the balance sheet date, the financial position is such that the book is closed and sold to another willing partner.
2. On the balance sheet date, the financial position is such that it will be able to hold its position during the following 12 months and also that during this period there will be an adverse scenario emerging so that on the balance sheet day 12 months later, the book will be closed and transferred to another willing partner.
3. The going-concern approach.

In the Swiss proposal (see Section 6.8), the solvency requirement is thought of as consisting of two parts above the technical provisions: a risk margin reflecting a run-off situation plus a margin reflecting a going-concern approach.

An extensive discussion on solvency and the capital requirements is given in IAA (2004, chap. 3).

In the literature there have been other terms used as synonyms of solvency; e.g., in Campagne (1961) the term *dynamic solvency* is used as a synonym for the going-concern approach and *static solvency* is used for the break-up situation (see also Kastelijn et al., 1986, p. 8, footnote).

Some other related concepts are discussed in Kastelijn et al. (1986, chap. 1.5); e.g., the *guarantee fund* in the EU solvency directives is one third of the minimum solvency margin and is the *absolute minimum capital* for a company if it is continuing to trade business. The term *minimum free reserve* can either mean solvency margin or guarantee fund. In some reviews the term *solidity* is used. See Kastelijn et al. (1986) for more terms used.

Summary

The *available solvency margin* is a capital buffer of free assets covering the liabilities. The buffer should be positive and consist of "good quality" assets. Its relative size depends on the time horizon. You could either define its size according to an immediate liquidation (run-off approach) or a situation where all payments are done as the debts mature (going concern approach).

Part A

Past and Present: A Historical Review and Different Approaches to Solvency (Chapters 3–6)

In Chapters 3 to 6 we discuss the models, assessment, and regulations from a historical perspective.

In Chapters 3 and 4 we consider the solvency regulation and accounting within EU. We briefly present the theory behind the non-life and life directives, the accounting directive, and the works of Professor Campagne, which are central to this area.

Chapters 5 and 6 present a smorgasbord[1] of different approaches to modeling and assessment.

Chapter 5 describes different organizations and their approaches. First we discuss the 1988 Basel Accord for banks and its credit risk assessment. We also look at the new Basel II Accord and its three-pillar system, including its risk charges for credit and operational risks.

This is followed by the new accounting standard proposed by the International Accounting Standard Board, where we move from an institutional view (insurance companies) to a functional view (insurance contracts). Fair valuation is also discussed.

This chapter and Appendix E also include a summary of the insurance principles and guidelines presented by the International Association of Insurance Supervisors. The fifth principle deals with capital adequacy and the solvency regime.

The International Actuarial Association (IAA) presented in 2004 a report on insurance solvency assessment. A brief summary of this report is given, where different risk categories are discussed.

In the final section we introduce the first phase of the European Union Solvency II project. A comparison is made betweem the Basel II project and the Lamfalussy procedure. We also give brief summaries of the KPMG report

---

[1] Smorgasbord: Swedish *smörgåsbord*, from *smörgås* ("open sandwich") + *bord* ("table") (approximately 1919); 1: luncheon or supper buffet offering a variety of foods and dishes (as hors d'oeuvres, hot and cold meats, smoked and pickled fish, cheeses, salads, and relishes); 2: heterogeneous mixture (*Webster's Ninth New Collegiate Dictionary*).

and the three supervisory reports on life insurance, non-life insurance, and the Sharma report on failures, as well as their causal chain and diagnostic and preventive tools.

In Chapter 6 we study different models and assessments used in a number of countries. We have chosen certain issues and characteristics that can inspire new models.

- *Australia*: The ideas are similar to those behind Solvency II. Liability valuation, risk categories, a factor-based prescribed method, and internal models.
- *Canada*: A factor-based system. Risk categories, the minimum capital test, dynamic capital adequacy testing, and minimum continuing capital and surplus requirements on ratings.
- *Denmark*: Fair valuation and a traffic light test system.
- *Finland*: A risk theoretical transition model and equalization reserve.
- *The Netherlands*: Fair valuation and minimum solvency and continuity analysis.
- *Singapore*: Valuation of assets and liabilities, risk categories, and two requirements in a risk-based system.
- *Sweden*: Valuation of assets and liabilities, risk categories, and a simple model.
- *Switzerland*: Valuation of assets and liabilities, risk categories, standard model, scenario tests determining the target capital, and internal models.
- *U.K.*: A twin peaks' approach under pillar I, individual capital adequacy standards under pillar II, and risks.
- *U.S.*: Risk-based capital model, correlation structure, and different intervention levels.

The German and Norwegian approaches are also touched upon.

3

The European Union: Solvency 0 and Accounting

The first non-life and life directives of EU, at that time the European Economic Community (EEC), were published July 24, 1973, and March 5, 1979, respectively (see EEC, 1973, 1979). These two directives marked the first steps toward the establishment of the free market in insurance within the European Community (see, e.g., Pool, 1990). Included in the directives are the requirements that the companies within the EEC should be able to meet in order to fulfill the solvency assessment. The works by Campagne (1961) are the main base for these requirements. We will therefore start this chapter with a summary of his proposals before we look closer at the directives. A very good description of the early works that were made within different organizations on the solvency assessment is given in Daykin (1984). He notes that at the time of signing the Treaty of Rome in 1957, OEEC[1] had already initiated discussions to harmonize the controls on international insurance operations. This discussion included not only the founder member states of the European Community, but also such countries as the U.K., Sweden, and Switzerland (cf. also Schlude,[2] 1979). The disagreement whether an insurance undertaking should be allowed to carry on both life and non-life business is the main reason for the lag of nearly 6 years between the first two directives.

Reserving and solvency assessment in different EU countries are discussed and compared in Wolthuis and Goovaerts (1997).

In Section 3.6 we will discuss the first accounting directive on insurance.

---

[1] At that time OEEC, the Organization for European Economic Cooperation, now OECD, the Organization for Economic Cooperation and Development.
[2] Administrator in the Division Insurance of the EEC, Brussels.

3.1   The Works of Campagne

In 1948, Professor Campagne published a report on solvency assessment for life insurance companies. It was based on data from 10 Dutch life companies for the years 1926 to 1945 ("Contribution to the Method of Calculating the Stabilization Reserve in Life Assurance Business"; see Kastelijn and Remmerswaal, 1986).

On the request of the OEEC Insurance Committee, Professor Campagne[3] presented a report on solvency in 1957 ("Minimum Standards of Solvency for Insurance Firms"; see Daykin, 1984). In the report he recognized, as Teivo Pentikäinen (1952) had done earlier, that in assessing the solvency position of a company, risk theoretical considerations should be made. The report made some simplifying assumptions about the distribution function underlying the solvency. He did not claim that the model should give any information about the solvency position of a company, but only provide an early warning system (Daykin, 1984). The data used in this first non-life report were taken from 10 insurance companies operating in Switzerland during 1945 to 1954. In this and the second non-life study, Campagne proposed that the probability of ruin over 3 years should be 1/1000, taken as approximately 3/10,000 in 1 year. Because expenses and commissions, in an average account, stand for 42% of the retained premiums, some 58% were available for claims (in relation to retained premiums). This is illustrated in Figure 3.1. From this and the model used, it was recommended that a solvency margin of 25% of the retained premiums was enough to meet the requirement of avoiding ruin. To this end, it was suggested that an additional 2.5% of the ceded reinsurance premiums should be added to cover against the risk of reinsurance failure.

The OEEC Insurance Committee set up a working party (WP) chaired by Professor Campagne. The WP consisted of 14 members from 10 countries, and Professors de Mori and Grossmann compiled the data based on a questionnaire. In the report to the OEEC, Campagne (1961) used data for the years 1952–1953 to 1957 from eight European countries for the non-life insurance industry and from five countries for the life insurance industry.

3.1.1   Campagne's Non-Life Approach

Campagne's approach is simple in its nature (see Campagne, 1961; Kastelijn and Remmerswaal, 1986, pp. 32–33; de Wit et al., 1980, p. 138). Let the net retained premium be 100%. From this we deduct a constant fraction equal to the average expense ratio of each country. The remaining part is what remains for claims payment. Calculate the *value at risk* of the loss ratio distribution (VaRLR) in each country and add this to the difference between

---

[3] Professor Campagne was the chairman of the *Verzekeringskamer* in the Netherlands.

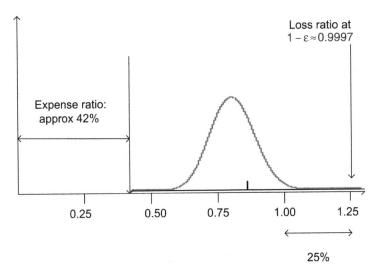

FIGURE 3.1
Illustration of Campagne's non-life approach. In the combined ratio the expense ratio is assumed constant and equal to 42%, and the loss ratio follows a beta distribution. At the 0.9997
percentile the loss ratio is approximately 83%. Thus, the combined ratio will be 125%. The
company needs 25% of the premiums during 1 year to meet the requirement.

100 and the expense ratio. The part that is above 100 would constitute a
solvency margin expressed in percent of the net premium income according
to this approach.

Let X_i be the net claims incurred during a year (for a country), P_i the net
retained premium income during the same period, and E_i the net operating
expenses during the same period. The ratios $ER_i = E_i / P_i$ and $LR_i = X_i / P_i$
are the net expense ratio and the net loss ratio, respectively.

We also define, for each country, the average net expense ratio

as $\overline{ER} = \dfrac{1}{n} \displaystyle\sum_{i=1}^{n} ER_i$. Campagne assumed that the net loss ratios are distributed

according to a beta distribution (see below). For each country the VaRLR is
calculated. The solvency margin in percent of the net premium income is

now defined as $[\text{VaRLR} + \overline{ER} - 100]$.

The beta distribution, with parameters α and β is

$$f(x;\alpha,\beta) = \frac{x^{\alpha-1}(1-x)^{\beta-1}}{B(\alpha,\beta)} \quad \text{for } 0 < x < 1$$

$$= 0 \text{ otherwise}$$

with $B(\alpha,\beta) = \int_0^1 u^{\alpha-1}(1-u)^{\beta-1}\,du$, i.e., the beta function.

Here x equals the loss ratio. Hence, these ratios cannot be larger than 100% in this approach. The maximum net loss ratio observed by Campagne was 97%. De Wit et al. (1980) used a beta distribution for $x/1.5$, as their maximum LR was 130%.

Campagne assumed that the probability that the sum of the average net expense ratio and the loss ratio is larger than a solvency margin is equal to a small value ε, i.e.,

$$P\left(\overline{ER} + LR > 1 + msm\right) = \varepsilon$$

where msm is the minimum solvency margin in terms of the net premium income.

This probability could be written as $P\left(LR > msm + 1 - \overline{ER}\right) = \varepsilon$. Letting $msm = LR_\varepsilon - (1 - ER)$, we get the value at risk, VaRLR, defined as the ε-quantile LR_ε, which is the smallest value satisfying $P[LR > LR_\varepsilon] = \varepsilon$, where LR is beta distributed. The VaRLR is computed as the empirical VaR based on the empirical distribution.

The parameters of the beta distribution were estimated by the method of moments. From the observed net loss ratios we can calculate its mean (m) and variance (s^2). The mean of the beta distribution is $\mu = \dfrac{\alpha}{\alpha+\beta}$ and the variance $\sigma^2 = \dfrac{\alpha\beta}{(\alpha+\beta)^2(\alpha+\beta+1)}$. If we calculate the mean and variance from the observed net loss ratios and let $z = (1-m)/m$, then the parameters of the beta distribution are

$$\alpha = \frac{z - s^2(1+z)^2}{s^2(1+z)^3} \text{ and } \beta = z\alpha$$

Some of the results presented by Campagne (1961, p. 60) are
As stated by Kastelijn and Remmerswaal (1986, p. 33,) Campagne used these results to propose a minimum solvency margin for the EEC of 25% of the net retained premium (and in addition 2.5% of premium ceded; see also Campagne, 1961, p. 59), because such a margin would not lead to unacceptable high ruin probabilities and most companies would be able to meet this standard, particularly if hidden reserves are taken into account. In addition to this, a concept of a minimum margin in absolute monetary units was

$\varepsilon = 0.000$	Denmark	France	Germany	U.K.	Italy	The Netherlands	Sweden	Switzer-land
NRR%	100	100	100	100	100	100	100	100
$\overline{ER}$	35	38	35	41	44	53	32	42
$\overline{LR}$	51	49	44	50	43	43	61	46
VaRLR	74	97	68	72	83	78	90	83
$\overline{LR} + VaRLR$	109	135	103	113	127	131	122	125
msm	9	35	3	13	27	31	22	25

Notes: NRR% = net retained premium ratio, in %; $\overline{ER}$ = mean net expense ratio, in %; $\overline{LR}$ = mean net loss ratio, in %; VaRLR = value at risk of the loss ratio distribution; msm = minimum solvency margin

introduced. This amount was set to 250,000 u/c A.M.E. (European Monetary Agreement units of account) (see Campagne, 1961, p. 73). One u/e A.M.E. was equivalent to one U.S. dollar.

The minimum solvency margin was thus proposed to be:

- 25% of the net retained premiums
- 2.5% of the ceded premiums
- 250,000 u/c A.M.E

De Wit et al. (1980, p. 142) used Dutch data from 71 companies in the years 1976 to 1978 to update the results. The following table is based on their results.

Probability of ruin	0.01	0.001	0.0003
NRR %	100	100	100
$\overline{ER}$	30	30	30
VaRLR	115	126	130
$\overline{ER} + VaRLR$	145	156	160
msm	45	56	60

Notes: NRR% = net retained premium ratio, in %; $\overline{ER}$ = mean net expense ratio, in %; VaRLR = value at risk of the loss ratio distribution; msm = minimum solvency margin

As a comparison, we can look at Swedish data for the years 1996 to 2003. Companies that are in run-off are excluded. Captives and what is called labor market companies are also excluded. Looking at individual companies, there are data from 116 companies (658 data points), and in the case of groups of companies, there were 17 companies (groups) (92 data points). The maximum LR for the individual companies was 1.82 and for the groups 1.37. In the first case, we divided all loss ratios by 2, and in the second case, by 1.5. This has been done arbitrarily, as in de Wit et al. (1980).

Swedish Data, 1996–2003	Individual Companies	Groups and Individual Companies
Number of data points	658	92
Loss ratio (LR)	LR/2.0	LR/1.5
Mean	0.363	0.558
Variance	0.020	0.025
Max LR	0.910	0.916
Max LR, not divided	1.820	1.374

This gives us the following minimum solvency margin as percent of the net retained premium.

	I.C.	I.C.	I.C.	G.C.	G.C.	G.C.
Probability of ruin	0.01	0.001	0.0003	0.01	0.001	0.0003
NRR %	100	100	100	100	100	100
$\overline{ER}$	34	34	34	28	28	28
VaRLR	142	160	167	132	141	143
$\overline{ER} + VaRLR$	176	194	201	160	169	171
msm	76	94	101	60	69	71

Note: I.C. = individual companies; G.C. = group and individual companies. NRR% = net retained premium ratio, in %; $\overline{ER}$ = mean net expense ratio, in %; VaRLR = value at risk of the loss ratio distribution; msm = minimum solvency margin

The Dutch data show an increase in the VaRLR, or *estimated* maximum loss ratio, from 78 to 130 during the 20 years from the first study and from 90 to 167 in Sweden during the last 40 years. We will not use the data for any deeper analysis, but they could be an indication that the impact of the investments yield has increased.

De Wit et al. (1980) points out the arbitrariness in dividing the loss ratios with a number (above the maximum registered loss ratio) and weakness of this method when the loss ratios are higher than 1. This is also one of the main remarks against this method in the criticism made by Ramlau-Hansen (1982). Ramlau-Hansen also points out that the observations cannot be assumed to be identical and independently distributed (i.i.d.), and the fact that the variation between companies increases with the number of companies involved. In the study by de Wit et al. (1980), a comparison is made by estimating the VaRLR using a Weibull distribution, and in Ramlau-Hansen (1982) a credibility approach is used (cf. also Kastelijn and Remmerswaal, 1986).

3.1.2 Campagne's Life Approach

In the report to the OEEC, Campagne (1961) used the same approach as in his study at the end of the 1940s. As the risk on investments is the most important factor for life insurance companies, and as the technical provisions (tp) are the most important invested amount, Campagne considers a minimum solvency margin (msm) as given by a percentage of the tp. He also discusses other possibilities, e.g., the msm as a percentage of the sum insured or of the sum at risk (Campagne, 1961, pp. 20–21; Kastelijn and Remmerswaal, 1986, p. 27). In the first case, the method of calculation can differ from one country to another and its meaning as a criterion is not clear. The sum at risk could be difficult to calculate, and it is not obvious how a negative sum at risk should be taken into account.

One main objection of the approach used by Campagne is that the more prudence there is in the technical provisions, the higher the msm will be. In other words, *the more prudent a company is, the more it has to pay for the solvency.*

From Campagne (1961, p. 53) we have the following table of characterization ratios from five European countries (1952 to 1957). The three ratios are the free assets (A) in relation to the technical provisions (A/tp), to the sum at risk (A/sr), and to the sum insured (A/si).

Ratios	France	Germany	Italy	The Netherlands	Sweden	Mean
A/tp = FR	32.4	3.5	46.1	11.5	13.6	21.4
A/sr	2.6	0.6	6.4	2.2	5.4	3.4
A/si	2.3	0.5	5.5	1.8	3.8	2.8

A loss ratio, LR, is defined as the loss (L) in a year as a percentage of the technical provisions (tp), LR = L/tp. A profit is a negative loss, and as cited by Kastelijn and Remmerswaal (1986, p. 28), "on the whole changes in the capital position of the company have been listed as profit or loss, the respective book-profits or book-losses have been considered without going into detail" (from the 1948 report of Campagne). The LRs are assumed to be i.i.d. for different years and companies.

The free reserve ratio of the technical provisions (FR = A/tp) must be such that

$$P(LR > FR) \leq \varepsilon$$

This can be defined as the value at risk of LR, VaRLR, defined as ε – quantile LR_ε, which is the smallest value satisfying $P(LR > LR_\varepsilon) = \varepsilon$, where LR is distributed according to a Pearson type IV distribution (see Campagne, 1961, p. 62).

For the data that were used, this led to the following frequency function (estimation is made by the method of moments), $x = LR$ (Kastelijn and Remmerswaal, 1986, p. 28):

$$f_x(x) = 31.73\left(1+\frac{x^2}{5.442^2}\right)^{-4.850} e^{2.226 \, arctg\left(\frac{x}{5.442}\right)}$$

The VaRLR was calculated both as the margin for 1 year and as margins that should be sufficient to keep the probability so that after 2, 3, 5, and 10 years, the total loss in this period exceeds the VaRLR is less than ε. Campagne used two different methods to calculate the VaRLR. The first was made by convolution of the 1-year frequency function and the second by fitting the same distribution to the accumulated loss data (of 4 years) divided by the average reserve. For the 1-year period the two methods are identical.

The minimum solvency margins, as a percentage of the technical reserves, are given in the following table (Campagne, 1961, p. 65).

VaRLR	1 year	2 years	3 years	5 years	10 years
$\varepsilon = 0.001$	9	10	10	12	14
$\varepsilon = 0.01$	7	7	7.5	8	9
$\varepsilon = 0.05$	3.5	4	4	4	3
$\varepsilon = 0.1$	2.5	2.5	2	2	1

Campagne proposed $\varepsilon = 0.05$, and therefore a necessary minimum solvency margin of 4% of the technical reserves.

3.2   Other Steps toward the First Directives

The description of the developments on solvency in this chapter is mainly based on a paper by Daykin (1984), but also on the description made by Pool (1990). The establishment of the European Economic Community (EEC) in 1957 also started cooperation between the supervisory authorities. The Conference of EEC Insurance Supervisory Authorities[4] began to discuss the steps toward a free insurance market. From the very beginning they discussed the technical reserves, assets backing these reserves, and control over the assets. Discussions between the supervisors and the industry, with the aid of OECD, ended up with a plan to pursue the work of Campagne. The working group that was set up, the Study Commission, gave its report to OECD in 1963. A member of Campagne's working group and this Study Commission, Professor de Mori, summarized the developments in a paper published in 1965 (de Mori, 1965).

The Study Commission developed the work started by Campagne and proposed alternative criteria for the minimum solvency margin based on three ratios, or yardsticks, according to Schlude (1979):

---

[4] From 2004 this organization was named CEIOPS (Committee of European Insurance and Occupational Pensions Supervisors).

- Free assets to premiums received during the last year
- Free assets to average incurred claims over the last 3 years
- Free assets to technical reserves

Data from five countries for a period of 10 years (1951 to 1960) were used. It was assumed that the gross loss ratio (gross claims paid divided by gross premiums earned) was following a normal distribution. The pragmatic solution for the solvency margin was to calculate it as $\bar{x} + 3s$ (mean + 3 standard deviations). The results from different countries were weighted proportionally to their market share. This procedure gave the following standards of solvency margin (see, e.g., de Mori, 1965; Kastelijn and Remmerswaal, 1986):

- 24% of gross premiums written
- 34% of incurred claims
- 19% of technical reserves

These percentages were also calculated for different branches as transport, cars, and others.

The method of using the technical reserves presupposes that they are set up in a uniform manner in various countries. However, as this was not the case, this approach was not used further. As claims payment fluctuates between years, it was proposed that an average over the last three financial years should be used.

Some further work by Professor de Mori was carried out for CEA (the Comité Européen des Assurances) in the mid-1960s. The discussions were also going on in two working groups set up in 1965 by the OECD. One considered additional guarantees of security (the de Florinier group) and one the estimation of technical reserves (the Homewood group). Reports from these two working groups were published in 1969 and 1976.

The OECD framework was built on solvency margins for insurance companies operating on an international market, and many countries wanted the solvency margin to be applicable for companies operating only on domestic markets. In 1976 the Conference of EEC Insurance Supervisory Authorities set up a working group to study the harmonization of technical reserves (the Angerer group). A report from this group was published in 1979.

3.3   The Non-Life Directives (First, Second, and Third)

Some countries thought that the proposed percentages for solvency made by the Study Commission were too high, and some countries thought that they were only sufficient; a compromise was eventually achieved (Schlude,

1979). The (required) solvency margin should be the higher of two indices, viz., the *premium index* and the *claims index*.

For the premium index:

- 18% of gross premiums up to 10 million units[5]
- 16% of gross premiums in excess of 10 million units

For the claims index:

- 26% of gross average incurred claims up to 7 million units
- 23% of gross average incurred claims in excess of 7 million units

The average in the claims index is usually taken over the last 3 years (7 years for certain risks, such as storms and hail). The result is reduced for reinsurance by the ratio (net paid claims)/(gross paid claims), with a maximum reduction of 50%. The shift from the premium index to the claims index will normally take place when the loss ratio is approximately 69% ($18 / 26 \approx 16 / 23 \approx 69\%$).

As stated by Schlude (1979, p. 28), one objection against this system is that it does not take account of the structure of a company's losses. One could argue that the solvency requirement for a company with 10 claims each amounting to 100,000 units should be higher than the requirement for a company with 1000 claims each amounting to 1000 units.

It must also be decided which assets should be used to cover the solvency margin when it has been calculated. This is stated in the directive.

There are also two concepts of *guarantee funds*:

- The relative guarantee fund is one third of the minimum solvency margin and termed the *minimum guarantee fund*.
- The *absolute guarantee fund*, or minimum fund, is a fixed amount categorized according to branches of insurance, e.g., 300,000 units of account in the case where all or some of the risks included in one of the classes listed in point A of Appendix B under numbers 1 to 8 and 16 are covered.[6]

If the net worth of a company is less than its minimum solvency margin, the supervisory authority must ask the company to set up a plan to restore sound financial relations (the *solvency plan*). If the company's net worth is less than the absolute guarantee fund, the supervisory authority must require a *finance plan* for a short-term procurement of net worth. It is then

---

[5] Article 5(a) of the first non-life directive defines the units of account: "means that unit which is defined in Article 4 of the Statute of the European Investment Bank."

[6] If the insurer is authorized to only write property damage or legal expense insurance, the minimum was set to 200,000 units of accounts, and if it is authorized to write any class of liability, credit, or suretyship insurance, the minimum was set to 400,000 units of accounts.

also possible for the supervisory authority to restrict or suspend the company from disposal of its assets.

The purpose of this first directive was to remove restrictions on the opening of branches and agencies by insurance undertaking in other member states. In order to do so, it was essential to eliminate differences in national supervisory legislations and to coordinate provisions relating to financial guarantees. It was important to have a clear definition of an insurance undertaking, and this is done in this first directive. As noted by Pool (1990), there was no definition of insurance, a fact that produced difficulties later. The structure of this first non-life directive (EEC, 1973) is given below:

Title I: General provisions

    Articles 1 to 5

Title II: Rules applicable to undertakings whose head offices are situated within the community

    Section A: Conditions of admission

        Articles 6 to 12

    Section B: Conditions for exercise of business

        Articles 13 to 21

        Article 15: Technical reserves

        Article 16: Solvency (paragraph 1-4 is given in Appendix C)

        Article 17: Guarantee fund (see Appendix C)

        Article 20: The solvency plan and the financial plan

    Section C: Withdrawal of authorization

        Article 22

Title III: Rules applicable to agencies or branches established within the community and belonging to undertakings whose head offices are outside the community

    Articles 23 to 29

Title IV: Transitional and other provisions

    Articles 30 to 32

Title V: Final provisions

    Articles 33 to 38

Annex

    A. Classification of risks according to classes of insurance (see below)

    B. Description of authorizations granted for more than one class of insurance

    C. Ancillary risks

The calculation of solvency is stated in Article 16, paragraph 1-4, and the guarantee fund is defined in Article 17 (see Appendix C).

The last sentence of Article 16, paragraph 1 ("at the report of ..."; see Appendix C) was a nonharmonization item, which produced differences between the member states. The hidden reserves, mentioned in this first paragraph, are usually the difference between the current market value and the purchase price of investments. The solvency margin, which does not include the technical provisions (reserves), is a sort of free reserve held above the technical provisions.

The commission invited Groupe Consultatif (GC) to comment on a report on the experience of the first non-life directive in 1979. The 1-year-old GC did not, at that time, feel able to prepare a report (Henty, 2003). At its meeting in 1980 it was decided to establish a Solvency Margin Committee. A solvency report was submitted to the GC's meeting in 1986, but as it included sensitive issues, it was not published. The technical annexes were made available to the member associations, including a note on the shortcomings of the solvency margins as defined in the first directives (both non-life and life directives). In 1988 the solvency issues were taken over by GC's new Insurance Committee.

One main feature of the directive was that it called for close cooperation between the national supervisory authorities within the member states. The directive was not intended to deal with *freedom of services,* which means the right of an insurance undertaking established in one member state to cover risks and policyholders in another member state without making an establishment in that state. As early as 1975 the commission had put forward a proposal for a second non-life insurance directive. It proposed an early expression of the principle of *home country control* with mutual recognition of standards (Pool, 1990). This means that an insurer established in country A but wishing to cover risks in country B would have to apply to its own supervisory authority in A for getting permission. The freedom of services would mean that the insurer could insist that the law of his country should be used to interpret contracts. In the European Parliament the main objection in 1978 against this was that all policyholders, even the small ones, would find themselves having contracts that followed an unfamiliar law from another country. The proposal from 1975 was amended as the commission accepted this argument and decided that a distinction between *large and small risks* should be made. Large risks should follow the law in the country of the insurer, but small risks should follow the law of the country where the risk is situated. The negotiations about this proposal were discontinued in 1985.

In 1987 the council working party on economic questions met and discussed the proposed second non-life directive. This led to the adoption of the second non-life insurance coordination directive on June 22, 1988 (EEC, 1988). The solvency rules defined in the first directive were not changed.

The structure of this second directive is as follows:

Title I: General provisions (Articles 1 to 4)

   Definitions

Title II: Provisions supplementary to the first directive (Articles 5 to 11)

   Large risks are defined in terms of the risks classified in the annex
      of the first directive

Title III: Provisions peculiar to the freedom to provide services (Articles
   12 to 26)

Title IV: Transitional arrangements (Article 27)

Title V: Final provisions (Articles 28 to 35)

Annex 1: Matching rules

Annex 2A and B: Underwriting account

This was an important step toward the internal market in insurance, as
this was established for large risks, but the commission was not satisfied
with it, as it did not make the internal market complete for all insurance
risks. In 1989 it was announced that the policy of the commission was to
achieve freedom of services with home country control and using a single-
license concept. This means that an insurance undertaking within the community only needed to have one authorization, i.e., the one from the state
of its head office.

The structure of the third non-life directive followed a structure that had
already been adopted by the second banking directive and is as follows
(EEC, 1992a):

Title I: Definition and scope (Articles 1 to 3)

Title II: The taking up of the business of insurance (Articles 4 to 8)

Title III: Harmonization of conditions governing pursuit of business
   (Articles 9 to 31)

   Article 24 replaces Article 16(1) from the first directive; for the new
      article, see Appendix C

Title IV: Provisions relating to freedom of establishment and freedom
   to provide services (Articles 32 to 46)

Title V: Transitional provisions (Articles 47 to 50)

Title VI: Final provisions (Articles 51 to 58)

3.3.1 Equalization Reserves

According to the credit insurance directive (EEC, 1987), all insurance companies underwriting credit insurance have to set up an equalization reserve.
It cannot be used for calculation of the solvency margin. In the directive,
there are four methods for the calculation of the reserve that can be used by
the member states. A summary of the methods and how they are used in

EU countries is given in Wolthuis and Goovaerts (1997). The credit insurance and the equalization reserve were introduced in the third non-life directive in Article 18 on the amendment of Article 15a from the first non-life directive.

Article 15a

1. Member States shall require every insurance undertaking with a head office within their territories which underwrites risks included in class 14 in point A of the Annex (hereinafter referred to as "credit insurance") to set up an equalization reserve for the purpose of offsetting any technical deficit or above-average claims ratios arising in that class in any financial year.

2. The equalization reserve shall be calculated in accordance with the rules laid down by the home Member State in accordance with one of the four methods set out in point D of the Annex, which shall be regarded as equivalent.

3. Up to the amount calculated in accordance with the methods set out in point D of the Annex, the equalization reserve shall be disregarded for the purpose of calculating the solvency margin.

4. Member States may exempt insurance undertakings with head offices within their territories from the obligation to set up equalization reserves for credit insurance business where the premiums or contributions receivable in respect of credit insurance are less than 4% of the total premiums or contributions receivable by them and less than ECU 2,500,000.

Equalization reserves are also used for other lines of business (LOBs) in different countries. In some countries the reserve is accepted by the tax authority, meaning that it is tax deductible; in others they are not allowed according to the local tax authorities. In Germany we have the *Schwankung-rückstellung*, and in Finland (see Section 6.4), the equalization reserve has been the main solvency pillar since the beginning of the 1950s. In Sweden there is a similar system that is prescribed by the supervisory authority and accepted by the tax authority as tax exempted.

3.4   The Life Directives (First, Second, and Third)

The two first non-life and life directives have a similar structure, reflecting the same approach to the problem they dealt with. The solvency approaches were also both mainly based on the Campagne (1961) proposals. As the member states had differing positions on both the composition and the amount of the solvency margin, the final result was a collection of compromises (Pool, 1990, p. 36). The first two articles of the directive defines, in a broad sense, what insurance activities should be defined as life insurance, e.g., Article 1(1)(d): "the type of insurance existing in Ireland and the United Kingdom known as permanent health insurance not subject to cancellation." It was up to the member states to decide if pension funds should be included in the directive.

The basic formula for the (required) solvency margin for life insurance companies was set up in the first life directive (EEC, 1979), mainly as:

First result: 4% of the mathematical reserves (gross of reinsurance)

+

Second result: 0.3% of the capital sum at risk

According to Kastelijn and Remmerswaal (1986, p. 30), this base formula could be used and accepted for:

- *Reinsurance*: The maximum allowance can be 15% of the mathematical reserves and 50% of the capital sum at risk.
- *Short-term temporary insurance*: A reduction of the 0.3% margin is possible.
- *Supplementary insurance and accidental death benefit*: The coverages of the non-life margins apply (see Section 3.3).

Zillmer adjustments of beyond 3.5% of the capital sum at risk are not allowed (see below).

According to Article 18 of the directive, the solvency margin shall consist of:

1. The assets of the undertaking (free of all liabilities, less intangibles), especially:
   - The paid-up share capital (or the paid-up amount of the mutual's fund)
   - One half of the unpaid-up share capital/fund once 25% of such capital/fund is paid up
   - Statutory reserves and free reserves not corresponding to the liabilities
   - Any carry forward of profits
2. Profit reserves, appearing in the balance sheet
3. With the agreement of the supervisory authority:
   - An amount equal to 50% of future profits: a factor (<10) times the estimated annual profit. The factor represents the average remaining duration left for the policies. The estimated annual profit is the average over the last 5 years for some activities (see Article 1)

- The difference between a non- or partially zillmerized reserve and a zillmerized reserve at a rate equal to the loading for acquisition costs included in the premium; the rate must not be >3.5%
- Hidden reserves

The minimum solvency margin defined above (and in Article 19) is, however, not a wind-up barrier, but an *early warning signal*. The guarantee fund (see Article 20) gives the wind-up barrier.

The structure of the first life directive (EEC, 1979) is given below. Note that the calculation of solvency is stated in Articles 18 and 19, and the guarantee fund is defined in Article 19; see Appendix D for the full text of these articles.

Title I: General provisions

    Articles 1 to 5

    Articles 1 and 2: The diversity of operations

Title II: Rules applicable to undertakings whose head offices are situated within the community

    Section A: Conditions of admission

    Articles 6 to 14

    Article 13: The separation of life and non-life insurance undertakings

    Article 14: Separate accounting for life and non-life insurance undertakings

    Section B: Conditions for carrying on activities

    Articles 15 to 25

    Article 17: Technical reserves

    Articles 18 and 19: Solvency (see Appendix D)

    Article 20: Guarantee fund (see Appendix D)

    Article 24: The solvency plan and the financial plan

    Section C: Withdrawal of authorization

    Article 26

Title III: Rules applicable to agencies or branches established within the community and belonging to undertakings whose head offices are outside the community

    Articles 27 to 32

Title IV: Transitional and other provisions

    Articles 33 to 37

Title V: Final provisions
  Articles 38 to 42
Annex
  Classes of risks

In the compromise in defining the solvency margin, the commission had to decide whether the margin should be represented by explicit items, implicit items, or a mixture of both (Pool, 1990, p. 36). *Explicit items* means the same type of items making the solvency margin for non-life insurance undertakings, and *implicit items* should reflect the future performance of the life insurance undertakings. This was not a new discussion, as an OECD report in 1971 favored the implicit item approach. This study was chaired by Mr. Buol from Switzerland (the Buol report[7]) (Pool, 1990, p. 36; Kastelijn and Remmerswaal, 1986, p. 31).

In the Buol report the margin is calculated on the basis of the difference between reserves based on *strengthened* (SR) and *unstrengthened* (UR) interest rate bases.

The unstrengthened interest rate is based on the interest rates of the past 20 years:

$$UR = 2/3 \text{ (minimum interest rate over the past 20 years) +}$$
$$1/3 \text{ (90\% of the most recent rate)}$$

The strengthened interest rate is defined as SR = 80% UR. The working group also considers 85% as a possibility.

The difference between the UR-based reserve and the SR-based reserve can for different types of insurance be expressed as a percentage of the UR-based reserve in combination with a percentage of the capital sum at risk. The Buol report suggested that for an average portfolio, the result could be expressed as

$$9\% \text{ (UR reserve) + 6\% (sum insured at risk)}$$

Short-term contracts, which lead to little or no reserving, were considered separately by a subgroup to Buol's working group.

Most member states considered explicit items to be essential because it was easy to understand what was involved. The compromise that was reached was that explicit items should be taken into account and that implicit items may be relied on in accordance with certain rules and up to certain limits (cf. Article 19).

The commission waited until the second non-life directive was on its way before it started to work on a corresponding proposal for life insurance

---

[7] Buol et al. (1971): Les garanties financieres requises des enterprises d'assurance vie. OECD.

undertakings (EEC, 1990). Experience from the non-life directives made it clear that the commission had to proceed in two stages.

In the first step (the second directive), there was a separation between home country control and destination state control. In life insurance the nature of such control has a more fundamental effect upon the products that an undertaking can put on the market than in non-life insurance business. The solvency rules defined in the first directive were not changed.

The structure of the second directive is as follows:

Title I: General provisions (Articles 1 to 3)
   Definitions

Title II: Provisions supplementary to the first directive (Articles 4 to 9)

Title III: Provisions relating specifically to the freedom to provide services (Articles 10 to 25)

Title IV: Transitional provisions (Articles 26 to 27)

Title V: Final provisions (Articles 28 to 32)

Annex: Statements

As for non-life business, the commission was not satisfied with this directive, as it did not include a single authorization (single license), and a third life directive was adopted (EEC, 1992b). The solvency rules defined in the first directive were not changed.

Title I: Definition and scope (Articles 1 and 2)

Title II: The taking up of the business of life insurance (Articles 3 to 7)

Title III: Harmonization of conditions governing pursuit of business (Articles 8 to 31)

   Article 25 replaces Article 18 (second subparagraph, point 1) from the first directive; for the full text of the new article, see Appendix D

Title IV: Provisions relating to right of establishment and freedom to provide services (Articles 32 to 44)

Title V: Transitional provisions (Articles 45 to 46)

Title VI: Final provisions (Articles 47 to 52)

Annex I: Matching rules

Annex II: Information for policyholders

3.5 Calculating the Solvency Margin for Non-Life Insurance Business

In this section, we will show the procedure to go from the directives to calculating the solvency margin. For simplicity, we use the non-life directives to set up a spreadsheet table for the calculations. The *available solvency margin* is calculated according to Article 16(1) in the first non-life directive (with changes according to Article 24 in the third directive), and the *(required) solvency margin* according to Article 16(3). The amounts are calculated in euros.

A similar approach was made by the Insurance Conference (now CEIOPS) to make it easier to compare statements between different member states. The spreadsheet table is based on the directives and the proposal made by the Insurance Committee. The Insurance Committee also made a similar table for the life insurance solvency calculation.

In the following summary the notions of the two tables for the available solvency margin and the solvency margin are used.

Summary

I. Total of components of the margin (available solvency margin)	=	ASM
of which components of A (Total A)		
of which components of B (Total B)		
{Solvency ratio: E/D	=	ASM/max[PI, CI]}

II. Amount of the margin to be established (solvency margin)		
A. First result (premium index)	=	PI
B. Second result (claims index)	=	CI
II. The minimum solvency margin: max[A, B]	=	max[PI, CI]

I. Components of the Margin (Available Solvency Margin)

Items under B are allowed on request and subject to proof by the undertaking and after agreement by the supervisory authorities of the home member state.

A.

(1) Paid-up share capital or the effective initial fund + members' account	+	
(2) 50% (unpaid share capital or initial fund if at least 25% of this capital has been paid up)	+	
(3) Reserves		
(3a) Statutory reserves	+	
(3b) Free reserves	+	
(4) Profits brought forward (carry forward of profits)		
(4a) Carry forward from previous period	+	
(4b) Profits not appropriated after distribution of the last balance	+	
(5) 50% of the possible calls for variable supplementary contributions within the financial year (subject to a limit of 50% of the margin)	+	
(6) Preferential share capital and subordinated loan capital; maximum 50% of the available solvency margin, no more than 25% of subordinated loan (see Article 24 of the third non-life directive); insofar as national legislation permits it as own funds	+	
(7) Other items; list the other items	+	
(8) Intangible items	–	
Total A: Available solvency margin	=	Total A

B.

(9) Hidden reserves resulting from the underestimation of assets	+	
Total B:	=	Total B

I. Total of the components	=	ASM = Total A + Total B
of which components in B as % of Total A + Total B	=	%

II. Amount of the Margin to be Established (Solvency Margin)

A. First result (premium basis):		
(10) Premium income (gross); all direct business last financial year	+	
(11) Premium income (gross); accepted reinsurance last financial year	+	
(12) Cancelled premiums in (10), also taxes pertaining to (10)	–	
(13) Sum of premium income: (10) + (11) – (12)	=	
(14) If (13) < 10,000,000 ecu	0.18*(13)	
(15a) If (13) ≥ 10,000,000 ecu		1,800,000
(15b) 16% of that part of (13) that is above 10,000,000 ecu	0.16*((13) – 10,000,000)	
(a) The sum (14) or (15a) + (15b)	=	
(16) The amount of claims, net of reinsurance	=	
(17) The amount of claims, gross	=	
(b) The claims retention ratio, (16)/(17)	=	
(c) The largest of 0.50 and (b)	=	(0.5)
A. First result: the premium index (a)*(c)	=	**PI**
or for health insurance practiced on a basis similar to that of life insurance: (a)*(c)/3	=	(PI)
B. Second result (claims basis):[a]		
(18) Claims paid; direct business	+	
(19) Claims paid; accepted reinsurance	+	
(20) Provisions or reserves for outstanding claims; both direct business and accepted reinsurance; at the end of the last financial year	+	
(21) Claims paid during the periods specified in Article 16(2)	–	
(22) Provisions or reserves for outstanding claims; both direct business and accepted reinsurance; at the beginning of the reference period	–	
(23a) The sum (18) + (19) + (20) – (21) – (22)	=	
(23b) The average (23a)/3 or (23a)/7, depending on the reference period	=	
(24) If (23b) < 7,000,000 ecu	0.26*(23b)	
(25a) If (23b) ≥ 7,000,000 ecu		1,820,000
(25b) 23% of that part of (23b) that is above 7,000,000 ecu	0.23((23b) – 7,000,000)	
(d) The sum (24) or (25a) + (25b)	=	
B. Second result: the claims index max[0.50, (b)] d	=	**CI**
or for health insurance practiced on a basis similar to that of life insurance: (b)*(c)/3	=	(CI)

[a] The reference period is usually the last three financial years, except for the risks of credit, storm, hail, and frost, where 7 years is used (cf. the Credit Directive 87/343/EEC, EEC, 1987).

3.6   The Insurance Accounting Directive (IAD)

In the fourth directive, from 1978, on annual accounts for certain types of companies,[8] it was laid down that the member states need not apply the provisions of it to insurance companies, banks, and other financial institutions. The seventh directive on consolidated accounts is also of importance (EEC, 1983). Some years later, a directive dealing with the annual accounting of banks and credit institutions was proposed. The bank accounting directive was adopted in 1986.

3.6.1   The Importance of Disclosure

At this time, the information that was available to the public, both private and institutions, about the financial position of the insurance companies varied from one country to another. In some, there were detailed rules about the layout and contents of the accounts, but in others, such rules were more or less absent. As the commission was striving toward a single market in insurance, it proposed a directive on the annual accounts of insurance companies.

There is not only a public interest in the financial business of insurers, but also an interest for brokers, risk managers, financial journalists, etc. Thus, it was important that the accountings in different countries were coherent.

One main reason for exclusion of the insurance companies from the 1978 accounting directive was that they would require special rules because of their nature (Pool, 1990, p. 83). The opinion of most member states was that the public is entitled to get a high level of financial information about the insurance undertakings and their activities. Therefore, the accounting information from the insurance companies had to be at least equivalent (never less) to that required under the 1978 accounting directive.

During the drafting of the Insurance Accounting Directive (IAD), it became apparent that the comparability is closer in non-life undertakings in different countries than in life undertakings (Pool, 1990). These differences are also shown in the directive.

A directive was proposed by the commission in December 1986, but was not adopted until December 19, 1991 (EEC, 1991). The text itself was not self-sufficient, as it states that both the fourth and seventh directives on accounts and consolidated accounts are applied except where the insurance accounting directive provides otherwise.

When someone buys insurance, he pays a premium in advance before payments are made as claims or settlements. This means that in *the balance sheet* the investments on the asset side and the technical provisions on the liability side are the main items. In *the profit and loss account* there are mainly

---

[8] EEC, 1978.

two types of income: premiums and return on investments. The profit and loss account is divided into a technical and a nontechnical part. The technical part is also separated for non-life and life insurance. The IAD is mainly concerned with rules relating to investments (including valuation rules) and rules relating to technical provisions.

The IAD and the fourth account directive have the same structure; i.e., in Section 3 we have the layout of the balance sheet and in Section 5 the layout of the profit and loss account. Section 7 discusses the valuation rules, and in the eighth section the contents of the notes are laid out. The fourth directive was essentially applied to companies having share capital, but as the IAD was much connected with the creation of an internal common market in insurance, it could not exclude some of the most important players, i.e., the mutuals and cooperatives.

3.6.2 Balance Sheet (Section 3, Article 6)

Most of the assets of insurance undertakings were investments that did not fit into the categories of fixed assets and current assets in the fourth directive, and therefore they were abandoned in favor of a single concept of investments. Because they are especially important to insurance undertakings in meeting future liabilities, there are detailed rules about the way they are shown.

The technical provisions are the amounts set aside for the liabilities and to meet the insurer's commitments under the contracts. The provisions can be split up into:

- *Non-life insurance*[9] (also for short-term life business)
  - Provisions for unearned premiums (that part of the premium that has already been received for the period that has not expired)
  - Provisions for claims (represents the estimated cost of settling claims that has been incurred or is believed to have incurred but not reported (IBNR))
- *Life insurance*
  - Mathematical provisions built up on the actuarial basis for the contract

There were two difficult areas to treat: the treatment of the equalization reserve and reinsurance. For the equalization reserves the directive seeks full disclosure and clarity in the presentation. For reinsurance the question was whether the technical provisions should be shown gross or net of reinsurance. It was decided that the gross amount, the reinsurance deduction, and the net value were all to be disclosed.

---

[9] The treatment of the technical provisions in non-life insurance follows a report from the Insurance Conference (now CEIOPS): the Angerer report (Pool, 1990).

The main layout of the balance sheet is given below:

Liabilities	Assets
A. Capital and reserves	A. Subscribed capital unpaid
B. Subordinated liabilities	B. Intangible assets
C. Technical provisions	C. Investments
D. Technical provisions for life insurance policies where the investment risk is borne by the policyholders	D. Investments for the benefits of life insurance policyholders who bear the investment risk
E. Other provisions	E. Debtors
F. Deposits received from reinsurers	F. Other assets
G. Creditors	G. Prepayments and accrued income
H. Accruals and deferred income	H. Loss for the financial year
I. Profit for the financial year	

The *equalization provisions* and the different provisions in terms of the gross value, the reinsurance value, and the net value are given under C (technical provisions). Note that the term used here is not equalization reserves, as in the third non-life directive.

3.6.3 Profit and Loss Account (Section 5, Article 34)

The profit and loss account is divided into a *technical account*, showing the results of the insurance activities in the classes of direct insurance and corresponding classes of reinsurance, and a *nontechnical account*. For the *non-life undertakings* the technical account should reflect the result of the *underwriting activity* before the investment income is taken into account. For the *life undertakings* most of the *investment income* is used for the benefit of the policyholders, and thus this is reflected in the technical account.

3.6.4 Valuation Methods (Section 7, Articles 45 to 62)

Investments made by the insurance undertakings are almost exclusively held for the purpose of meeting future liabilities. Therefore, their valuation is of vital importance.

The technical provisions of the *liabilities* are to be valued in such a way that the amount at all times is such that an undertaking can meet any liability arising out of insurance contracts as far as can reasonably be foreseen (Article 56). This is discussed in more details in Articles 57 to 62 and means that there is a valuation that includes "pillows" and prudence.

On the other hand, the assets may be valued in different ways (cf. the table below). It was thought that only *historical cost*, as a sole method, would be inadequate in some circumstances. Another possible valuation technique could be to use *current value*, which in some circumstances can be hard to establish. By the directive it is possible to use one, but the other value must be given in a note on the account. The reader of the accounts will thus have access to both figures.

Valuation Method	Assets/Investments (Articles 45 to 55)
Purchase price	Yes, Article 45 and 46
Market price/current value	Yes, Articles 46 to 49
Fair value[a]	Yes, Article 46(a)(2): Assets D

[a] Directive 2003/51/EC amending the four directives mentioned in the text (COM, 2003a; CONSLEG, 2003).

Some member states have chosen to have a conservative accounting approach, as they use the prudent valuation on the liability side and historical value (or purchase prise) on the asset side. Other countries, such as the U.K. and Sweden, have chosen a mismatch approach, as the liabilities are valuated with prudence and the assets at market value.

Some amendments to EEC (1991) were done in 2003 (COM, 2003a). A consolidated text on the annual accounts and consolidated accounts of insurance undertakings was published in 2003 (CONSLEG, 2003).

4

The European Union: Solvency I

During the process with the third directives, the council discussed the possibility of reviewing the provisions concerning the solvency margin. But in order not to delay the completion of the insurance single market, it was decided to do so later. In agreement with the commission, the council included articles in the third directives to oblige the commission to present a report to the Insurance Committee (IC) within 3 years of the implementation of the two directives on the need for further harmonization of the solvency margin: Article 25 in the third non-life directive and Article 26 in the third life directive.

At the IC's meeting in April 1994, the question about a solvency review was raised. The IC agreed to ask the European supervisory authorities[1] (the conference) to establish a working group to look into solvency issues in a broad sense. Dr. Helmut Müller, from the German insurance supervisory authority,[2] chaired the group. The report that the working group presented in 1997 will be named the Müller report (1997). A review of these discussions is given in EC (1997).

The Müller group had used a questionnaire to its members, and a similar questionnaire was sent out by the commission to three European organizations: Groupe Consultatif (GC), representing the European actuaries; Comité Européen des Assurances (CEA), representing the European insurance industry;[3] and Association des Assureurs Coopératifs et Mutualistes Europèens (ACME), representing mutual and cooperative insurance undertakings.

For GC it was decided in 1995 to reconvene its solvency working party, as it was asked to consider the operation of the current regime and to make recommendations in light of any proposals that might be made by the commission (Henty, 2003, pp. 23–24). A general report on the current solvency regime was submitted to the commission in December 1996. Some of its members presented a report at the International Congress of Actuaries

---

[1] The Conference of Insurance Supervisory Departments of European Community Member States (from 2004 CEIOPS).

[2] Bundesaufsichtsamt für das Versicherungswesen, BAV.

[3] Rantala and Vesterinen (1995) prepared a report for CEA based on the Finnish solvency system.

in 1998 (Horsmeier et al., 1998), including the GC's 1996 report to the commission (Part I) and the response to the questionnaire (Part II).

The commission's work on solvency developed in two parts:

- Solvency I: A review of the current regime, based on the Müller report
- Solvency II: A fundamental new approach (see Section 5.5)

It was considered in the Müller report that the current solvency margin requirement had proved satisfactory. The insurance industry and the EU supervisory authorities were in favor of a simplification in the calculations for the premium index and claims index for the non-life insurance business. The Müller group identified cases and analyzed the causes of deficiencies of insurance undertakings in countries within the European economic area during the past 20 years. Only a few cases of deficiencies had been observed. Most of the deficiencies could have been remedied through capital increase or takeover by other companies.

The Müller report (p. 3) concludes:

> It was found that even if the solvency rules had been applied and observed more strictly, and even if they had contained stricter requirements than they do at present, a number of the economic collapses that happened could not have been prevented. The solvency margin as a rule fulfils its warning and safety function but it does not at all replace an effective company analysis and even less a prudent establishment and coverage of the technical provisions.

The Müller report pointed out some specific cases of deficiencies that could have been avoided by more accurate solvency margin regime. They occurred in relation to long-tail risks in non-life insurance, investment and asset–liability mismatches, rapidly growing enterprises, and inadequate reinsurance taken.

The commission considered that the principles governing the operation of the present regime should be maintained (EC, 1997, pp. 7, 10). This did not mean that some detailed arrangements could not be adjusted and supplemented. The commission also stated, "Every effort should then be made to avoid any additional cost for the industry, except in the case of specific risk situations for which the current level of requirement proves to be inadequate" (EC, 1997, p. 10).

The commission had the view that further work should be done to improve the solvency margin regime and harmonize the provisions (EC, 1997, p. 11). The commission proposed that a new working group chaired by the commission's services and consisting of government experts should be set up. The results presented in the Müller report and the answers to the questionnaires given by the insurance industry and the GC should also be considered by this new working group.

The working group should look at the composition and calculation of the solvency margin, investments as cover for the solvency margin, measures available to the supervisory authorities, and the level of harmonizations.

The work of this group resulted in the Solvency I proposal.

As the questionnaire from the Müller group included a list of risks to be discussed, and the report also discussed solvency requirements and the components of solvency margin, this will be discussed in Section 4.1. Comments by the Groupe Consultatif are discussed in Section 4.2. In Section 4.3 the new directive will be discussed and reviewed. A calculation spreadsheet, as for the non-life insurance for Solvency 0 (see Section 3.5), will be given in Section 4.4.

4.1 The Müller Report

The Müller group took the view that the subdivision of the solvency system into minimum guarantee fund, guarantee fund, and solvency margin should be maintained.

The amount of the minimum guarantee fund is proposed to be increased considerably, at least to an amount compensating for the inflation since 1973 (non-life) and 1979 (life). Special regulations are to be provided for small life but not for non-life undertakings.

For non-life insurance the group proposed the use of at least three indices: the premium and claims indices, as in the first non-life directive, complemented with a *provision index* in order to take account of the long-tail business. The index should be applied either alternatively or additively. The index resulting in the highest margin should be the decisive one. A minority of the members of the working group believed that the investment risk would not be taken account of in an adequate way. Introducing a fourth index could do this: the *investment index*. It should be applied additively, and the yardstick for this index should be the weighted assets of the insurance companies, in a way similar to that of the U.S. risk-based capital (RBC) system or the EU banking regulation.

For life insurance, the group proposed that the solvency margin should not be increased. The technical risk should, as it is in the first life directive, be taken into account by the second result (3‰ of the capital at risk). A majority of the group was in favor of the present regulation in taking care of the investment risk by the first result (4% of the mathematical provisions). A minority suggested a solution as for non-life insurance, based on the U.S. RBC system or the EU banking regulation.

The Müller report suggested that the new regulations should refer not only to the solvency margin, but also to the composition of the margin and the guarantee fund. The definition of own funds must be reviewed. The admissible own fund listed in the catalog of the directive should be used to cover the margin and the guarantee fund in the future, but to a lower degree.

It was also proposed that the instruments available to the supervisory authorities should be reviewed. The working group wished that the supervisory authorities should have the right to intervene even if the requirements regarding the technical provisions and solvency are still being met.

4.1.1   Risks for Insurance Undertakings

The working group identified three groups of risks with 20 risk categories. The three groups were classified as technical risks, investment risks, and nontechnical risks. We here give a short review of all these risks and risk categories.

4.1.1.1   Technical Risks

Current risks

- *Risk of insufficient tariffs*: Miscalculation, if it was made deliberately, could be classified as a management risk.
- *Deviation risk*: Risk factors changing subsequently claims frequency and extent, mortality, morbidity, price and wage levels, cancellation probability, legislation, and falling interest rates.
- *Evaluation risk*: The risk that the technical provisions are insufficient.
- *Reinsurance risk*: The risk of nonpayment by the reinsurer and poor quality of reinsurance.
- *Operation expenses risk*: The risk that the amount for operating expenses is insufficient.
- *Major losses risk* (only non-life): The risk due to the size and number of major losses.
- *Accumulation or catastrophe risk*: Risks due to single events, e.g., earthquakes, storms, etc.

Special risks

- *Growth risk*: Excessive growth, uncoordinated growth.
- *Liquidation risk*: The risk that the existing funds of the undertaking are insufficient to meet the liabilities.

4.1.1.2   Investment Risks

- *Depreciation risk*: Investments losing their value due to credit, nonpayment, and market risks.
- *Liquidity risk*: Risks due to investments not being able to be liquidated at the right time and in a proper manner.

- *Matching risk*: The risk that the assets are poorly matched to the liabilities.

- *Interest rate risk*: Risk of changing interest rates, including reinvestment risk.

- *Evaluation risk*: The risk that an investment has been evaluated at too high a value.

- *Participation risk*: Risk due to the undertakings holding shares in other undertakings.

- *Risks related to the use of derivative financial instruments*: Specific market, credit, and liquidity risks. Untrained staff of the undertaking is also a risk.

4.1.1.3 Nontechnical Risks

- *Management risk*: Incompetent or criminal intentions of the management.

- *Risks in connection with guarantees in favor of third parties*: Risk that the economic capital of the undertaking is strained.

- *Risk of the loss of receivables due from insurance intermediaries*: Risk that external third parties do not meet their obligations.

- *General business risks*: Risk of change in general legal conditions, e.g., tax laws and regulations.

The report gives a list of financial difficulties of insurance undertakings in Annex 1. The RBC system of the U.S. is presented in Annex 2 (see also Section 6.10) and resilience tests for life insurance undertakings in Annex 4.

4.2 Comments from Groupe Consultatif

The Groupe Consultatif (Horsmeier et al., 1998) states that from an actuarial point of view the current solvency regime is valuable and is becoming generally accepted. The GC also states that the EU directives do not govern all balance sheet items and that the non-life technical provisions are not subject to a proper appraisal of risks. *Solvency is a combination of technical provisions and a solvency margin.* The GC also lists risks that it saw as affecting the solvency and reserving. They were divided into quantifiable and more or less quantifiable risks (see below).

4.2.1   Non-Life Risks

Quantifiable risks

- *Asset liability management* (ALM): Two important issues: the current and expected future yield of the investments and the current and expected future of the investments. The yield is important with a view to the time value of money that is incorporated in both premium and technical provisions. The value of investments plays an important role when liquidation is required for claim settlement. The valuation of assets should be in line with the valuation of liabilities.
- *Risk assessment*: The systems used for premiums and technical provisions.
- *Expenses*: The premium should be sufficiently loaded with expense allowances.
- *Premium growth*: Fast-growing and fast-declining portfolios may jeopardize the undertaking.
- *Reinsurance*: The reinsurance program and the reinsurers involved.

Risks hard to quantify or measure a priori

- *Management:* Failings of management.
- *Major business decisions*: A downside risk.
- *Underwriting*: Failings within the underwriting department.
- *Claims handling*: Failings within the claims handling department.

4.2.2   Life Risks

Quantifiable risks

- *Investment risk*: Default, liquidity, inadequate investments return, reinvestment risk.
- *Currency risk*: The matching of assets and liabilities in one specific currency.
- *Asset liability management* (ALM): Two important issues: the current and expected future yield of the investments and the current and expected future of the investments. The yield is important with a view to the time value of money that is incorporated in both premium and technical provisions, given also guaranteed interest rates. The value of investments plays an important role when liquidation is required for claim settlement in case of lapses and at the maturity date. The valuation of assets should be in line with the valuation of liabilities.

- *Mortality, disability, and morbidity:* Longevity risk, disability, and morbidity.
- *Lapse/surrender risk:* The policyholders initiate these risks.
- *Expense risk:* Expenses may have a major impact according to the long-term nature of life business.
- *Taxation risk:* Changed taxation rules.
- *Inflation risk:* Inflation may have a major impact according to the long-term nature of life business.
- *Options/guarantees:* A major risk due to the long-term nature of life business.
- *Business risk:* Faulty business conditions, administration systems, fraud.
- *Catastrophes:* Risks like AIDS.

Risks hard to quantify or measure a priori

- *Management:* Failings of management.
- *Major business decisions:* A downside risk.
- *Underwriting:* Failings within the underwriting department.
- *Claims handling:* Failings within the claims handling department.

GC also proposed that the risks that are quantifiable should be covered by the technical provisions and used in the calculation of the solvency margin. This is done for life insurance but not for non-life (cf. the Müller report's provision index). There should not be double counting, so if a risk is well provisioned, it should not be taken into account once again in the solvency margin. The factors and numbers to be applied should be the same for all undertakings.

GC also preferred to have the requirements differentiated according to classes of insurance and a distinction made between short-term and long-term insurance.

4.3 The Solvency I Directives

Following the Müller report (1997) and the commission's report (EC, 1997), there were four expert meetings during 1997 and 1998. The commission also consulted the European insurance industry (represented by, e.g., Groupe Consultatif, CEA, ACME, and AISAM[4]). In its first meeting it was decided to make simulations in order to analyze the financial impact of (1) the third index in non-life insurance based on the technical provisions and (2) an

---

[4] Association Internationale des Sociétés d'Assurance Mutuelle.

increase in the minimum guarantee fund (MGF). The results of the simulations were presented in September 1998 (EC, 1999). The conclusion made was that the efficiency of the provision index was doubtful, as it did not seem to affect the targeted insurance undertakings. With regard to the MGF, very large increases were estimated for small undertakings.

The working document (EC, 1999) concluded that the future EU legislation should be more flexible in order to incorporate developments in the financial services industry faster. Analysis has shown that the solvency margin system had functioned satisfactorily, but this did not guarantee that it would function effectively in the future. Some jurisdictions had chosen to adopt a risk-based capital system, like the U.S. RBC, for the solvency margin. It is stated in the report that "at some future date it may be desirable to undertake a wider review of the EU solvency margin system and to consider whether more explicit recognition of the different risks is required."

The working document sets out draft proposals for an improvement of the system and also some sketches on further developments that could be considered (i.e., a future Solvency II system).

The work led to the proposal for new life and non-life undertaking directives in 2000 (COM, 2000a, 2000b). On March 5, 2002, the European Parliament adopted the new solvency directives for life and non-life insurance undertakings (COM, 2002a, 2002b). The new life insurance directive was repealed by COM (2002c). The directive was optional for small mutual companies (premium income less than 5 million euro), and the guarantee fund was set to 3,000,000 euro for life and liability and 2,000,000 euro for other non-life business.

4.3.1   The Solvency I Non-Life Directive

The Solvency I non-life directive, adopted March 5, 2002 (COM, 2002a), has the following structure:

Article 1: Amendments to Directive 73/237/EC

> Point 2 replaces Article 16 of this first directive: Available solvency margin
>
> Point 3 inserts a new article, Article 16a: Required solvency margin
>
> Point 4 replaces Article 17 of the first directive: Guarantee fund
>
> Point 5 inserts a new article, Article 17a: On the amounts in euro

Article 2: Transitional period

Article 3: Transposition

Article 4: Entry into force

Article 5: Addressees

The new articles, 16, 16a, 17, and 17a, are given in Appendix C.

The guarantee fund was raised from 300,000 to 2,000,000 euro, except for risk classes 10 to 15, where the increase is from 400,000 to 3,000,000 euro. These amounts should be reviewed annually (from September 20, 2003). The new Article 16 discusses the available solvency margin and what it can consist of, and the new Article 16a discusses the required solvency margin.

The premium income for classes 11 to 13 should be increased by 50%. The premium basis should be calculated by using the higher of the gross written premiums and gross earned premiums. A spreadsheet is given in Section 4.4 to illustrate the calculation of the solvency margin for a non-life business. This could be compared to the Solvency 0 calculation in Section 3.5. If the claims reserve is discounted, a correction should be made. The amount calculated in the first result (premium index) shall be divided into two portions: the first portion extending up to 50 million euros (earlier 10 million) and the second comprising the excess. The level of the second result (claims index) is increased from 7 million to 35 million euros.

There is also a 20% higher solvency margin for all liabilities, except motor third-party liability.

4.3.2 The Solvency I Life Directive

The Solvency I life directive was approved March 5, 2002 (COM, 2002b), and was repealed by COM (2002c), which is a consolidated directive. The consolidated directive has 8 titles and 74 articles. Title III is the most interesting from a solvency perspective. Categories of authorized assets are stated in Chapter 2 (Articles 20 to 26), and the available solvency margin (Article 27), required solvency margin (Article 28), and guarantee funds (Article 29) are found in Chapter 3.

The main changes from Solvency 0 are that the available solvency margin covering the technical provisions must be of good quality, and that for unit-linked contracts, a solvency margin of 25% of the last financial year's net administrative expenses is introduced. The authorized assets are listed in Article 23, and the solvency margin for undertakings that bear no investment risk (such as unit-linked contract) is given in Article 28(7)(c). The guarantee fund is now EUR 3 million (Article 29).

Note that Article 69.5 states:

> Not later than 1 January 2007 the Commission shall submit to the European Parliament and to the Council a report on the application of Articles 3(6), 27, 28, 29, 30 and 38 and, if necessary, on the need for further harmonisation. The report shall indicate how Member States have made use of the possibilities under those articles and, in particular, whether the discretionary powers afforded to the national supervisory authorities have resulted in major supervisory differences in the single market.

The structure of the consolidated directive and Articles 27 to 30 on the solvency and guarantee funds is given in Appendix D.

4.4 Calculating the Solvency Margin for Non-Life Insurance Business

Now we will show the procedure how to go from the directives to calculating the solvency margin. For simplicity, we use the non-life directives to set up a spreadsheet for the calculations. The *available solvency margin* is calculated according to the new Article 16 in the Solvency I non-life directive, and the *required solvency margin* according to the new Article 16a. The amounts are calculated in euro. For comparisons with the Solvency 0 calculation, see Section 3.5.

In the following summary the notions of the two spreadsheets for the available solvency margin and the required solvency margin are used.

Summary

II. *Amount of the margin to be established* **(solvency margin):**		
A. Solvency margin (g or k), comparison with last year	=	**g or k**
B. Guarantee fund, Article 17	=	**GF**
C. The minimum solvency margin: max[A, B]	=	**max[g or k, GF]**
I. *Total of components of the margin* **(available solvency margin):**		
D. Available solvency margin	=	**ASM**
Requirement on the ASM		
C. The minimum solvency margin	=	**max[g or k, GF]**
E. Minimum guarantee fund, C/3	=	**MGF = max[g or k, GF]/3**
F. Guarantee fund	=	**GF**
G. max[E, F]	=	**max[MGF, GF]**

I. Components of the Margin (Available Solvency Margin)

"A assets"

(1) Paid-up share capital or the effective initial fund + member's account	+	
(2) Reserves		
(2a) Statutory reserves	+	
(2b) Free reserves	+	
(3) Profits brought forward (carry forward of profits)		
(3a) Carry forward from previous period	+	
(3b) Profits not appropriated after distribution of the last balance	+	
(4) Preferential share capital and subordinated loan capital; maximum 50% of the lesser of the available solvency margin and the required solvency margin, no more than 25% of subordinated loan	+	
(5) Other items; list the other items	+	
Sum of A assets	=	**Total A**

"Reducing assets"

(6) The amount of own shares directly held by the insurance undertaking	+	
(7) The difference between undiscounted technical provisions and the discounted technical provisions	+	
(8) Intangible items	+	
Reducing assets	=	**Reducing**

"B assets"

(9) 50% (unpaid share capital or initial fund if at least 25% of this capital has been paid up)	+	
(10) Mutuals: 50% (maximum contributions – contributions called in); maximum 50% of the available solvency margin	+	
(11) Hidden reserves resulting from the underestimation of assets	+	
Sum of B assets	=	**Total B**

Available solvency margin

I. Total of the components	=	**ASM = Total A – Reducing + Total B**
Of which components in B as % of ASM	=	%

II. Amount of the Margin to be Established (Required Solvency Margin)

A. First result (premium basis):		
(12a) Premium income (gross); all direct business last financial year except classes 11–13	+	
(12b) Premium income (gross); direct business in classes 11–13 last financial year		
(12c) (12b) 1.50	=	
(13) Premium income (gross); accepted reinsurance last financial year	+	
(14) Cancelled premiums in (12a) and (12b), also taxes pertaining to (12a) and (12b)	−	
(15) Sum of **gross premium written**: (12a) + (12c) + (13) − (14)	=	**GPW**
(16a) Premium earned (gross); all direct business last financial year except classes 11–13	+	
(16b) Premium earned (gross); direct business in classes 11–13 last financial year		
(16c) (16b) 1.50	=	
(17) Sum of **gross premium earned**: (16a) + (16c)	=	**GPE**
(18) max[(15), (17)]	=	max[GPW, GPE]
(19) If (18) < 50,000,000 euro	0.18 (18)	
(20a) If (18) ≥ 50,000,000 euro		9,000,000
(20b) 16% of that part of (18) that is above 50,000,000 euro	0.16 ((18) − 50,000,000)	
(a) The sum (19) or (20a) + (20b), premium index	=	**PI**
(21) The amount of claims, net of reinsurance	=	
(22) The amount of claims, gross	=	
(b) The claims ratio, (21)/(22)	=	
(c) The largest of 0.50 and (b)	=	(0.5)
(23) Adjusted premium index: (a)(c)	=	**API**
B. Second result (claims basis):		
(24a) Claims paid; direct business except classes 11–13	+	
(24b) Claims paid; direct business classes 11–13		
(24c) (24b) 1.50	=	
(25) Claims paid; accepted reinsurance	+	
(26a) Provisions or reserves for outstanding claims; both direct business and accepted reinsurance; at the end of the last financial year; except classes 11–13	+	
(26b) Provisions or reserves for outstanding claims in classes 11–13; both direct business and accepted reinsurance; at the end of the last financial year		
(27c) (27b) 1.50	=	
(28a) Recoveries except classes 11–13	+	
(28b) Recoveries classes 11–13		
(28c) (28b) 1.50	=	
(29a) Provisions or reserves for outstanding claims; both direct business and accepted reinsurance; at the beginning of the reference period; except classes 11–13	−	

(29b) Provisions or reserves for outstanding claims in classes 11–13; both direct business and accepted reinsurance; at the beginning of the reference period		
(29c) (29b) 1.50		
(30a) The sum (24a) + (24c) + (25) + (26a) + (26c) – (28a) – (28c) – (29a) – (29c)	=	
(30b) The average (30a)/3 or (30a)/7, depending on the reference period	=	
(31) If (30b) < 35,000,000 euro	0.26 (30b)	
(32a) If (30b) ≥ 35,000,000 euro		9,100,000
(32b) 23% of that part of (30b) that is above 35,000,000 euro	0.23 ((30b)– 35,000,000)	
(d) The sum (31) or (32a) + (32b); claims index	=	CI
(33) A(21) — The amount of claims, net of reinsurance	=	
(34) A(22) — The amount of claims, gross	=	
(b) The claims ratio, A(21)/A(22)	=	
(c) The largest of 0.50 and (b)	=	(0.5)
(35) Adjusted claims index: (d) (c)	=	ACI

Required solvency margin — comparison with last year's solvency margin

(23) First result (adjusted premium index)	=	API
(35) Second result (adjusted claims index)	=	ACI
e. max[(23), (35)]	=	max[API, ACI]
f. The largest of (23) and (35) last financial year	=	
g. Solvency margin: if e > f, then e is taken to the summary		max[API, ACI]
If f > e		
h. Provisions or reserves for outstanding claims; both direct business and accepted reinsurance; at the end of the last financial year	=	
i. Provisions or reserves for outstanding claims; both direct business and accepted reinsurance; at the beginning of the reference period	=	
j. The ratio h/i	=	h/i
k. Solvency margin: last year's solvency margin multiplied by the ratio j:(g)(j)	=	(g)(j)

5

Steps toward Solvency II: 1

Chapters 5 and 6 are a *smorgasbord*,[1] with different ideas in modeling solvency.

In this chapter we will first discuss influences from different organizations, such as the banking supervisory system (Basel II), proposals for a new insurance accounting system (International Accounting Standards Board (IASB)), frameworks from the International Association of Insurance Supervisors (IAIS), and the International Actuarial Association's (IAA) work on a global solvency assessment system. We then look closer at the first steps made by the European Commission toward a new solvency system.

5.1 Bank for International Settlements (BIS): The New Basel Capital Accord

In a presentation of the proposed work on Solvency II in 2001 (MARKT, 2001b), the commission described the New Basel Capital Accord as "an interesting basis for incorporating the latest thoughts of banking supervisory into the Solvency II project: while the banking prudential system is necessarily different from that for insurance ..., it may nevertheless serve as a source of inspiration as regards design." In another note from the commission (MARKT, 2001c), the question of whether banking rules are relevant for the insurance sector was raised. The paper also briefly described the New Basel Accord, usually known as *Basel II*.

In 1988 the Basel Committee on Banking Supervision introduced global standards for regulating the capital requirements for banks. This accord is usually referred to as Basel I. The accord established a minimum requirement of 8% of capital to risk-weighted assets. The requirements of Basel I have

---

[1] Smorgasbord: Swedish *smörgåsbord*, from *smörgås* ("open sandwich") + *bord* ("table") (approximately 1919); 1: luncheon or supper buffet offering a variety of foods and dishes (as hors d'oeuvres, hot and cold meats, smoked and pickled fish, cheeses, salads, and relishes); 2: heterogeneous mixture (*Webster's Ninth New Collegiate Dictionary*).

been adopted in more than 100 countries and are a cornerstone in banking supervisory regulations.

The business of banking, its risk management, and the whole financial market had changed in a way that the Basel I requirements were inadequate. In June 1999 the Basel Committee proposed a first version (Consultation Paper 1, CP1) of a proposal for a New Capital Accord, referred to as Basel II (BIS, 1999). The third version of the new system, CP3, was published in April 2003 (BIS, 2003) for a comment period of 3 months, and the New Basel Capital Accord, Basel II, was adopted in June 2004 and should be in force from 2006 (BIS, 2004).

Before looking closer at Basel II and the credit and operational risks, we will look at the 1988 Accord to see the definition of the capital charge for credit risk.

5.1.1   1988 Capital Accord

The Basel Committee on Banking Supervision set up a working committee to "secure international convergence of supervisory regulations governing the capital adequacy of international banks" (BIS, 1988). After a proposal from the committee in December 1987 (BIS, 1987) and a consultative process, the first capital accord, 1988 Accord or Basel I, was adopted by the publication of BIS (1988). The framework was designed to establish minimum levels of capital mainly in relation to credit risk, or the risk of a counterparty failure.

5.1.1.1   *The Capital Base*

The core capital, i.e., equity capital and disclosed reserves, is the only element of capital that is common to all countries' banking systems. For supervisory purposes the capital was defined in two tiers. The first tier (tier 1) consists of the core element comprised of equity capital and disclosed reserves from posttax retained earnings, and the second tier (tier 2) of supplementary capital such as undisclosed reserves, asset revaluation reserves, general provisions/general loan loss reserves, hybrid (debt/equity) capital instruments, and subordinated debt. The sum of tier 1 and tier 2 capital will be eligible for inclusion in the capital base, subject to some limits and restrictions, such as tier 1 capital of >50% of the capital base and tier 2 capital of <50% of the tier 1 capital (see Appendix A in BIS, 1988).

5.1.1.2   *The Minimum Capital Requirement*

The capital base is required to be higher than 8% of a sum of risk-weighted assets or off-balance-sheet exposures. The framework of weights was kept as simple as possible and only five weights were assumed: 0, 10, 20, 50, and 100%. The committee believed that a risk ratio of this type had advantages over simpler gearing approaches, because it provides a fairer basis for making international comparisons between banking systems and it allows off-balance-sheet exposures to be incorporated more easily. The risk ratio,[2] or target standard ratio, can be written as

$$\frac{Capital\ Base}{\sum_j r_j A_j} \geq 0.08 \qquad (5.1)$$

where r_j denotes the risk weights taking on values 0, 0.1, 0.2, 0.5, and 1, and A_j is on-balance-sheet or off-balance-sheet j. The core capital will be at least 4% of the capital base. The minimum capital requirement (MCR) can now be written as

$$MCR = 0.08 \sum_j r_j A_j = \sum_j w_j A_j \qquad (5.2)$$

where $w_j = 0.08 r_j$ is the risk factor applied to asset j. Risk weights and on-balance-sheet assets are given in Table 5.1.

For off-balance-sheet items there are predefined credit conversion factors that would be multiplied by the risk weights applicable to the category of the counterparty for an on-balance-sheet transaction, see BIS (1988, Annex 3). As an example, certain transaction-related contingent items, e.g., performance bonds, bid bonds, warranties, and standby letter of credit related to particular transactions, have a conversion factor of 50%.

The importance of collateral in reducing credit risk is recognized, but it was not possible to develop a basis for recognizing collateral in the weighting system. The limited recognition is only applied to loans secured against cash or against securities issued by OECD central governments and specified multilateral development banks (MDBs). These will have the weight given to the collateral, i.e., zero or low weight. Loans that are partially collateralized by these assets will also have the equivalent low weight on the part of the loan that is collateralized.

The transitional period for the 1988 Accord was 4.5 years. In 1996 the Basel Committee on Banking Supervision published an amendment to the 1988 Accord to incorporated market risks (BIS, 1996). As we are not particularly interested in the market risk,[3] we will only briefly show how it is incorporated in the above model. Corresponding to the risk-weighted assets there is a *measure of market risk* (MMR), as defined in BIS (1996). The MMR is divided by 0.08, or multiplied by 12.5, to ensure consistency in the calculation

---

[2] "The Committee as a whole has not endorsed any precise indicative figure at this stage but the present view of those ten countries wishing to promulgate a figure now as a basis for consultation is that the target standard ratio of capital to weighted risk assets should be 8 per cent (of which the core capital element should be at least 4 per cent)" (BIS, 1987, p. 18).
[3] Of course market risk per se is interesting, but we will handle that in another way in Appendix A.

TABLE 5.1

Risk Weights and On-Balance-Sheet Assets

Risk Weights r	On-Balance-Sheet Assets
0%	Cash (includes gold)
	Claims on sovereigns and their central banks denominated in national currency and funded in that currency
	Other claims on OECD central governments and central banks
	Claims collateralized by cash of OECD central government securities or guaranteed by OECD central governments
0, 10, 20, or 50% at national discretion	Claims on domestic public sector entities, excluding central government, and loans guaranteed by such entities
20%	Claims on multilateral development banks and claims guaranteed by or collateralized by securities issued by such banks
	Claims on banks incorporated in the OECD and loans guaranteed by OECD-incorporated banks
	Claims on banks incorporated in countries outside the OECD with a residual maturity of up to 1 year and loans with a residual maturity of up to 1 year guaranteed by banks incorporated in countries outside the OECD
	Claims on nondomestic OECD public sector entities, excluding central government, and loans guaranteed by such entities
	Cash items in process of collection
50%	Loans fully secured by mortgage on residential property that is or will be occupied by the borrower or that is rented
100%	Claims on the private sector
	Claims on banks incorporated outside the OECD with a residual maturity of over 1 year
	Claims on banks incorporated outside the OECD (unless denominated in national currency and funded in that currency)
	Claims on commercial companies owned by the public sector
	Premises, plant, equipment, and other fixed assets
	Real estate and other investments (including nonconsolidated investment participations in other companies)
	Capital instruments issued by other banks (unless deducted from capital)
	All other assets

Source: BIS, International Convergence of Capital Measurement and Capital Standards, Basel Committee on Banking Supervision, July 1988.

of the capital requirements for the credit and market risks. The risk ratio is now written as

$$\frac{Capital\ Base}{12.5MMR + \sum_{j} r_j A_j} \geq 0.08 \tag{5.3}$$

or if we rewrite this in terms of the MCR we get

$$MCR = MMR + 0.08 \sum_{j} r_j A_j \tag{5.4}$$

A tier 3 capital was also defined to solely support market risks.

5.1.2 Basel II

In June 1999 the Basel Committee on Banking Supervision published a framework report on a new capital accord (BIS, 1999), the first consultative paper (CP1). The new framework consists of three pillars: minimum capital requirement, a supervisory review process, and effective use of market discipline. In the construction of the new capital framework it is important to continue to recognize the *minimum (regulatory) capital requirements*. This is the first pillar. A financial institution's capital adequacy and internal assessment process is the second pillar, and the third is the need for greater market discipline (Figure 5.1).

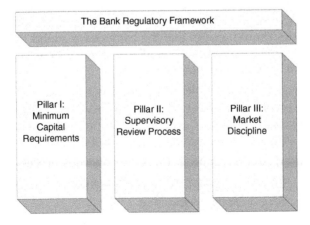

FIGURE 5.1
The three-pillar system of Basel II.

The regulatory framework is based on three mutually reinforcing pillars:
- **Pillar I: Minimum capital requirements**
  - The minimum capital requirements are composed of three fundamental elements: a definition of regulatory capital, risk-weighted assets, and the minimum ratio of capital to risk-weighted assets. Different risk types are proposed as contributing to the capital charge of banks (BIS, 2004, p. 27):
    - Credit risk
    - Market risk
    - Operational risk
    - Liquidity risk
    - Legal risk
  - Operational risk is new. For the credit and operational risks Basel II proposes one or more standardized approaches and also internal models based on the bank's own models; e.g., the credit risk standard approach is supported by external credit assessments, and the internal approach is supported by the bank's internal rating system (explicitly approved by the banking supervisor). We will look closer at credit and operational risks below.
  - The liquidity risk, which is highly correlated to the credit risk, and the legal risk will be addressed in pillar II. Liquidity is vital for an ongoing concern and its capital positions can affect their ability to obtain liquidity, e.g., in a crisis.
  - The supervisory process in pillar II will monitor interest rate risk.
- **Pillar II: Supervisory review process**
  - The second pillar concerns the prudential supervision of banks' capital adequacy and internal risk management. The local authorities should ensure that the banks have sound internal processes in place to take care of all existing and potential risks and capital adequacy requirements.
  - There are four principles of pillar II (cf. also, e.g., PWC, 2003):
    - *Principle 1*: Banks should have a process for assessing their overall capital adequacy in relation to their risk profile and a strategy for maintaining their capital levels.
    - There are five main features of a rigorous process:
      - Board and senior management oversight
      - Sound capital assessment
      - Comprehensive assessment of risks
      - Monitoring and reporting
      - Internal control review

- *Principle 2*: Supervisors should review and evaluate banks' internal capital adequacy assessments and strategies, as well as their ability to monitor and ensure their compliance with regulatory capital ratios. Supervisors should take appropriate supervisory action if they are not satisfied with the result of this process.
- *Principle 3*: Supervisors should expect banks to operate above the minimum regulatory capital ratios and should have the ability to require banks to hold capital in excess of the minimum.
- *Principle 4*: Supervisors should seek to intervene at an early stage to prevent capital from falling below the minimum levels required to support the risk characteristics of a particular bank and should require rapid remedial action if capital is not maintained or restored.

- **Pillar III: Market discipline**
  - The market participants should be able to assess key information about a bank's risk profile and level of capitalization. This procedure will increase disclosure.

It was concluded by the committee that the 1988 Accord risk weighting of assets was resulting, at best, in a crude measure of economic risk, because degrees of credit risk exposure are not calibrated to adequately differentiate between borrowers' differing default risks.

The committee also examined the capital treatment of a number of important *risk mitigation techniques*, as the use of the 1988 Accord discouraged the use of such techniques. Another factor of interest is the *maturity of a claim*. Two more consultative papers were published in 2001 (in fact, 11 different papers) and 2003 (BIS, 2001, 2003), and in June 2004 the new capital accord, Basel II, was published.

All risk categories are of interest for the insurance solvency assessment regime, but credit risk and the new one introduced in Basel II, operational risk, are of special interest.

5.1.2.1 Credit Risk

As we have seen above, the Basel Committee on Banking Supervision recognized that the credit risk measure used was crude in measuring the economic risk as the degrees of credit risk exposure are not calibrated to adequately differentiate between borrowers' risk of default. In the committee's first consultative paper (BIS, 1999), some main and very important issues were discussed. One was the use of *portfolio credit risk models* and another the use of *risk mitigation techniques*. In using internal models, the concepts of probability of default (PD), loss given default (LGD), exposure at default (EAD), and maturity were highlighted in the consultative papers.

A risk mitigation technique is one used by banks to reduce the credit risk on a loan or other exposures. This could be, e.g., taking collateral or obtaining a third-party guarantee.

In the New Basel Capital Accord there are three approaches to measure the minimum capital requirements. The first is the *standardized approach*, which makes use of predetermined risk weights. These risk weights vary with external credit assessments. As a proxy for risk judgment, ratings from of rating agencies are used. In the second approach, the *internal ratings-based approach* (IRB), the proxy for risk judgments, is the bank's internal credit rating. However, other factors, such as the claims severity, will be based on standard factors. In the third approach, the *advanced securitization framework*, the banks can use their own data to determine additional risk components.

5.1.2.1.1 The Standardized Approach

In the first accord of 1988 the banking book exposures were multiplied by constant factors, e.g., 0.50, irrespective of the riskiness of the assets. A change from this old fixed system came with the release of the second consultative papers (package) (BIS, 2001). As an example, loans issued to similar counterparts, e.g., private firms or sovereigns, will require different capital charge depending on their intrinsic risk. These intrinsic risks are evaluated by external rating agencies.

In the 1988 Accord, as seen above, the amount of capital required by a 100-eurocent loan to a private firm was 8 eurocents. In the new system, this 8% is risk weighted depending on the ratings. A high-rated firm will be risk weighted by a factor of 0.20, i.e., $0.20 \times 0.08 = 0.016$, and a low-rated firm by 1.50, i.e., $1.50 \times 0.08 = 0.16$. In the first case we have a decreased risk charge to 1.6 eurocents, and in the last case an increase to 16 eurocents.

Risk parameters of each loan are turned into a capital charge. The risk parameters behind the models are the four mentioned above: PD, LGD, EAD, and maturity.

The final capital charge for the credit risk is the sum of all individual requirements. Subadditivity (see Section 7.4.1) is not admitted.

In Table 5.2 the new risk weights are summarized and compared to the 1988 Accord's weights. The minimum capital requirements for the credit risk can be written as

$$MCR = 0.08 \sum_{j,c} r_{jc} A_{jc} = \sum_{j} w_{jc} A_{jc} \qquad (5.5)$$

where r_{jc} is the risk weight according to Table 5.2[4] and j the asset category and c the rating of the exposure and the risk factors $w_{jc} = 0.08 r_{jc}$.

In Table 5.3 we illustrate the risk factors for rated and unrated corporate claims, including claims on insurance companies. The local supervisor

---

[4] The ratings are only an example and are based on Standard & Poor's.

TABLE 5.2

A Shortened Table with Old and New Risk Weights

On-Balance-Sheet Assets	1988 Accord Risk Weights	Ratings Basel II Risk Weights, r					
		AAA to AA−	A+ to A−	BBB+ to BBB−	BB+ to B−	Below B−	Unrated
Claims on							
Sovereigns	0%	0%	20%	50%	100%	150%	100%
Non-central government PSEs	0–50% on national discretion	On national discretion according to option 1 or 2 for banks					
MDB	20%	On external credit assessments as set out under option 2 for banks					
		Option 1: Risk weight one category less than that assigned on sovereigns					
		AAA to AA−	A+ to A−	BBB+ to BBB−	BB+ to B−	Below B−	Unrated
		20%	50%	100%	100%	150%	100%
		Option 2a: Risk weight based on external credit assessment					
Banks and securities firms	20%	AAA to AA−	A+ to A−	BBB+ to BBB−	BB+ to B−	Below B−	Unrated
		20%	50%	50%	100%	150%	50%
		Option 2b: For short-term claims under option 2					
		AAA to AA−	A+ to A−	BBB+ to BBB−	BB+ to B−	Below B−	Unrated
		20%	20%	20%	50%	150%	20%

-- continued

TABLE 5.2 (CONTINUED)

A Shortened Table with Old and New Risk Weights

On-Balance-Sheet Assets	1988 Accord Risk Weights	Ratings Basel II Risk Weights, r				
		AAA to AA–	A+ to A–	BBB+ to BB–	Below BB–	Unrated
Corporates, including insurance firms	100%	20%	50%	100%	150%	100%
Regulatory retail portfolios	100%	75% and following four criteria: orientation, product, granularity, and low value of individual criteria				
Secured by						
Residential property	50%	35% if it is or will be occupied by the borrower or is rented				
Commercial real estate	100%	100%				
Past-due loans	100%	100–150% depending on specific provisions				
Higher-risk categories	100%	150–350% depending on risk				
Other assets	100%	100%				
Off-balance-sheet assets	CCFs	These items will be converted into credit exposure equivalents through the use of credit conversion factors (CCFs)				

Note: PSE = public sector entities; MDB = multilateral development banks.

Source: BIS, International Convergence of Capital Measurement and Capital Standards, Basel Committee on Banking Supervision, July 1988; BIS, International Convergence of Capital Measurement and Capital Standards, A Revised Framework, Basel Committee on Banking Supervision, Bank for International Settlements, June 2004.

should increase the standard risk weights (see Table 5.2) for unrated claims where they judge that a higher risk weight is warranted by its default experience.

In Gordy (2003) a rigorous presentation of the model used to generate the capital requirements against credit risk is presented. Two quick introductions to the model are given by Finger (2001) and Resti (2002). The internal ratings-based approach is also discussed in Gordy (2004).

5.1.2.1.2 Credit Risk Mitigation

Banks are using different methods to mitigate the credit risk; e.g., exposures may be collateralized by first-priority claims with cash or securities (cf. the 1988 Accord above). A third party may guarantee a loan exposure or a bank may buy a credit derivative to offset a credit risk. If these credit risk mitigation (CRM) techniques meet the requirements for legal certainty, they may be recognized for regulatory capital purposes. The banks may use either a *simple approach*, similar to the one in the 1988 Accord, or a *comprehensive approach*.

In the simple approach the risk weighting of collateral is substituted by the risk weighting of the counterparty for the collateralized portion of the exposure, subject to a 20% floor.

The comprehensive approach allows fuller offset of collateral against exposures by reducing the exposure amount by the value ascribed to the collateral. Using *haircuts*,[5] the banks are required to adjust both the amount of the exposure to the counterparty and the value of any collateral received in support of that counterparty to take account of possible future fluctuations by market movements in the value of either.

TABLE 5.3

Risk Factors for Claims on Corporates, Including Insurance Companies

On-Balance-Sheet Assets	1988 Accord Risk Factor	Ratings Basel II Risk Factors, w				
		AAA to AA-	A+ to A-	BBB+ to BBB-	Below B+	Unrated
Corporates, including insurance firms	0.08	0.016	0.04	0.08	0.12	0.08

Source: BIS, International Convergence of Capital Measurement and Capital Standards , Basel Committee on Banking Supervision, July 1988; BIS, International Convergence of Capital Measurement and Capital Standards, A Revised Framework, Basel Committee on Banking Supervision, Bank for International Settlements, June 2004.

---

[5] Haircuts: Usually a discount on the market value of a bond. It may also be any discount or deduction from the normal value (*Dictionary of International Business Terms*, Financial World Publishing, Canterbury, Kent, UK, 2001), or it could be seen as the difference between the value of a loan and the value of the collateral securing that loan. Generally, a term used for any of a wide variety of spreads and margin.

The bank may use either of the two approaches in the banking book, but only the comprehensive approach in the trading book.

The exposure amount after risk mitigation is under the comprehensive approach, calculated as

$$
\begin{aligned}
A^* &= \max\left\{0,\left[A(1+He)-C(1-Hc-Hfx)\right]\right\} \\
&= \max\left\{0,\left[(A-C)+AHe+CHc+CHfx\right]\right\}
\end{aligned}
\tag{5.6}
$$

where A^* is the exposure after risk mitigation, A the current value of the exposure, He the haircut appropriate to the exposure, C the current value of the collateral received, Hc the haircut appropriate to the collateral, and Hfx the haircut appropriate for currency mismatch between the collateral and exposure.

The exposure amount, after risk mitigation, is multiplied by the risk weight of the counterparty to obtain the risk-weighted assets (RWAs):

$$
MCR = 0.08 \sum_{j,c} r_{jc}^* A_{jc}^* = \sum_{j} w_{jc}^* A_{jc}^*
\tag{5.7}
$$

The asterisks (*) indicate collateralized risk weights (exposure after risk mitigation) or standard risk weights (exposure before risk mitigation).

If a collateral is a basket of assets, the haircut of the basket is a weighted sum of haircuts, with weights equal to the weight of the asset as measured by units of currency. Standard supervisory haircuts are given in Table 5.4. Banks are permitted to use their own internal haircuts on approval of the supervisor.

Adjustments of the haircut for different holding periods are allowed for some transactions, depending on the nature and frequency of the revaluation and remargining provisions. In the framework for collateral haircuts distinctions are made between repo-style transactions (5 business days; daily remargining), secured lending (20 business days; daily revaluation), and other capital market transactions (10 business days; daily remargining). When the frequency of remargining or revaluation is longer than minimum, the haircuts will be scaled up depending on the actual number of business days:

$$
H = H_M \sqrt{\frac{N_R + (T_M - 1)}{T_M}}
$$

where H is the haircut and H_M is the haircut under the minimum holding period. N_R is the actual number of business days between remargining for capital market transactions or revaluation for secured transactions. T_M is the minimum holding period for the type of transaction. If a bank calculates the

TABLE 5.4

Standard Supervisory Haircut

Issue Rating for Debt Securities	Residual Maturity	Sovereigns	Other Issuers
AAA to AA–/A–	≤ 1 year	0.005	0.01
	>1 year, ≤ 5 years	0.02	0.04
	>5 years	0.04	0.08
A+ to BBB– unrated bank securities	≤ 1 year	0.01	0.02
	>1 year, ≤ 5 years	0.03	0.06
	>5 years	0.06	0.12
BB+ to BB–	All	0.15	
Main index equities (including convertible bonds) and gold		0.15	
Other equities (including convertible bonds) listed on a recognized exchange		0.25	
UCITS/mutual funds		Highest haircut applicable to any security in which the fund can reinvest	
Cash in the same currency		0	

Note: UCITS = undertakings for collective investments in transferable securities

Source: BIS, International Convergence of Capital Measurement and Capital Standards, A Revised Framework, Basel Committee on Banking Supervision, Bank for International Settlements, June 2004, p. 33.

volatility on a T_N day holding period that differs from the minimum holding period T_M, then the haircut under the minimum holding period is

$$H_M = H_N \sqrt{\frac{T_M}{T_N}}$$

H_N is the haircut based on the holding period T_N.

Effects of bilateral netting arrangements covering repo-style transactions are recognized on a counterparty-by-counterparty basis. Equation 5.6 is used to calculate the capital requirements for transactions with netting agreements:

$$A^* = \max\left\{0, \left[\left(\sum A - \sum C\right) + \sum AsHs + \sum AfxHfx\right]\right\} \qquad (5.8)$$

where A^* is the exposure value after risk mitigation, A the exposure before, C the value of the collateral received, As the absolute value of the net position in a given security, Hs the haircut appropriate to As, Afx the absolute value of the net position in a currency different from the settlement currency, and *Hfx* the corresponding haircut.

As an alternative to standard (or own estimate) haircuts, banks are allowed to use value-at-risk (VaR)[6] models to reflect price volatility of the exposure and collateral for repo-style transactions. The approach applies to repo-style transactions covered by bilateral netting arrangements on a counterparty-by-counterparty basis. It is only banks that have received supervisory recognition for an internal market risk model that can use the VaR approach. The exposure value of the risk mitigation is calculated as

$$A^* = \max\left\{0, \left[\left(\sum A - \sum C\right) + VaR\text{-}output \times multiplier\right]\right\} \qquad (5.9)$$

VaR-output is the value-at-risk output from the internal market risk model. And according to paragraph 180 in BIS (2004), the multiplier ranges from 1 in a green zone, to 1.13 to 1.28 in a yellow zone, to 1.33 in a red zone.

For more details, see BIS (2004, Part II).

The internal ratings-based approach (IRB) is outlined in BIS (2004, Part III).

5.1.2.2   Operational Risk

Operational risk is defined as the risk of loss resulting from inadequate or failed internal processes, people, and systems or from external events.

The Basel II framework presents three methods for calculating operational risk capital charges in a continuum of increasing sophistication and risk sensitivity:

1. The basic indicator approach (BIA)
2. The standardized approach (SA)
3. Advanced measurement approaches (AMA)

The banks are encouraged to move along the spectrum of available approaches as they develop more sophisticated operational risk measurement systems and practices. Basel II also defines qualifying criteria for the standardized and advanced measurement approaches. A bank will not be allowed to choose to revert to a simpler approach without supervisory approval once it has been approved for a more advanced approach.

5.1.2.2.1   The Basic Indicator Approach

Banks using the basic indicator approach must hold capital for operational risk equal to a fixed percentage (denoted α) of the average annual gross

---

[6] See Section 7.4.2.

income over the previous 3 years. The charge may be expressed as follows (BIS, 2004, p. 137):

$$K_{BIA} = \frac{\alpha}{n^+} \sum_{i=1}^{3} GI_i^+$$

where K_{BIA} is the capital charge under the basic indicator approach, GI_i^+ is the annual, positive gross income over the previous 3 years, n^+ is the number of previous years for which the gross income is positive, and α is 15%, which is set by the committee.

Gross income is defined as net interest income plus net noninterest income. It is intended that this measure should:

- Be gross of any provisions (e.g., for unpaid interest)
- Be gross of operating expenses, including fees paid to outsourcing service providers
- Exclude realized profits/losses from the sale of securities in the banking book
- Exclude extraordinary or irregular items as well as income derived from insurance

5.1.2.2.2 *The Standardized Approach*

A bank's activities are divided into eight business lines in the SA: corporate finance, trading and sales, retail banking, commercial banking, payment and settlement, agency services, asset management, and retail brokerage. The business lines are defined in detail in BIS (2004).

The gross income is an indicator that serves as a proxy for the scale of business operations (within each line of business).

The capital charge for each business line is calculated by multiplying gross income by a factor β assigned to that business line.

The beta serves as a proxy for the industry-wide relationship between the operational risk loss experience for a given business line and the aggregate level of gross income for that business line.

The total capital charge is calculated as the 3-year average of the regulatory capital charges across each of the business lines (the positive part of each line). The total capital charge may be expressed as

$$K_{SA} = \frac{1}{3} \sum_{i=1}^{3} \max \left[\sum_{j=1}^{8} \left(GI_j \times \beta_j \right), 0 \right]$$

where K_{SA} is the capital charge under the standardized approach; GI_j is the annual gross income in a given year, as defined above in the basic indicator approach, for each of the eight business lines ($j = 1, ..., 8$) (see below); and β_j is a fixed percentage, set by the committee, relating the level of required capital to the level of the gross income for each of the eight business lines ($j = 1, ..., 8$) (see below).

Business Lines	Beta Factor
1. Corporate finance	0.18
2. Trading and sales	0.18
3. Retail banking	0.12
4. Commercial banking	0.15
5. Payment and settlement	0.18
6. Agency services	0.15
7. Asset management	0.12
8. Retail brokerage	0.12

There is also an alternative standardized approach (ASA) that can be used if the bank can prove to the supervisory authority that this approach is superior to the ordinary SA (BIS, 2004, p. 139, footnote 97).

5.1.2.2.3 Advanced Measurement Approaches

Under the advanced measurement approaches, the regulatory capital requirement will be equal to the risk measure generated by the bank's internal operational risk measurement system using the quantitative and qualitative criteria for the AMA discussed in the Basel Accord (BIS, 2004, p. 140).

5.2 IASB: Toward a New Accounting System

In 1997 the International Accounting Standards Committee (IASC) started a project with the objective of developing an International Accounting Standard (IAS) for insurance. Later IASC was transformed into a board, the International Accounting Standards Board (IASB). New IASs are now described as International Financial Reporting Standards (IFRS).

At the start of the project, there was a wide range of accounting standards used by insurers, and these standards often differed from accounting standards for other enterprises in the same country. The objective of the project was "to produce a single set of high quality, understandable and enforceable global accounting standards that require high quality, transparent and comparable information in financial statements."

In the first stage of the IASC's project a committee produced two volumes of issues papers in December 1999 (IASC, 1999). The papers got *general support* from the actuarial profession, from the U.K., Australia, Canada, and parts of Scandinavia, and *general opposition* from the industry and regulators in the rest of Europe, the U.S., and Japan. One of the most substantial responses came from the International Actuarial Association (IAA, 2000). IASB had earlier started to develop a new standard for financial instruments, and this influenced the work on insurance standards. As it was expected that the valuation of financial instruments should be based on *fair value*, it was assumed that the insurance standards should also be based on the same valuation approach (see Chapter 8). The fair value concept used for insurance contracts should be consistent with accounting principles for other financial sectors, such as banking and the securities industry, but since these are not at the moment in a position to move to fair value, other valuation concepts have been proposed.

At its meeting on July 17, 2000, the European Council of Finance Ministers endorsed the commission's proposal from June of the same year that all listed EU companies, including insurance undertakings, should prepare their consolidated accounts in accordance with the IAS by 2005. The IAS regulation came into force July 19, 2002 (COM, 2002e). *The introduction of the new accounting system for the insurance industry is much more than a technical issue and will result in fundamental changes to the way the industry reports and does business.*

A steering committee, established by the IASC, finalized a report to the new IASB based on the comments on the issues papers: a Draft Statement of Principles (DSOP) for an IAS for insurance (IASB, 2001). The steering committee observed at an early stage that insurance contracts are not traded in a deep, liquid market as other assets (or liabilities). Hence, the determination of fair value of insurance liabilities gives rise to difficult conceptual and practical issues. The view of the steering committee was that a fair value approach should be based on assumptions that an independent marketplace participant would make in determining the *charge* it would make to acquire the liability. The steering committee also discussed a *nonfair valuation* based on a company's own assumptions and expectations. This approach is referred to as the *entity-specific valuation*.

It was required that an asset and liability reporting approach, a *full balance sheet approach*, was to be used, instead of a deferral and matching reporting approach (Abbink et al., 2002). The reasons for this were that an asset and liability approach will:

- Provide greater transparency
- Produce accounts that are more understandable
- Make it easier for users to make comparisons between different sets of accounts

> The new accounting standard applies to insurance contracts (a functional view) and not to insurance companies (an institutional view).

A contract should be unbundled into its insurance and noninsurance (financial or service) parts, and the accounting will be according to three different standards: IAS 18 for service contracts, IAS 39 for financial instruments, and IFRS 4 for insurance contracts. The definition of insurance contracts is crucial.

In May 2002 IASB decided to split the work on standards for insurance accounting into two phases. Phase I is an interim measure, and one of the main objectives for the first phase was the definition of insurance contracts, as opposed to financial instruments. Full implementation will occur with its phase II.

In 2003 IASB produced an exposure draft on insurance contracts, ED 5 (IASB, 2003), and with its publication "IFRS 4 Insurance Contracts" in March 2004, IASB ended phase I (IASB, 2004).

In IASB (2004, "Basis for Conclusions") tentative conclusions for phase II are given (BC = "Basis for Conclusions"):

BC6

The Board sees phase I as a stepping stone to phase II and is committed to completing phase II without delay once it has investigated all relevant conceptual and practical questions and completed its due process. In January 2003, the Board reached the following tentative conclusions for phase II:

(a) The approach should be an asset-and-liability approach that would require an entity to identify and measure directly the contractual rights and obligations arising from insurance contracts, rather than create deferrals of inflows and outflows.

(b) Assets and liabilities arising from insurance contracts should be measured at their fair value, with the following two caveats:

(i) Recognising the lack of market transactions, an entity may use entity-specific assumptions and information when market-based information is not available without undue cost and effort.

(ii) In the absence of market evidence to the contrary, the estimated fair value of an insurance liability shall not be less, but may be more, than the entity would charge to accept new contracts with identical contractual terms and remaining maturity from new policyholders. It follows that an insurer would not recognise a net gain at inception of an insurance contract, unless such market evidence is available.

(c) As implied by the definition of fair value:

(i) An undiscounted measure is inconsistent with fair value.

(ii) Expectations about the performance of assets should not be incorporated into the measurement of an insurance contract, directly or indirectly (unless the amounts payable to a policyholder depend on the performance of specific assets).

(iii) The measurement of fair value should include an adjustment for the premium that marketplace participants would demand for risks and mark-up in addition to the expected cash flows.

(iv) Fair value measurement of an insurance contract should reflect the credit characteristics of that contract, including the effect of policyholder protections and insurance provided by governmental bodies or other guarantors.

(d) The measurement of contractual rights and obligations associated with the closed book of insurance contracts should include future premiums specified in the contracts (and claims, benefits, expenses, and other additional cash flows resulting from those premiums) if, and only if:

(i) Policyholders hold non-cancellable continuation or renewal rights that significantly constrain the insurer's ability to re-price the contract to rates that would apply for new policyholders whose characteristics are similar to those of the existing policyholders; and

(ii) Those rights will lapse if the policyholders stop paying premiums.

(e) Acquisition costs should be recognized as an expense when incurred.

(f) The Board will consider two more questions later in phase II:

(i) Should the measurement model unbundle the individual elements of an insurance contract and measure them individually?

(ii) How should an insurer measure its liability to holders of participating contracts?

From January 1, 2005, quoted companies in the European Community were required to report according to the new international accounting standards.

5.3 IAIS: Insurance Principles and Guidelines

The International Association of Insurance Supervisors (IAIS) has as one of its current initiatives the development of a global framework for insurer capital requirements. IAIS is publishing principles, standards, and guidances for supervisory authorities. The proposals and guidances should hold in all jurisdictions. That means that the text in some cases can be very general.

The IAIS has not fully adopted the three-pillar system of the Basel II project (see Section 5.1), but instead has created a new supervisory framework, which is compatible with the Basel II system. A draft of this framework was released for consultation in October 2004 (IAIS, 2004c). The framework includes three levels reflecting three different responsibilities, but also three blocks of issues: financial issues, governance issues, and market conduct issues (Figure 5.2).

There are two sets of basic conditions that have to be in place before an effective insurance supervision is functioning (level 1: preconditions). They are the environments for the supervisory authority and for the insurance industry, but also the insurance supervision. Level 2 is the regulatory requirements. It consists of three blocks of issues:

- *The financial block*: Pertains to the field of solvency and capital adequacy, i.e., valuation and adequacy of the technical provisions, forms of capital, investments, and financial reporting and disclosure.

- *The governance block*: Refers to governance processes and controls, fit and proper testing (directors and management), internal controls such as risk management, shareholder relationships, and governance risks posed by group structures.

- *The market conduct block*: Includes areas such as dealing with customers in selling and handling insurance policies, and disclosure of relevant information both to the market and to policyholders.

At the third level, the supervisory action process, it is understood that the supervisors should have access to the insurer's risk profile, controls, and available support. The common structure and standards for the assessment of insurer solvency are an important part of the framework.

In February 2005 IAIS released a draft paper on eight key elements or *cornerstones* for the formulation of regulatory financial requirements (IAIS, 2005). The paper was mainly built on the framework paper (IAIS, 2004c) and outlines a more precise view on some issues and acts as a conceptual guide rail for further work.

In its "Insurance Core Principles" (IAIS, 2003c) capital adequacy is described as an area that has to be addressed in legislation or regulation. In January 2002 (IAIS, 2002) a principles paper on capital adequacy and solvency was published (see below). At the same time, IAIS posed to the International Actuarial Association (IAA) a question about conducting an investigation into solvency assessment from an actuarial perspective (see Section 5.4). In 2003 a guidance paper on solvency control levels was published (IAIS, 2003b).

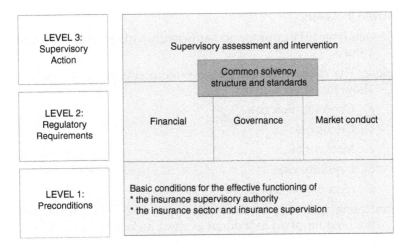

FIGURE 5.2
Outline of the framework for insurance supervision. (From IAIS, A New Framework for Insurance Supervision, Towards a Common Structure and Common Standards for the Assessment of Insurer Solvency, IAIS, October 2004c.)

In the "Road Map for the Development of Future Work" (MARKT, 2004c), the European Commission seeks advice from the new European Supervisory Committee, CEIOPS. Among the references in this road map are IAIS's insurance core principles.

In Appendix E there is a list of the principles, standards, and guidances that have been published by IAIS.[7]

As the principles on capital adequacy and solvency (principle 5) are of interest for the solvency assessment framework, we will list the 14 principles and the short comments given by IAIS. To each principle there is a general discussion and also suggestions. We use the abbreviation CASR for *capital adequacy and solvency regime* in the text by IAIS (it is our notation).

"An insurance company is solvent if it is able to fulfil its obligations under all contracts under all reasonable foreseeable circumstances. Insurance regulatory authorities require insurers to maintain assets or surplus capital in excess of liabilities, that is, a solvency margin."

Guidances have been or will be published for each of the principles.

Principle 1: Technical provisions

> "Technical provisions of an insurer have to be adequate, reliable, objective and allow comparison across insurers."

Principle 2: Other liabilities

> "Adequate provisions must be made for all other liabilities insofar as they are not included in the technical provisions."

---

[7] The published principles and guidelines can be found at www.iaisweb.org.

Principle 3: Assets

"Assets have to be appropriate, sufficiently realisable and objectively valued."

Note: The framework may impose requirements on the assets to allow for items such as:

a. Concentration risk

b. Credit risk

c. Market risk

d. Liquidity risk

e. Liquidation risk

In dealing with concentration risk, supervisory rules may prohibit the holding of an individual asset or a class of assets in excess of a certain level. If concentrated holdings are permitted, the capital and solvency regime should ensure that only a certain level of this holding is counted for in the CASR.

Principle 4: Matching

"CASRs have to address the matching of assets with liabilities."

Note: The CASR should address the risk of loss arising from mismatch of the assets and liabilities in the following:

a. Currency

b. Timing of cash flows

c. Amount of cash flows

Principle 5: Absorption of losses

"Capital requirements are needed to absorb losses that can occur from technical and other risks."

Note: The undertakings must be able to evaluate the risks that they underwrite and to establish an adequate level of premiums. Among the risks that the CASR should include are:

a. Other current technical risks (e.g., deviation risk, risk of error, evaluation risk, reinsurance risk, operating expenses risk, and risks associated with major or catastrophic losses)

b. Special technical risks (e.g., liquidation risk and the risk of excessive or uncoordinated growth)

c. Operational, market, organizational, and conglomerate risks

d. Investment risk (e.g., depreciation, liquidity, matching, interest rate, evaluation, and participation risks)

Principle 6: Sensitivity to risk

"CASRs have to be sensitive to risk."

Note: The CASR comprises the:

a. Valuation of liabilities (including the technical provisions)

b. Requirements on assets (including requirements for valuation of assets)

c. Definition of appropriate forms of capital

d. Required solvency margin

Principle 7: Control level

"A control level is required."

Note: See Guidance 6.

The authorities have to establish a control level, or a series of control levels, that triggers intervention by the authority when the available solvency falls below this control level. The level should be set sufficiently high to allow an intervention at an early stage.

Principle 8: Minimum capital

"A minimum level of capital has to be specified."

Principle 9: Definition of capital

"CASRs have to define the suitable form of capital."

Note: In determining the form of suitable capital, insurance regulators should consider the extent to which the capital element:

a. Represents a permanent and unrestricted investment fund

b. Is freely available to absorb losses

c. Does not impose any unavoidable charges on the earnings of the insurer

d. Ranks below the claims of policyholders and other creditors in the event of the insurer being wound up

Principle 10: Risk management

"CASRs have to be supplemented by risk management systems."

Note: The required solvency margin should be considered as the last resort after all other measures taken by the insurer to secure its financial stability have failed. The insurer must have risk management systems in place to be comprehensive and cover all risks to which the insurer is exposed.

Principle 11: Allowance for reinsurance

"Any allowance for reinsurance in a CASR should consider the effectiveness of the risk transfer and make allowance for likely security of the reinsurance counterparty."

Note: Reinsurance arrangements are a primary tool for risk transfer.

Principle 12: Disclosure

"The CASR should be supported by appropriate disclosure."

Principle 13: Solvency assessment

"Insurance supervisory authorities have to undertake solvency assessment."

Principle 14: Double gearing

"CASRs have to address double gearing and other issues that arise as a result of membership in a group."

5.3.1   Guidance 6: Solvency Control Levels Guidance Paper

This guidance paper addresses sub-principle 7 of principle 5 as a component of a solvency regime that it is assumed satisfies principle 6 and 8. Guidance with respect to principle 6 and 8 will be the subject of future papers.

The purpose of the guidance paper is to discuss the importance of setting solvency control levels, identifying key factors in setting the levels, and discussing possible supervisory actions when the control level is breached.

Three general issues that supervisors should take into consideration when setting a solvency control level are risk coverage issues, supervisory and jurisdictional issues, and other issues. These are discussed in the guidance.

5.4   IAA: A Global Framework for Solvency Assessment

In early 2002 the IAIS addressed to the International Actuarial Association (IAA) a question about making an investigation on solvency assessment from an actuarial perspective. IAA's Insurance Regulation Committee formed a working group, the Insurer Solvency Assessment Working Party (WP), to prepare a report. The report was published in May 2004 (IAA, 2004). The IAA is also assisting the IASB in determining a standard approach to actuarial principles and methods for the determination of actuarial reserves in accordance with the new accounting standard.

5.4.1   Short Summary

The short summary given here is mainly based on the "blue book's" executive summary (IAA, 2004, chap. 2). We will also look closer at the WP's classification of risks.

The WP approved the Basel II approach of a three-pillar regime for a successful implementation of the global framework proposed in the report (cf. Section 5.1). The *three-pillar approach*, similar to that used by banks, is seen as extremely useful given the common features shared by the two financial sectors and that many insurance supervisors are part of an integrated financial supervisory authority and are well acquainted with the Basel Capital Accord.

The WP believes that a proper assessment of an insurer's true financial strength for solvency purposes requires appraisal of its *full balance sheet*, depending on realistic values and consistent treatment of both assets and liabilities that does not generate hidden surplus or deficit.

Capital requirements, by themselves, cannot prevent failures. Therefore, the WP considered the role of rating agencies in assessing insurance undertakings, but also the link between the degree of protection and time horizon.

In determining the current financial position of an undertaking, a reasonable *time horizon* is about 1 year. It should not be confused with the need to consider the full term of all of the assets and obligations of the insurer. There must be a high level of confidence, say, 99%, of meeting all obligations for the 1-year time horizon as well as the present value at the end of the time horizon of the remaining future obligations (e.g., best estimate value with a moderate level of confidence such as 75%). For a long-term risk, the valuation for its lifetime could be made using a series of consecutive 1-year tests with a very high level of confidence (99%), or with a single equivalent with a lower level of confidence for the entire assessment time horizon (e.g., 90 or 95%).

It was recommended, in principle, that all type of risks are considered (both implicitly and explicitly). Some could be hard to quantify, and therefore could only be supervised under pillar II. The types of risks that ought to be addressed under pillar I are:

a. Underwriting risk
b. Credit risk
c. Market risk
d. Operational risk

That is, these are the insurance-specific underwriting risk and the three risk categories proposed in Basel II (see Section 5.1).

A satisfactory *risk measure* for solvency assessment will exhibit a variety of desirable properties, such as consistency. It is difficult for one risk measure to convey all the information needed for a particular risk. One risk measure that exhibits several desirable properties for many risks is the *tail value at risk*[8] (see Section 9.1). The TailVaR is often more suited to insurance than value at risk (VaR), as the risk event distribution is often skewed.

The solvency assessment method should recognize the impact of *risk dependencies*, *concentration*, and *diversification*. The tail dependency could be assessed by techniques like copulas, which is discussed in the report's Appendix H.

---

[8] Also called TVaR, TailVar, conditional tail expectation, CTE, or policyholder's expected shortfall.

The impact of risk transfer or risk-sharing mechanisms used by the insurer should be recognized by the solvency assessment method. The ways that an insurer can *manage its risk*, beyond prudential claim management, include risk reduction, integration, diversification, hedging, transfer, and disclosure. Stress testing, which can provide significant insight for company management into the impact of the risks they are faced with, is a part of the risk management.

The report discusses and presents suggestions on both standardized approaches (Chapter 6) and more company-specific advanced approaches (Chapter 7).

The WP sets out a framework for *capital requirements* and risk oversight that could be applied in almost all jurisdictions to suit the circumstances of each jurisdiction. In the development of capital requirement, it is desirable to consider not only the *target capital requirement* (TCR), but also a *minimum capital requirement* (MCR). The TCR refers to the appropriate amount of capital to be held in consideration of the risks assumed by the insurer, and MCR is a final threshold requiring maximum supervisory measures in the event that it is not fulfilled. The WP uses the term *free surplus* (FS) for the assets exceeding those needed to cover the liabilities and the TCR requirement.

Regulatory capital (i.e., the TCR) has aspects of both the *going concern* and *run-off* situation. The TCR can neither be seen strictly on a going-concern basis nor on a run-off basis.

5.4.2   Risk Categories

In modeling risks, special attention should be paid to the following key components of risk:

- *Volatility*: This is the risk of random fluctuation in either the frequency or the severity of a contingent event. This risk is *diversifiable*; i.e., the relative volatility declines as the number of independent insured risks increases.

- *Uncertainty*: This is the risk that the models used to estimate the claims or other processes are misspecified (model risk) or that the parameters within the models are misestimated (parameter risk). The structural risk includes changes in the parameter structure over time or for other reasons.

- *Extreme events*: They can be described as high-impact, low-frequency events for a company as a whole. One or more extreme events can cause fluctuations much greater than can be expected to arise from normal volatility or uncertainty fluctuations. The extreme events need special consideration.

The WP suggested (IAA, 2004, chap. 5) that insurer risk could be categorized under four major headings, each containing several more specific risks:

1. *Underwriting*: Specific insurance risks through the insurance contracts it sells
2. *Credit* (see Appendix E in IAA, 2004): Generally present in other financial institutions, e.g., change in the credit quality of issuers of securities
3. *Market* (see Appendix D in IAA, 2004): Generally present in other financial institutions, e.g., changes in the values of invested assets
4. *Operational*: Includes both internal and external nonunderwriting risks

It was also recommended that other risks, such as strategic risk and liquidity risk, should be examined within pillar II and the supervisory review.

5.4.2.1   Underwriting Risk

Risks, headed by the underwriting risk, are associated both with the perils covered by the specific line of business (fire, motor, accident, liability, death, windstorm, etc.) and with the specific processes associated with the conduct of the insurance business. Appendices A, B, and C of IAA (2004) provide detailed descriptions of the considerations involved in assessing underwriting risk in life, non-life, and health insurance. This is done by case studies.

The more generic risks considered by the WP are the following (cf. IAA, 2004, p. 29):

- *Underwriting process risk*: The risk from exposure to financial losses related to the selection and approval of risks to be insured.
- *Pricing risk*: The risk that the prices charged by the company for insurance contracts will ultimately be inadequate to support the future obligations arising from those contracts.
- *Product design risk*: The risk that the company faces exposure under its insurance contracts that was unanticipated in the design and pricing of the insurance contract.
- *Claims risk* (for each peril): The risk that many more claims occur than expected or that some claims that occur are much larger than expected, resulting in unexpected losses. This includes both the risk that a claim may occur and the risk that the cost of the claim might develop adversely after it occurs.
- *Economic environment risk*: The risk that social conditions will change in a manner that has an adverse effect on the company.

- *Net retention risk*: The risk that higher retention of insurance loss exposures results in losses due to catastrophic or concentrated claims experience.

- *Policyholder behavior risk*: The risk that the insurance company's policyholders will act in ways that are unanticipated and have an adverse effect on the company.

- *Reserving risk*: The risk that the provisions held in the insurer's financial statements for its policyholder obligations (also claim liabilities, loss reserves, or technical provisions) will prove to be inadequate.

Among the special considerations in life insurance is the importance of modeling the mortality, lapse, and expense risks, but also the long-term nature of the majority of the business, as the importance of, e.g., participating or with-profits policies, etc.

Mortality risk is divided into the key components discussed above (IAA, 2004, Section 6.3.1): volatility, by comparing normal power (NP) approximation and simulations, catastrophe risks (severe epidemic, natural catastrophes, and terrorist attack), and level and trend uncertainty.

5.4.2.2   Credit Risk (see also IAA, 2004, Appendix E)

Credit risk is the risk of default and change in the credit quality of issuers of securities, counterparties, and intermediaries, to which the company has an exposure. The WP includes the following risks (IAA, 2004, pp. 29–30):

- *Direct default risk*: The risk that a firm will not receive the cash flows or assets to which it is entitled because a party with which the firm has a bilateral contract defaults on one or more obligations.

- *Downgrade or migration risk*: The risk that changes in the possibility of a future default by a debtor will adversely affect the present value of the contract with the debtor today.

- *Indirect credit or spread risk*: The risk due to market perception of increased risk (i.e., perhaps because of the business cycle or perceived credit worthiness in relation to other market participants).

- *Settlement risk*: The risk arising from the lag between the value and settlement dates of securities transactions.

- *Sovereign risk*: The risk of exposure to losses due to the decreasing value of foreign assets or increasing value of obligations denominated in foreign currencies.

- *Concentration risk*: The risk of increased exposure to losses due to concentration of investments in a geographical area or other economic sector.

- *Counterparty risk*: The risk of changes in values of reinsurance, contingent assets, and liabilities (i.e., such as swaps that are not otherwise reflected in the balance sheet).

The credit risks are separated into type A and type B risks (IAA, 2004, Section 6.6):

- *Type A risk*: Risk relating to actual assets held and the insurer's ability to manage its credit loss position. In capturing the type A risk the WP recommends that the Basel II requirements for credit risk should be considered under a standard approach.
- *Type B risk*: Credit risk involved with future reinvested assets. The development of standard approaches for capturing type B risks is difficult, and the WP recommends that the supervisor should encourage companies to develop internal models. The WP also proposes what can be included in standard approaches; e.g., if it is possible to directly compute the present value of future liability cash flows, provision for type B credit risk can be made directly through use of credit risk spread.

5.4.2.3   Market Risk (see also IAA, 2004, Appendix D)

Market risks come from the level or volatility of market prices of assets. They involve the exposure to movements in the level of financial variables, e.g., stock prices, interest rates, exchange rates, and commodity prices. The WP considered the following risks (IAA, 2004, p. 31):

- *Interest rate risk*: The risk of exposure to losses resulting from fluctuations in interest rates.
- *Equity and property risk*: The risk of exposure to losses resulting from fluctuation of market values of equities and other assets.
- *Currency risk*: The risk that relative changes in currency values decrease values of foreign assets or increase the value of obligations denominated in foreign currencies.
- *Basis risk*: The risk that yields on instruments of varying credit quality, liquidity, and maturity do not move together, thus exposing the company to market value variation that is independent of liability values.
- *Reinvestment risk*: The risk that the returns on funds to be reinvested will fall below anticipated levels.
- *Concentration risk*: The risk of increased exposure to losses due to concentration of investments in a geographical area or other economic sector.

- *Asset/liability management risk*: The risk that the insurer fails to manage its net asset liability management (ALM) position.
- *Off-balance-sheet risk*: The risk of changes in values of contingent assets and liabilities, such as swaps that are not otherwise reflected in the balance sheet.

Market risk can only be measured appropriately if the market values of assets and liabilities are measured adequately. Because of the lack of real market values of insurance liabilities, the concept of *replicating portfolio* may be useful in this context.

The market risks are also separated into type A and type B risks (IAA, 2004, Section 6.7):

- *Type A risk*: Some portions of the liabilities may have durations (e.g., terms) comparable to readily available liquid assets. Then it is possible to match these liabilities with assets; i.e., a replicating portfolio of assets is available in the market. In this case market risk will be focused on the volatility of the market value of the actual assets held and the market value of the replicating portfolio of assets and the ability of the insurer to manage the volatility.
- *Type B risk*: The long-term duration of some liabilities requires consideration of long-term reinvestment, as there will be no feasible replicating portfolios. Measuring these risks will include considerable uncertainty about the replicating portfolio and the manner of its reinvestment to mature the underlying cash flows. Long-term contracts can contain complex options or guarantee schemes for which replicating portfolios do not exist.

5.4.2.4 Operational Risk

This risk has primarily emerged from the banking industry and was initially defined as "all risks other than market or credit." Operational risk for capital purposes is now defined as "the risk of loss resulting from inadequate or failed internal processes, people, systems or from external events." (See also Section 5.1.)

The WP recommends that operational risk should be addressed in pillar I. It may be reasonable to offer a Basel II type of approach (see Section 5.1), but the WP recommends that the supervisors, the insurance industry, and the actuarial profession work together to develop methods to measure operational risk.

5.4.2.5 Liquidity Risk

Liquidity risk is exposure to loss due to insufficient liquid assets being available. This can occur when a company has to borrow unexpectedly or sell assets for an unanticipated low price. The WP goes on and defines possible sources of liquidity. The WP recommends that the risk will be subject to pillar II assessment, as it is triggered by events that are difficult to predict.

5.5 EU: Solvency II — Phase I

The EU working document (EC, 1999) sketched out some further reflections that could be considered. These reflections later became the Solvency II project. At the 23rd meeting of the Insurance Committee (now EIOPC), it was agreed that a more fundamental and wide-ranging review of the overall financial position of an insurance undertaking should be done. This review should also include risk classes that have not been considered yet, i.e., investment risk (market and credit), operational risk, liquidity risk, and ALM risk.

In a paper entitled "The Review of the Overall Financial Position of an Insurance Undertaking (Solvency II Review)" (MARKT, 1999), the commission services outlined the new project for the first time. They found six key aspects to discuss further:

1. *Technical provisions*: There is inadequate harmonization of non-life technical provisions. The Manghetti group of the Insurance Committee (IC) had a crucial role to play in harmonizing the provisions (Manghetti report, 2001). Subjects to discuss were included discounting the provisions, setting up equalization reserves, and incurred but not reported (IBNR)/incurred by not enough reported (IBNER) reserves. The situation seemed more satisfactory in life insurance. A description of the reserving approaches made in nine EU countries is given in Wolthuis et al. (1997).

2. *Assets/investment risk*: Many member states are using stress tests (or resilience tests). Also, the Müller report identified investment risk is important. One reinsurer had indicated that 60 to 85% of the variation in the financial results could be attributed to volatility in investments.

   Another issue to discuss is asset valuation rules. Increasing transparency could diminish prudence.

3. *Asset liability management*: Certain ALM features are already established in the existing rules. It would be interesting to examine how alternative risk transfer (ART),[9] derivative products, catastrophe bonds, etc., could be used to control and transfer risks. Liquidity requirements are also of interest.

---

[9] The commission services presented a study on ART conducted by Tillinghast Towers-Perrin (EC, 2000).

4. *Reinsurance*: No explicit account is taken of the difference in the quality of reinsurance arrangements. A report on financial interest and ART has been commissioned, and this result and a more general study on a possible EU framework for reinsurance supervision are of vital importance.

5. *Solvency margin requirement — methodology*: A methodology that more accurately reflects the true risks encountered is of interest. It should remain simple, feasible, robust, and transparent, but also reflect the reduction in risk due to portfolio diversification.

   It would be of interest to review risk models in other jurisdictions and also investigate risk capital requirements in other financial sectors. This is essential for financial conglomerates.

6. *Accounting system*: The accounting system links the previous five key factors. The accounting liabilities can have fiscal repercussions. IASC (now IASB) has started its insurance accounting project (see Section 5.2). Relevant issues are the focus on insurance contracts (and not undertakings) and the approach to valuing liabilities and assets.

At the end of the 1990s the Insurance Conference (now CEIOPS) set up a working group to continue the work done by the Angerer group (cf. Section 3.2) on technical provisions in non-life insurance. The group, which was chaired by Mr. Giovanni Manghetti, was focused on the concept of ultimate cost[10] and discounting. They presented a report in 2001: the Manghetti report (2001).

The commission services also discuss several external sources, such as IASB, IAIS, Comité Européen des Assurances (CEA), etc. At the end of December 2000 the commission contracted KPMG Deutsche Treuhand Gesellschaft (hereafter KPMG) to conduct a study on background knowledge and updates on market developments in relation to its Solvency II project. A report was published in May 2002 (KPMG, 2002; also see Section 5.5.3).

In May 2001 it was decided to split the project into two phases. The first phase, which ended 2002–2003, should produce a decision on the general design of the system, and the second, more technical phase should focus on developing detailed rules; see Chapter 13, where a summary of the work done is given.

During 2001 the commission services produced three notes to the Solvency subcommittee on (1) a presentation of the proposed work (MARKT, 2001a), (2) banking rules' relevance for the insurance sector (MARKT, 2001b), and (3) a risk-based capital system (MARKT, 2001c). In the notes different questions were asked; e.g., if internal models and risk-based capital (RBC) systems are to be used, are the banking rules (Basel system) and the corresponding RBC system of interest? The services also summa-

---

[10] See Article 28 of the IAD (EEC, 1991).

rized the Basel project and the RBC systems of the U.S., Australia, and Canada.

The outline of proposed work in MARKT (2001a) is summarized by headlines below. The commission papers on different subjects are given within parentheses.

1. Qualitative criteria for a solvency system
2. Risk modeling adapted to each firm (internal models) (MARKT, 2002a)
3. RBC solvency systems (MARKT, 2001c)
4. European solvency system and supplementary rules
5. Solo solvency requirements and insurance groups
6. Study of the banking prudential system (MARKT, 2001b)
7. Accounting developments
8. Study of the risks to be taken into account
9. Non-life technical provisions (MARKT, 2002f; see Section 5.5.5)
10. Life technical provisions (MARKT, 2002e; see Section 5.5.4)
11. Rules on the admissibility of assets covering technical provisions
12. Investment risk
13. Asset–liability management (life) (MARKT, 2002e; see Section 5.5.4)
14. Reinsurance
15. Qualitative rules

In its review of the work (MARKT, 2002b), the commission services stated that the project has two phases: the first phase consists of examining the issues relating to the general design of the system (and to gather knowledge), while the second phase focuses on the detailed arrangements for taking account of risks in the new system. In notes to the IC Solvency Subcommittee the services discuss current work done by IAIS and IAA/Groupe Consultatif (MARKT, 2002c; cf. also Sections 5.3 and 5.4) and the Lamfalussy procedure (MARKT, 2002d).

5.5.1 Lamfalussy Procedure

In 2001 the European Council endorsed the so-called Lamfalussy procedure for regulation and supervision of the European securities markets. The approach is described in a report of the Committee of Wise Men, chaired by Baron Alexandre Lamfalussy, on the regulation of European security markets.

The procedure is to get a more flexible and efficient regulatory approach and permit more rapid decision making and improved supervisory convergence at the EU level.

In the Solvency II project there was an agreement to seek a solvency margin regime that better reflects the true risks and is easier to change when the financial environment changes. This would call for a more detailed regulation that could not be adopted by a directive or a regulation (the primary level), but it could be implemented under a comitology regime (the secondary level).

The Lamfalussy procedure is a four-level approach:

Level 1: The commission adopts a proposal for a directive or a regulation containing framework principles. Once the parliament and the council agree on the framework, the detailed implementing measures are developed in level 2.

Level 2: After consulting the level 2 committee, the European Insurance and Occupational Pensions Committee (EIOPC), the commission will request a level 3 committee on advice, the Committee of European Insurance and Occupational Pensions Supervisors (CEIOPS). The CEIOPS prepare this advice in consultation with market participants, e.g., Groupe Consultatif and CEA, and submit it to the commission. A formal proposal is then made by the commission and submitted to EIOPC, which must vote on the proposal within 3 months. After that, the commission adopts the measure.

Level 3: The CEIOPS works on joint interpretation recommendations, consistent guidelines, and common standards (e.g., IAIS or IAA Standards). It should also undertake peer review and compare regulatory practice to ensure consistent implementation and application.

Level 4: The commission checks the member states' compliance with the EU legislation and may take legal action.

Two working groups within the Solvency subcommittee, one on life insurance and one on non-life insurance, published reports to the Insurance Committee (MARKT, 2002e, 2002f; see also Sections 5.5.4 and 5.5.5, respectively).

The Conference of European Insurance Supervisors contributed to the first phase of the project by setting up a working group to study recent insolvencies and near insolvencies. This London group, or Sharma group, after its chairman, published a report (the Sharma report) in 2002 (Sharma, 2002; see Section 5.5.6).

A summary and a winding up of the first phase of the project is given in MARKT (2002h; see also MARKT, 2002g).

5.5.2 Summary of Phase I

The winding-up paper (MARKT, 2002h) consists of three parts. The first is a recapitulation of the first phase, the second draws lessons, and the third outlines the future prudential supervisory system.

The Solvency subcommittee of the EU Insurance Committee started four different projects during the first phase. All these projects can be seen as parts in a learning process.

- The learning project was done by KPMG (see Section 5.5.3).
- The life group looked at rules for calculating mathematical provisions and asset–liability management methods (see Section 5.5.4).
- The non-life group looked at rules for calculating technical provisions (see Section 5.5.5).
- The Insurance Conference, now CEIOPS, was initiated to study insolvency and near insolvency within EU (see Section 5.5.6 on the London group, or the Sharma group).

The Insurance Conference (now CEIOPS) set up a working group to draw up principles for insurance undertakings and supervisory authorities in assessing an undertaking's internal control systems (the Madrid group). A report was published at the end of 2003 (Madrid report, 2003).

The working documents for the project were circulated widely, and comments from member states and organizations like CEA and Groupe Consultatif were taken on board in the commission's work.

It was learned that there are three meanings of the term *solvency*. One is the direct one relating to the *solvency margin*, i.e., a set of rules for calculating a minimum capital requirement and the available solvency capital. The next concept relates to the set of rules intended to ensure that a company is *financially sound*. The last concept, known as *overall solvency*, corresponds to a company's financial soundness, taking account of the external environment and the conditions under which it operates. The concept of solvency is discussed in Section 2.2.

The risks were also discussed. It was stated that the capital requirement in itself could not be the sole measure of the risk exposure, but that a solvency system must include other rules for measuring and limiting the risks. The capital requirement could serve as a binding minimum threshold for a company to remain in the market, and the minimum margin could serve as an early-warning system. Another approach could be to determine what capital is required to provide against business fluctuations.

Harmonization within the European market was seen as important, and the adoption of the Lamfalussy procedure was one step in this direction. International standards were also seen as important parts in the work, e.g., the new accounting standards by IASB and new standards from IAIS and IAA.

A new idea for the future solvency system was the introduction of the concept of the *target capital* (or desirable capital), which would replace the solvency margin concept. At this stage it was seen as a soft threshold. Later the concept was changed to solvency capital requirement.

One of the main issues that the commission borrowed from the bank sector was the three-pillar system as proposed in Basel II (see Section 5.1).

The *first pillar* would contain the system's quantitative requirements, at least rules on provisions and investments and the latter's capital. The two working groups on life and non-life provisions reviewed these topics (see Sections 5.5.4 and 5.5.5).

For the investment rules, the third directives lay down the principle of prudent financial management (*prudent man principle*[11]). This principle should be further clarified by quantitative rules on the diversification and spread, the asset–liability management, and possible extension of the coverage rules on items other than technical provisions on the liability side of the balance sheet. The prudent person rule, as it is called in EU directives, is clarified in the directive on supervision of institutions for occupational retirement provisions (COM, 2003b; see also Appendix I).

The capital requirement in a prudential system can be viewed as

- The target capital (for maintaining an acceptable probability of default)
- An early-warning system
- The absolute minimum capital

These three concepts give rise to different margin requirements and different kinds of intervention rules (by the supervisory authorities).

The *second pillar* would contain the supervisory review process as set out within the Basel II project. The review process includes internal controls and rules for sound risk management. The harmonization and introduction of the Lamfalussy procedure give a common framework for prudential supervision, which would include:

- A common framework for assessing corporate governance
- Compilation of common statistics
- Harmonization of the main early-warning indicators
- Devising reference scenarios (stress tests)
- On-site inspection
- A common validation framework for internal models

---

[11] The prudent man principle or rule was stated by Judge Samuel Putnum in 1830: those with responsibility to invest money for others should act with prudence, discretion, intelligence, and regard for the safety of capital as well as money. A prudent man is one that does not place all of his risks into one investment, nor does he chose to unwisely pay too much. With this principle an investment decision should be made on the basis of a standardized code of conduct.

- A device for sharing information and coordination in crisis

The *third pillar* would contain market discipline, i.e., greater transparency and harmonization of accounting rules. The main factors contributing to market discipline are the financial markets and rating agents.

There should be a close link between the third pillar measures and those of the first and second pillars.

5.5.3   The KPMG Report (KPMG, 2002)

The KPMG report was published in May 2002. Its purpose is to sum up background knowledge and update developments in relation to the Solvency II project. The following summary is based on its executive summary.

The current system is based on three interconnected pillars:

- Assets
- Technical provisions
- The solvency margin (based on fixed ratios)

The main limitations of the approach are the narrow scope of risks considered and the insensitivity of capital requirements with respect to company-specific risk profiles.

There is a need for a level playing field across the global financial sector and a trend toward convergence of prudential rules for different sectors. Changes in the international accounting system provide further pressure on the European solvency system.

Key risks to the financial position of the insurance undertakings are:

- Insurance risk (underwriting risk and technical provisions)
- Asset risks (market values, interest rates and inflation, exchange rate, and commodity risk)
- Liability risk factors
- Credit risk (mainly in relation to reinsurer security and bond portfolios)
- Liquidity risk
- ALM risk
- Operational risks

It is also important to recognize the interaction between these risks.

The development in the banking sector is helpful in getting insights. Internal models must be a basis for decision making, and it must be possible to quantify the risks involved and provide a value as a result. The models have to be validated.

Technical provisions are discussed. There is a difference between the existing prudent valuation of the provisions and the fair valuation discussed in the IFRS. Decisions regarding methods and approaches have not been made yet. The disclosure of the technical provisions consists of, e.g., the disclosure of used methodologies and assumptions made, the sensitivity of the calculations to changes in assumptions, and details of run-off development. Additional safety margins as equalization reserves are studied.

The study also looks at stress tests (resilience tests) on the assets side of the balance sheet.

Various risk reduction techniques are discussed, e.g., reinsurance, ALM, and portfolio diversification.

Other topics that are studied are the new accounting rules from IASB, the role of rating agencies, and solvency margin methodologies. In the latter case, there is a comparison between the current EU system and risk-based capital systems used in other parts in the world.

Risk categories in a risk-based system should, as a minimum, include:

- Underwriting risk (exposures less reinsurance)
- Market risk
- Credit risk
- Operational risk
- Asset–liability mismatch risk

KPMG proposes the use of the banking three-pillar system:

- Pillar I: Financial resources
- Pillar II: Supervisory review
- Pillar III: Market discipline

5.5.4 The Life Report (MARKT, 2002e)

In 2001 the Insurance Committee Solvency Subcommittee decided to set up two working groups, of which the life working group should study the following two issues:

- Rules for calculating mathematical provisions
- Asset–liability management methods

The working group was composed of member state experts and one representative of the Groupe Consultatif. To acquire common knowledge, the members described the characteristics of their different markets. By this knowledge they tried to find common concerns and also common European solutions. The questions they discussed were:

- Guaranteed interest rates
- Annuities and mortality risk
- Profit-sharing clauses
- Unit-linked products
- Options embedded in the contracts

For each question the members discussed principles and methods for mathematical provisions, but also the principle of premium sufficiency. The discussion on asset liability management methods followed the same line and the group discussed ideas for improvement.

The group believed that the directives contain the most necessary prudence principles, but that other types of principles could be created or strengthened in the directives. Therefore, they suggested that two supplemented principles should be developed:

- A principle of prudence in the choice of mortality table (corresponding to the existing prudence principle in choice of interest rate)
- A principle of asset diversification applied to unit-linked products

Other prudence principles suggested by the group were:

- A principle aimed at protection of policyholders and fair conduct of business:
  1. With-profit products: Perhaps a general principle of fair sharing of the profits should be established.
  2. Unit- or index-linked products: Necessary with disclosure principles to ensure that policyholders are aware of the risks in these products.
- Principles regarding risk management and supervisory review: One way would be to introduce a requirement for companies to use appropriate prospective tools for their ALM, which could be used as the supervisory basis.

One method to use in improving the calculation of the mathematical provisions would be to focus on the interest rate used in the calculations. One way to go would be to make reference to current market interest rates, and another way would be to introduce a resilience provision.

If the last principle above is laid down at the EU level, then it will be necessity to harmonize or at least coordinate the supervisory methods. First, the supervisors need to have *benchmarks*, e.g., maximum interest rate, reference mortality tables, and, for ALM, reference adverse scenarios. Second, the supervisors need to have a role in *monitoring*, e.g., by exchanging indicators and statistical data, and third, they need to have supervisory *powers* defined.

5.5.5 The Non-Life Report (MARKT, 2002f)

The non-life working group set up by the Insurance Committee Solvency Subcommittee in 2001 should study the major issue: rules for calculating technical provisions — outstanding claims and equalization.

As the corresponding life group, it was composed of member state experts and one representative of the Groupe Consultatif. The report was to be seen as a complement to the Manghetti report on technical provisions in non-life insurance (Manghetti report, 2001; KPMG, 2002 — see Section 5.5.3).

5.5.5.1 *Provisions for Outstanding Claims*

The group explored the possibility of comparing levels of prudence of provisions for outstanding claims using statistical indicators (quantitative approach), but also of having discussions with transnational insurance groups, which have experience of different provisioning (qualitative approach), to get a better understanding of these issues. They found that the level of prudence in provisions is a more complex issue than is usually presented.

The group found that the supervisors lack a common set of data for the analysis of claims run-off, and therefore it would be useful to require companies to provide statistical data according to a common layout. Such a database would be the basis for a common supervisory review process (monitoring tool).

The group believed that it was better to let the provisioning converge toward a common level of prudence and, in this respect, set up principles and guidance for sound claims management and provisioning practices.

5.5.5.2 *Provisions for Equalization*

Equalization provisions are used as a buffer for bad years or catastrophes, and they are an additional buffer to the solvency. The group believed that it would make sense to take these provisions into account when assessing the solvency position. It should nevertheless be noted that under IFRS, equalization reserves do not exist. If Solvency II and IFRS should be coordinated, it must be decided whether such reserves will be permitted to smooth the results over time or for tax reasons.

There is a large diversity in the size of equalization provisions. One way to promote further harmonization would be to explicitly link the provisions to the volatility of the business written.

5.5.6 The Sharma Report (Sharma, 2002)

When the European Commission started its project on insurance regulation, the Solvency II project, in May 2001, the Insurance Conference (now CEIOPS) was asked to give input and make recommendations. It set up a working group of insurance supervisors, called the London Working Group (LWG), as Paul Sharma from the U.K. Financial Services Authority chaired it. Its report, which was published in December 2002, is usually called the Sharma report (see Sharma, 2002).

The goals of the LWG were to understand the risks to solvency of insurance undertakings and how better to monitor the undertaking's risk management. They followed four main lines in their work:

- Risk classification and causal chain mapping
- Surveys on actual failures and near misses from 1996 to 2001 — an update of the Müller report (see Section 4.1)
- Discussion of 21 detailed case studies
- Diagnostic and preventive tools questionnaire

All the case studies showed a *chain of multiple causes*, but the most obvious causes were inappropriate risk decisions, external trigger event, and resulting adverse financial outcomes. The study also showed that these causal chains started with *underlying internal causes*: problems with management, shareholders, or other external controllers. The problems included:

- Incompetence
- Operating outside area of expertise
- Lack of integrity or conflicting objectives
- Weakness in the face of inappropriate group decisions

The underlying internal problems led to inadequate internal controls and decision-making processes, which resulted in inappropriate risk decisions. The LWG concluded that supervision would be most effective when there are tools to tackle the full causal chain.

Risks can be described and categorized by their causes or their effects. The LWG confirmed the observation that the effects of a risk are usually more obvious than its causes. This led to a full cause–effect analysis to get a better idea of the real causes behind the observed effects.

Based on a theoretical cause-chain model, the LWG designed a risk map template (see Figure 3.2 in Sharma, 2002). Using this template, the group mapped their own risk classification onto it, giving us the detailed causal-chain risk map shown in Figure 5.3.

For each of the 21 case studies, the LWG made a risk map based on the template and a map similar to Figure 5.3. The risk models can be presented in three different levels:

- A summary level showing the broad categories

To get knowledge about the background to insolvency or near insolvency, the LWG used a cause–effect methodology. For a general introduction to causal inference, in a statistical sense, see, e.g., the classical work by Blalock (1961). In its simplest form, the methodology could be described as

CAUSE → EFFECT

If the cause is denoted X and the effect Y, we can write this simple causal chain as X ⇒ Y and measure the strength between the cause and effect by, e.g., correlation.

The correlation coefficient between Y and X describes their joint behavior. A partial correlation describes the behavior of Y and X_1, when X_2, ..., X_k are held fixed, i.e., when their effect is removed. The partial correlation is denoted $\rho_{X_1 Y \cdot X_2 .. X_k}$.

Assume the following simple causal chain model: X ⇒ Z ⇒ Y. The partial correlation between Y and X when the effect of Z is removed can be written as

$$\rho_{XY \cdot Z} = \frac{\rho_{XY} - \rho_{XZ}\rho_{ZY}}{\sqrt{(1 - \rho_{XZ}^2)(1 - \rho_{ZY}^2)}}$$

If $\rho_{XY \cdot Z} = \rho_{XY}$, then X and Y are both uncorrelated with Z.

Note: The concept of partial correlation is not used in the Sharma report.

- A detailed level applied to the specific cases (as in Figure 5.3)
- A detailed level showing the relevant supervisory tools and lessons learned (see Annex E in Sharma, 2002)

Based on the questionnaires on actual failures, near misses, and the case studies, the LWG identified 12 risk types according to similarities in the causal chain rather than the effects:

1. *Parent sets inappropriate policy in pursuit of group objectives* (strategic investments): See risk map at Annex E1 in Sharma (2002).

   The insurer's parent undertaking set an aspect of policy that had a detrimental effect on the insurance undertaking because it had objectives other than prudent management of the insurance undertaking. Group management overrode the local management.

2. *Parent sets inappropriate policy through poor understanding of insurance*: See risk map at Annex E2 in Sharma (2002).

   The insurer's parent undertaking had a noninsurance focus, and because it lacked a proper understanding of the insurance business,

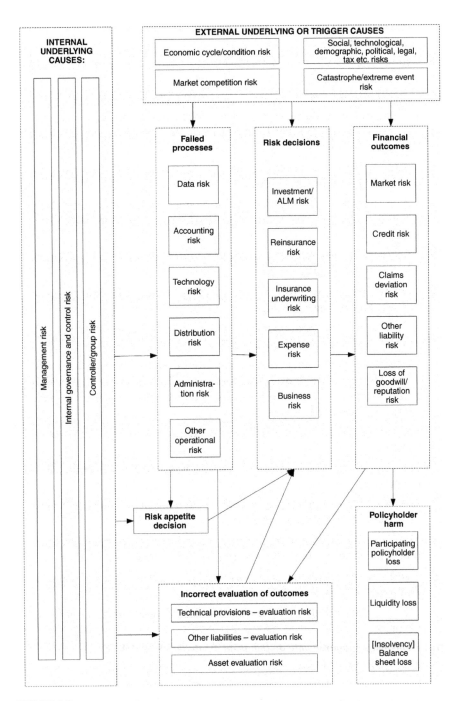

FIGURE 5.3
A risk map showing the cause–effect structure. (Adapted from Figure 3.3 of Sharma, Prudential Supervision of Insurance Undertakings, paper presented at Conference of Insurance Supervisory Services of the Member States of the European Union (now CEIOPS), December 2002.)

it set an aspect of policy that had a detrimental effect on the insurance undertaking.

3. *Mutual insurer faces conflicting objectives*: See risk map at Annex E3 in Sharma (2002).

   A mutual insurer's management may have social or other objectives besides prudent management of insurance business; e.g., in one case this led the management to invest in other activities for the benefit of the members.

4. *Business risk: large insurer faces merger integration issues*: See risk map at Annex E4 in Sharma (2002).

   Large composite firms can be hard to manage efficiently, particularly those that have grown up through a series of acquisitions and mergers.

5. *Cross-border management of insurance group*: See risk map at Annex E5 in Sharma (2002).

   Poor management centrally can affect the conduct of operations in more than one member state.

6. *Life insurer: high expectations/long-term interest rate guarantees*: See risk map at Annex E6 in Sharma (2002).

   Management sets policies that gamble on future economic conditions. Interest rate guarantees can contain long-term options that could be expensive to service or high expectations of discretionary bonuses with low reserves.

7. *Stagnating insurer diversifies*: See risk map at Annex E7 in Sharma (2002).

   Stagnating undertakings can move into noncore business with little experience.

8. *Underwriting risk: niche player with an evolving market*: See risk map at Annex E8 in Sharma (2002).

   Management takes naïve approach, ignoring developments in their market, changing the nature of the risks they take on.

9. *Insurer matches liabilities with correlated investments*: See risk map at Annex E9 in Sharma (2002).

   The correlation between the risk profiles of assets and liabilities is ignored.

10. *Firms have inappropriate distribution strategies*: See risk map at Annex E10 in Sharma (2002).

    Inappropriate strategy concerning agents and brokers can have adverse effects on the insurer, e.g., poor customer service or high distribution costs not linked to portfolio outcomes.

11. *Catastrophe/inadequate reinsurance planning*: See risk map at Annex E11 in Sharma (2002).

    An undertaking may find that it has insufficient reinsurance when catastrophic losses occur.

12. *Outsourcing of key functions*: See risk map at Annex E12 in Sharma (2002).

    An undertaking may outsource a key activity and fail to maintain proper control over it.

The LWG found that almost all of the 21 case studies shared the same root causes: poor or inexperienced management leading to inadequate decision making or internal controls.

Based on a questionnaire on diagnostic and preventive tools, the LWG set up tool kits for the supervision of the insurance undertakings. These tool kits are overlaid on the risk map structure (see Figure 5.1 in Sharma, 2002). The tool kits are:

Tool kit 1: Underlying causes — internal (management, governance, and ownership)

Tool kit 2: Underlying or trigger causes — external (general and insurance specific)

Tool kit 3: Inadequacy of failed internal processes, people, or systems

Tool kit 4: Risk decision and outcomes — investments, credit, ALM risks

Tool kit 5: Risk decision and outcomes — underwriting and technical provisions

Tool kit 6: Risk decision and outcomes — reinsurance risk

Tool kit 7: Risk decision and outcomes — other risks and general tools

Tool kit 8: Curative tools

For each of the first seven tool kits the LWG listed what they expect of firms, their supervisory aims, tools that have been used, examples of other tools in use, and ideas for new tools or uses of tools.

The LWG also gave general considerations on a new supervisory approach. There should be a right balance between supervisory control and prevention and maintaining an insurer's freedom of operations. In order to ensure the right mixture, a prudential regime should address risk in three main ways.

- Capital adequacy and solvency regime
- A broad range of tools needed to cover the full causal chain
- Internal factors

The LWG also listed future work for the Insurance Conference (now CEIOPS), e.g., harmonize common early-warning signals, improve focus on risk management and internal controls, and develop a framework and guidance on asset–liability matching.

6

Steps toward Solvency II: 2

In our steps toward a new solvency requirement for Europe we will look at different frameworks and different proposals made to enhance the solvency assessment throughout the world. The descriptions will be very brief, as a thorough description of each system would be too long and it is the framework of each system that is of interest. Therefore, components that are vital for the systems have been omitted and the interested reader is referred to the respective system provider for all details. The presentation should be seen as a *smorgasbord* with different ideas. We have picked out different issues and characteristics that could give the interested modeler some ideas:

- *Australia*: The ideas are similar to those behind Solvency II. Liability valuation, risk categories, a factor-based prescribed method, and internal models.
- *Canada*: A factor-based system. Risk categories, the minimum capital test, dynamic capital adequacy testing, and minimum continuing capital and surplus requirements on ratings.
- *Denmark*: Fair valuation and a traffic light test system.
- *Finland*: A risk theoretical transition model and equalization reserve.
- *The Netherlands*: Fair valuation and minimum solvency and continuity analysis.
- *Singapore*: Valuation of assets and liabilities, risk categories, and the requirements in a risk-based capital system.
- *Sweden*: Valuation of assets and liabilities, risk categories, and a simple model.
- *Switzerland*: Valuation of assets and liabilities, risk categories, standard model, scenario tests determining the target capital, and internal models.
- *U.K.*: A twin peaks' approach under pillar I, individual capital adequacy standards under pillar II, and risks.
- *U.S.*: Risk-based capital model, correlation structure, and different intervention levels.

The German and Norwegian approaches are also touched upon.

6.1   Australia

In the beginning of the 1970s, 16 Australian non-life insurers[1] defaulted and
caused financial loss to their policyholders and loss of confidence in the
market. Up until this time there were more or less no requirements for
companies that wished to conduct non-life business in Australia; the only
requirement was a deposit of $200,000.

The failures forced the authorities to introduce the 1973 Insurance Act,[2]
which sets out statutory minimum solvency requirements for non-life insurers or branches in Australia. The minimum requirement was that the insurer
must have total assets exceeding total liabilities by not less than the greater
of either $2 million or 20% of gross written premium income or 15% of
outstanding claims — a requirement similar to the requirement laid out in
the first non-life directive within EU (cf. Chapter 3). The Insurance Act also
imposed on the insurers that they have a reinsurance arrangement approved
by the authority. During the next 25 years the insurance industry rationalzed
and stabilized. There were nearly 20 non-life failures during this period, but
none resulting in major policyholder losses. The failures were due to causes
like fraud, rapid premium growth, underwriting and operating losses, poor
management, and poor asset spread. Assets were valued at net market value,
and the outstanding claims' liabilities were discounted, but the provisions
included actuarial margins. Prudential margins were not part of the Insurance Act at this stage.

In 1995 the Insurance and Superannuation commissioner raised the issue
of whether a measure of solvency of non-life insurers could or should be
disclosed. This raised the question as to whether the existing measures were
adequate, and the Institute of Actuaries of Australia (IAAust) established
two working groups. They reported to the new Australian Prudential Regulation Authority (APRA)[3] in the spring of 1999.

When the prudential supervision of non-life insurers transformed to
APRA, its board commissioned an assessment of the requirements with a
view to modernize and improve the non-life insurance regime. In September
1999, APRA issued three consultation papers — the September discussion
papers — for a dialogue with the industry and the IAAust (APRA, 1999a,
1999b, 1999c). These discussion papers laid out the structure of a new modern prudential supervisory require,[4] for example, taking account of risks
faced by non-life insurers: assets risks, liability risks, ALM risks, and operational risks. All financial institutions, which are supervised by APRA, will
have to carry out a fit and proper persons test for its directors, top manage-

[1] General insurers in Australia.
[2] For life insurance there is the Life Insurance Act 1995.
[3] The APRA was formed on July 1, 1998.
[4] Prudential regulation is about promoting prudent behavior by insurance companies and other
financial institutions, with the objective of reducing the likelihood of institutional insolvency
and consequential losses to policyholders.

ment, and other persons in a position to influence the conduct of the undertaking.

In APRA (1999b) the valuation of liabilities is discussed. The Insurance
Act required companies to have assets, at market value, in excess of reported
liabilities by at least the prescribed amount given above. The provisions for
liabilities have been accepted for prudential and solvency reporting purposes, but also for general financial reporting purposes.

The outstanding claims provision, which is the major part of the provisions, can typically be represented as

Central estimate + Prudential margin

The central estimate, or best estimate, can in terms of a probability distribution be thought of as the discounted mean of that distribution. The financial
reporting act was silent on prudential margins, which has given rise to
different interpretations and practices. In some classes of business the insurers have adopted prudential margins that vary from 0 to 30% for similar
portfolios.

According to APRA (1999b), the size of the prudential margin is driven
by factors such as:

- The assumed degree of uncertainty in the liability estimate
- The risk appetite of the insurer
- The amount of available capital
- The desire for consistency from one year to the next
- Market practice
- Advice from actuaries
- The impact on reported profit
- Taxation

They also conclude that it is inappropriate to allow some of the factors,
for example, the amount of available capital, to drive the reported liability.
Therefore, a solvency structure that starts with the balance sheet liabilities
needs a consistent approach to the valuation of the liabilities (Section 7.1).
As there is an absence of a deep market in insurance liabilities, it is only
possible to estimate the value of the liabilities suitably consistent with the
assets (at market value). The liabilities can be thought of as being comprised
by

Central estimate + Risk margin

In estimating the risk margin, it is assumed there is a proxy marketplace
consisting of hypothetical buyers and sellers. The basic principles, as presented in APRA (1999b, p. 4), are

- An arms-length buyer would like to be fairly certain that the settlement price is adequate. This may be reexpressed in terms of a requirement to expect a return for assuming the risk associated with the uncertainty of the outcome.

- An arms-length seller would be willing to pay a premium for transforming the risk associated with the uncertainty of the outcome to another party.

The expected return to an arms-length buyer (or equivalently, the premium to a seller) can be considered to be a function of:

- The level of uncertainty surrounding the estimate of the liability, in particular, the amount of downside risk

- The cost of capital required to support the business (statutory requirements and market forces impose capital requirements)

In April 2000, APRA published a new discussion paper based on the comments given regarding the September discussion papers (APRA, 2000a). APRA's preference, according to this paper, was to have a small number of standards, each outlining the key requirements for a particular prudential measure. The standards were proposed in respect to capital adequacy, liability valuation, and qualitative requirements for reinsurance arrangements, and operational risks. Under the proposed liability valuation standard the calculation of a risk margin, with respect to the liabilities, is required. Its intention is to secure the liabilities with high probability. "In APRA's view, a 75 per cent probability of sufficiency would be commensurate with prudently set technical provisions under the current regime" (APRA, 2000a, p. 18). In Collings and White (2001) the determination of the risk margin is discussed and modeled.

Later the same year, APRA published a discussion paper on a principles-based approach in harmonizing prudential standards for the financial sector under its supervision, such as banks and insurers (APRA, 2000b). In March 2001, APRA published a policy discussion paper (APRA, 2001a), which became the basis for the new prudential supervision of non-life insurers, that became effective on July 1, 2002. The new framework comprised a three-layered system of regulation: the General Insurance Reform Act 2001,[5] APRA (2001b), and prudential standards and guidance notes. The prudential standards,[6] specifying mandatory requirements, were:

- Prudential Standard: Capital Adequacy (GPS 110)
- Prudential Standard: Assets in Australia (GPS 120)
- Prudential Standard: Liability Valuation (GPS 210)

---

[5] The legislation at the top of the new framework is still the Insurance Act 1973, as amended by the General Insurance Reform Act 2001.

[6] The standards and guidances can be found on the following Web site: www.apra.gov.au.

- Prudential Standard: Risk Management (GPS 220)
- Prudential Standard: Reinsurance Arrangements (GPS 230)
- Prudential Standard: Transfer and Amalgamation of Insurance Business for General Insurers (GPS 410)
- Prudential Standard: Early Approvals of Auditors and Actuaries (GPS 900)

To most of these standards there are one or more guidance notes. For example, to GPS 110 there are five guidance notes labeled GGN 110.1 to GGN 110.5. These guidance notes allow judgments and approaches made by individual companies.

The liability valuation standard was designed to produce a consistent estimate of the value of insurance liabilities, and consistency between different classes of business and between different insurers. The standard requires a level of uncertainty, which the market would require in estimating a value of the liabilities. This has been set at the 75th percentile of the range of possible outcomes.[7,8]

In setting the capital adequacy standard, a default rate of around 0.5% over a 1-year time horizon was deemed to be an appropriate benchmark. In converting this benchmark into a measure of required capital, and to assist with comparability with the existing standards (the 1973 Insurance Act), the profitability was proposed to be measured as a percentage of each company's outstanding claims provisions, a percentage of the annual premium income, and a percentage of total assets. This examination was made both at the individual company level and at the group level.

According to the capital adequacy standard (GPS 110), an insurer may choose one of two methods for determining its *minimum capital requirement* (MCR):

1. The *prescribed method*, as outlined in the standard.
2. Those with sufficient resources are encouraged to develop a capital measurement model, referred to as the *internal model-based* (IMB) method. The IMB approach is discussed in GGN 110.2.

The use of the IMB has to be approved by APRA and the treasurer and will require the insurer to satisfy both qualitative and quantitative criteria. According to Sutherland-Wong and Sherris (2004), no insurer has yet adopted the IMB approach.[9] The results obtained by Sutherland-Wong and Sherris show that the IMB method produced a *higher* MCR than the prescribed method.

---

[7] If the hypothetical probability function is standard normal, a one-sided confidence level would be 0.67, i.e., $\mu + 0.67\sigma$.

[8] One argument for the 75% confidence level is given in Section 8.4.2.

[9] A dynamic financial analysis (DFA) approach is used by Sutherland-Wong and Sherris (2004) to calibrate the IMB approach.

Regardless of which approach a company uses, the MCR is determined with regard to three main risk categories:

- *Insurance risk*: The risk that the true value of net insurance liabilities is greater than the value determined under GPS 210: Liability Valuation.
- *Investment risk*: The risk of an adverse movement in the value of an insurer's assets or off-balance-sheet exposures.
- *Concentration risk*: The risk associated with an accumulation of exposures to a single catastrophic event.

A company using the IMB approach should address these risks, together with other relevant risks, in its calculation.

The insurer must, at all times, have *eligible capital* in excess of its MCR. The eligible capital is comprised of tier 1 and tier 2 capital, where the tier 1 eligible capital must exceed the tier 2 eligible capita (cf., for example, the Basel Accord in Section 5.1). There are slightly different requirements for foreign insurers authorized to operate in Australia as branches. *Disclosure* and *transparency* are important issues in the supervisory process.

6.1.1 The Capital Base

There are different matters that are relevant to whether a capital instrument is adequate for supervisory purposes. The capital instruments eligible for inclusion in the insurer's capital base are outlined in GGN 110.1: Measurement of Capital Base, broadly following the Basel Accord.

6.1.2 Prescribed Method

The prescribed method of MCR is the sum of the capital charges for:

C_{InsR}: Insurance risk; GGN 110.3: Insurance Risk Capital Charge

C_{InvR}: Investment risk; GGN 110.4: Investment Risk Capital Charge

C_{ConR}: Concentration risk; GGN 110.5: Concentration Risk Capital Charge

$$\text{i.e., } C_{PM} = C_{InsR} + C_{InvR} + C_{ConR} \tag{6.1}$$

The underlying assumption is that the three risk categories are fully dependent (cf. Section 9.4).

6.1.2.1 Insurance Risk in Equation 6.1

The insurance risk has two components:

- A charge for the outstanding claims' risk (in response to the risk that the true value of the net outstanding claims' liabilities is greater than the value determined under GPS 210)
- A charge for the premium's liability risk (in response to the risk that premiums relating to postcalculation date exposures, including premiums written after the calculation date, will be insufficient to fund liabilities arising from that business)

The total capital charge for the two components is the sum of them. X is the outstanding claims' liability and P the net premium's liability, i.e,

$$C_{InsR} = \sum_{j=1}^{15} \alpha_j X_j + \sum_{j=1}^{15} \beta_j P_j \qquad (6.2)$$

where the sum runs over the 15 different classes of business according to Table 6.1. The risk factors are given in Table 6.1.

6.1.2.2 Investment Risk in Equation 6.1

The investment risk derives from different sources such as:

- Credit risk: The risk of an adverse movement of an asset owing to changes in the credit quality of the issuer of that asset, including default of the issuer.
- Market/mismatch risk: The risk of an adverse movement in the value of asset, which is not offset by a corresponding movement in the value of liabilities.
- Liquidity risk: The risk that the reported asset value will not be readily realized in certain circumstances.

The capital factors assigned to assets reflect broad judgments by APRA for capital adequacy purposes only. Each of an insurer's assets, and certain off-balance-sheet exposures, are assigned to one of nine categories. The capital charge is determined by multiplying the balance sheet value of each asset by an appropriate investment capital factor. Let A be an asset; then we have

$$C_{InvR} = \sum_{i=1}^{9} \omega_i A_i \qquad (6.3)$$

where the sum runs over the nine asset categories. The risk factors are briefly described in Table 6.2. The grading of 1 to 5 corresponds to a five grading

TABLE 6.1

The Risk Capital Factors Used for the Calculation of the Insurance Risk

Class of Business	Outstanding Claims' Risk Capital Factor, α	Premium's Liability Risk Capital Factor, β
Direct Insurance		
Householders		
Commercial motor	0.09	0.135
Domestic motor		
Travel		
Fire and ISR		
Marine and aviation		
Consumer credit	0.11	0.165
Mortgage		
Other accident		
Other		
CTP		
Public and product liability	0.15	0.225
Professional indemnity		
Employer's liability		
Inward Reinsurance		
Property		
Facultative proportional	0.09	0.135
Treaty proportional	0.10	0.150
Facultative excess of loss	0.11	0.165
Treaty excess of loss	0.12	0.180
Marine and aviation		
Facultative proportional	0.11	0.165
Treaty proportional	0.12	0.180
Facultative excess of loss	0.13	0.195
Treaty excess of loss	0.14	0.210
Casualty		
Facultative proportional	0.15	0.225
Treaty proportional	0.16	0.240
Facultative excess of loss	0.17	0.255
Treaty excess of loss	0.18	0.270

Source: GGN 110.3., Guidance notes 3 to Prudential Standard 110: Capital Adequacy, July 2003.

scales of counterparty grades. For example, grade 1 corresponds to Standard & Poor's AAA and Moody's Aaa.

The capital charge may be reduced by risk mitigation techniques, such as collateral and guarantees (see the GGN 110.4 for details).

6.1.2.3 Concentration Risk in Equation 6.1

The capital charge is based on the company's *maximum event retention* (MER) for catastrophe purposes. It is the responsibility of the board and senior management to ensure that the MER is set at a level consistent with the insurer's risk profile and its reinsurance program.

TABLE 6.2

The Risk Capital Factors Used for the Calculation of the Investment Risk

Asset	Investment Capital Factor, ω
Cash	0.005
Debt obligations of	
• The commonwealth government	
• An Australian state or territory government	
• The national government of a foreign country where	
• The security has a grade 1 counterparty rating or if not rated,	
• The long-term foreign currency counterparty rating of that country is grade 1	
Any debt obligation that matures or is redeemable in less than 1 year with a rating of grade 1 or 2	0.01
Cash management trusts with a rating of grade 1 or 2	
Any other debt obligation (that matures or is redeemable in 1 year or more) with a rating of grade 1 or 2	0.02
Reinsurance recoveries and other reinsurance assets due from reinsurers with a counterparty rating of grade 1 or 2	
Unpaid premiums due less than 6 months previously	0.04
Unclosed business	
Any other debt obligations with a rating of grade 3	
Reinsurance recoveries and other reinsurance assets due from reinsurers with a counterparty rating of grade 3	
Any other debt obligations with a counterparty rating of grade 4	0.06
Reinsurance recoveries and other reinsurance assets due from reinsurers with a counterparty rating of grade 4	
Any other debt obligations with a counterparty rating of grade 5	0.08
Reinsurance recoveries and other reinsurance assets due from reinsurers with a counterparty rating of grade 5	
Listed equity instruments	
Units in listed trusts	
Unpaid premiums due more than 6 months previously	
Direct holding of real estate	0.10
Unlisted equity instruments	
Units in unlisted trusts	
Other assets not specified elsewhere in this table	
Loans to directors of the insurer or directors of related entities (or a director's spouse)	1.00
Unsecured loans to employees exceeding $1000	
Assets under a fixed or floating charge	
Goodwill	0
Other intangible assets	
Future income tax benefits (deducted from tier 1 capital)	

Source: GGN 110.4, Guidance notes 4 to Prudential Standard 110: Capital Adequacy, July 2003.

The MER is the largest loss to which an insurer will be exposed due to concentration of policies after netting out any reinsurance recoveries. The probable maximum loss (PML) is the largest loss to which an insurer will be exposed due to concentration of policies without any allowance for reinsurance recoveries.

The capital charge for the concentration risk is given by

C_{ConR} = MER + the cost of one reinstatement premium for the catastrophe reinsurance cover   (6.4)

6.1.2.4 Total Capital Charge in Equation 6.1

The total capital charge, MCR, is given by combining Equations 6.2 to 6.4 into Equation 6.1:

$$C_{PM} = \sum_{j=1}^{15} \alpha_j X_j + \sum_{j=1}^{15} \beta_j P_j + \sum_{i=1}^{9} \omega_i A_i + MER + \text{the cost of one reinstatement}$$

premium for the catastrophe reinsurance cover   (6.5)

6.1.3   Disclosure

The following items should be published in a company's annual report:

1. The amount eligible tier 1 capital, with separate disclosure of each of the items specified in GGN 110.1
2. The aggregate amount of any deductions from tier 1 capital
3. The amount of eligible tier 2 capital, with separate disclosure of each of the items specified in GGN 110.1
4. The aggregate amount of any deductions from tier 2 capital
5. The total capital base of the insurer derived from items 1 to 4
6. The MCR of the insurer
7. The capital adequacy multiple of the insurer: item 5 divided by item 6

Because of a failure of an insurer (HIH), the recommendation was made by the HIH Royal Commission that further work is required to strengthen the prudential supervision framework for non-life insurance. A number of proposals were published in a discussion paper released in November 2003 by APRA (2003). This further work includes an extension of the prudential requirements to cover consolidated groups, improved disclosures, and strengthening of actuarial, supervisory, governance, and audit arrangements.

6.2 Canada

During the mid-1980s work on a risk-based capital system for life insurance companies began in Canada. The Canadian regulatory Office of the Superintendent of Financial Institutions' (OSFI)[10] standard came into effect in 1992: Guideline on Minimum Continuing Capital and Surplus Requirements (MCCSR). This risk-based system was seen as a retrospective system, and the Canadian Institute of Actuaries (CIA) had the view that this more or less static system was insufficient for the life undertaking's needs. Instead, the CIA wanted a forward-looking and dynamic system, and thus developed its dynamic solvency testing (DST). In the DST, a company's capital position was analyzed through projections of generally unfavorable scenarios. The DST was introduced in 1990, and it was suggested by CIA that the regulator OSFI should impose the requirements on the undertakings. The Insurance Companies Act introduced in 1992, saying that the superintendent may require actuaries to report on the company's expected future financial condition to management, provided this. The DST was later changed to dynamic capital adequacy testing (DCAT). A new revised guideline on MCCSR was published in October 2004 (OSFI, 2004). The OSFI considers each year whether changes are required to improve the risk measures (minimum and target capital levels), address emerging issues, and encourage improved risk management.

For non-life insurance companies there was a fixed ratio system, similar to the EU system, up until 2002. At that time the companies calculated a margin of admitted assets over the adjusted liabilities. This margin should be at least 10%. By admitted assets we mean the total assets adjusted for the difference between market value and book value of investments and deducting nonadmitted assets, such as deferred acquisition costs. By adjusted liabilities we mean the liabilities plus margins depending on claims and premiums. The DCAT was introduced for the non-life undertakings in 1998. A new risk-based capital requirement for the non-life undertakings took effect in 2003: the minimum capital test (MCT) (OSFI, 2003).

Continuing changes in the financial markets have been leading OSFI to review its supervisory practices to ensure their effectiveness. One such framework review was made during 1997 and 1999, resulting in OSFI (1999). Some of the key issues from this framework are:

- The supervisory process should be conducted on a consolidated basis, using information from other regulators where appropriate.
- Work performed should be focused on clearly identified risks or areas of concern.

---

[10] www.osfi-bsif.gc.ca/eng/.

- The level and frequency of scrutiny should depend on the risk assessment of the undertakings.
- The supervision should include reviews of major risk management control functions.

It is important to understand the environment in which an undertaking operates. OSFI has grouped the risks inherent in the activities of the companies. These are:

- Insurance risk
- Credit risk
- Market risk
- Operational risk
- Liquidity risk
- Legal and regulatory risk
- Strategic risk

6.2.1 Dynamic Capital Adequacy Testing

Before looking at the capital requirements, we will see what the DCAT process involves. In the process of conducting a DCAT, i.e., the report on expected future financial condition to the management and the board of directors, a copy is also sent to OSFI, who uses these reports in its supervisory role and in its work with CIA to enhance the quality of the reports. The key issues in the DCAT process are the development of a *base scenario*, the identification and examination of *threats* to solvency, the development of plausible *adverse scenarios*, the *projection and analyses* of capital adequacy, and, at last, the *reporting*.

The DCAT begins with a base scenario, usually the company's business plan. This must include projections of policies in force, but also future sales and results of noninsurance operations. The projections are made for a period of several years into the future. The effect of a scenario should fully develop, and therefore, in general, the projection period is 5 years for life insurance undertakings and 2 for non-life undertakings. The liabilities at the end of the projection period reflect that the scenario experience will continue in effect in the future.

The intention of DCAT is to identify and test the impact of adverse, but plausible, assumptions that affect the financial condition of the company. It is the company's actuary that decides the choice of scenarios to be tested.

Stress testing may be required to determine the potential threat. The CIA's standard of practice lists a number of sources of risk that should be tested. Some sources to risk to life undertakings are mortality, morbidity, persistency, cash flow mismatch, deterioration of asset values, and new business, and to non-life undertakings we have frequency and severity, pricing, mis-estimating of liabilities, inflation, interest rate, and premium volume. Common sources of risk are expenses, reinsurance, government and political actions, and off-balance-sheet items.

6.2.2 Minimum Continuing Capital and Surplus Requirements for Life Insurance Companies

Under the MCSSR there are two main triggers: the *minimum capital require-ment* and the *target capital requirement*. To determine these requirements (or levels), companies apply factors for each of five risk components to different on- and off-balance-sheet assets and liabilities and add them together. The risk components are:

- Asset default risk (RC1, our notation): Risk of loss from asset default (on-balance sheet), loss of market value of equities and reduction of income, etc.
- Mortality/morbidity/lapse risk (RC2, our notation): Risk that assumptions about mortality, morbidity, and lapse are wrong.
- Interest margin pricing risk (RC3, our notation): Risk of interest margin losses, with respect to investment and pricing decisions, on business in force, other than RC1 and RC4.
- Changes in interest rate environment risk (RC4, our notation): Risk of loss resulting from changes in interest rate environment, other than RC1 and RC3.
- Segregated funds risk (RC5, our notation): Risk of loss arising from guarantees embedded in segregated funds.

In defining the capital used for measuring the adequacy, there are some main considerations that are made. They are its permanence, its being free of mandatory fixed charges against earnings, and its subordinated legal position to the rights of policyholders and other creditors of the institution. The capital is defined as comprising two main tiers, the first being the core capital tier. The tire 1 comprises the highest-quality capital elements. The second tier comprises the supplementary capital. This is a capital element that falls short in meeting either of the first two properties listed above, but that contributes to the overall strength of the company as a going concern. If there is any doubt as to the availability of capital, it is classified as tier 2. The tier 2 is eventually split up into three subtiers: 2A, 2B, and 2C.

Foreign life insurers are required to maintain an adequate margin of assets in Canada over its liabilities in Canada. The *test of adequacy of assets in Canada and margin requirements* (TAAM) is the framework within which the super-intendent assesses whether a company operating in Canada on a branch basis is maintaining an adequate margin (OSFI, 2004, Section 7).

A MCCSR/TAAM ratio is calculated, comparing the available capital to the required capital, by applying factors for specific risks. The minimum MCCSR/TAAM requirement is set to 120%. It is higher than 100% because the calculation excludes other risks, e.g., operational, strategic, and legal risks.

In order to cope with volatility in markets and economic conditions, international developments, etc., each undertaking is believed to establish a target capital level (TCL). An adequate TCL provides additional capacity to absorb unexpected losses not covered by the minimum MCCSR/TAAM. Each company is expected to establish such a target total MCCSR/TAAM ratio no less than the supervisory target of 150%, i.e.,

$$\text{Target total MCCSR/TAAM ratio} \geq 150\%$$

As the core capital in tier 1 is the primary element of capital that helps underwritings to absorb possible losses during their ongoing activities, they should also calculate a tier 1 target ratio, which is no less than 70% of the supervisory target, i.e., no less than 105%.

Some of the factors, which are applied to components associated with participating policies and assets backing participating policies' liabilities, can be reduced. The reduction can only be achieved if the participating policies and the assets backing them are *qualifying participating policies*.[11]

To illustrate the application of applying factors for the risk components, we look at the asset default risk, which include both on- and off-balance-sheet risks. Those related to qualifying participating policies are tracked separately. These asset default factors are 50% of the regular ones. The risk for on-balance-sheet items covers losses from asset default and related loss of income and of market value of equities. Factors for bonds are one example that we use for illustration (see Table 6.3).

We end up with another example, illustrating the mortality risk. For the risk "insurance against death (including accidental death and dismemberment)," the measure of exposure (before reinsurance ceded) is *net amount at risk*, i.e., the total face amount of insurance less policy reserves (even if negative) for both direct and reinsurance acquired business. For example, for participating policies the factor is 0.0005 for groups and 0.0010 for all other policies.

6.2.3   Minimum Capital Test for Non-Life Insurance Companies

In 1995 the Canadian Council of Insurance Regulators initiated a project on harmonizing Canadian solvency tests, and a year later the project went on to develop a risk-based test. A prototype of the MCT was launched in 1997.

---

[11] Four criteria have to be met: the policies must pay meaningful dividend, the participating dividend policy must be publicly disclosed, its scale must be regularly reviewed, and the undertaking must demonstrate to OSFI that it follows its policy.

TABLE 6.3

Example of Factors for the Asset Default Risk Bonds/Loans/Private Placements

Factor		Bonds
Regular	**Qualifying Participating**	
0%	0%	Canadian and all provincial and territorial government bonds and bonds of OECD central governments and claims against OECD central bonds
		Bonds rated:
0.25%	0.125%	AAA, A++, or equivalent
0.5%	0.25%	AA, A+, or equivalent
1%	0.5%	A, A–, or equivalent
2%	1%	BBB, B++, or equivalent
4%	2%	BB, B+, or equivalent
8%	4%	B, C, or equivalent
16%	8%	Lower than B, C, or equivalent

Source: OSFI, Minimum Continuing Capital and Surplus Requirements (MCCSR) for Life Insurance Companies, Guideline A, Office of the Superintendent of Financial Institutions, Ottawa, Canada, October 2004.

During the years since then, the MCT has developed and the latest version was published in 2003 (OSFI, 2003).

Under the MCT (see OSFI, 2003), there are two main triggers, the minimum capital test and the target capital level. The risk-based capital adequacy framework assesses the risk of assets, policy liabilities, and off-balance-sheet exposures by applying factors. The companies are required to meet a *capital available-to-capital required test*. At a minimum, the companies are required to maintain an MCT ratio of 100%. OSFI also expects all undertakings to establish a target capital level and maintain ongoing capital, at no less than the supervisory target of 150% MCT.

The *capital available* consists of instruments with residual rights that are subordinate to the rights of policyholders and that will be outstanding over the medium term, and an amount that reflects changes in the market value of investments. The *capital required* is the sum of capital for on-balance-sheet assets (as listed in OSFI, 2003), margin for unearned premiums and unpaid claims, catastrophe reserves and additional policy provisions, an amount for reinsurance ceded to unregistered reinsurers, and capital for off-balance-sheet exposures. The same considerations on measuring capital adequacy are made as for the MCCSR.

As an example of risk factors we look at the risks for policy liabilities, reflecting the insurer's risk profile by individual classes of insurance and results in margin requirements on policy liabilities. The risk is divided into three parts: variation in claims' provisions (unpaid claims), possible inadequacy of provisions for unearned premiums, and occurrence of catastrophes. For personal property and commercial property, the margin on unearned premiums is 8% and the margin on unpaid claims 5%.

The MCT applies only to Canadian non-life undertakings. A similar test for branches of foreign undertakings is the branch adequacy and assets test (BAAT).

6.3   Denmark

Denmark, as one of the member states in the EU, started to move in the direction of Solvency II before its appearance.

Assets like stocks and property had been valued at market values. New rules were introduced to valuate fixed-rate assets, like bonds, at market value. In 1998 the Danish Ministry of Economic Affairs appointed a Market Value Committee to consider valuation principles for the technical provisions, as assets, like bonds, were valuated at market values. This was done as a measure to increase the efficiency on the financial market (see, e.g., Jørgensen, 2004). The committee's recommendations led the Danish Financial Supervisory Authority (FSA) to issue a new executive order regarding the annual financial reports: all assets must be reported as market value[12] from 2002 on. Because market discipline will require transparency, no hidden reserves should be included in the liabilities. The Danish FSA executive order also required the companies to report liabilities at market value from January 1, 2003 on.

According to the executive order, the FSA required a new decomposition of the liability side of the life and pension companies' balance sheet. A comparison of the old and new liability sides in the balance sheet is given in Figure 6.1.

- The old liability side: The *technical provision* was estimated on a policy-by-policy basis using the different technical insurance basis. Nondistributed profits were listed in an entity called *bonus smoothing reserves* (bonus equalization reserves).
- The new liability side: The technical provisions are split into three parts, and for each a market value must be reported. Two of the new components relate to the individual *bonus potential* of the policies (on respective paid premiums on future premiums). The third part is the *guaranteed benefits*. The bonus smoothing (equalization) reserves were renamed the *collective bonus (potential)*.

To illustrate the three market value parts of the old technical provisions, we can look at Figure 6.2. The variables A to D in Figure 6.1 are defined in Figure 6.2.

---

[12] The Danish insurance industry has not yet adopted the fair value terminology (see Jørgensen, 2004), which is our source for this description.

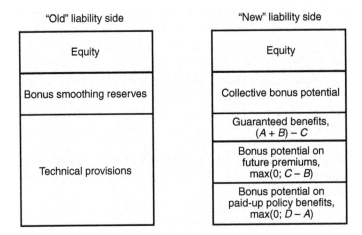

FIGURE 6.1
The old and new sides of the balance sheet. Only the most significant entries are listed. (From Jørgensen, P.L., *Scand. Actuarial J.*, 5, 372–394, 2004. With permission.)

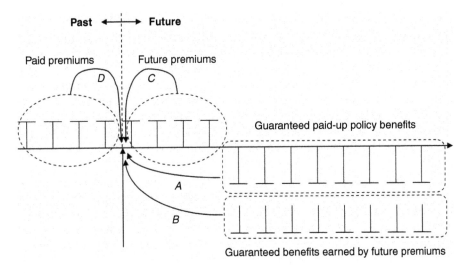

FIGURE 6.2
The cash flow of a life insurance policy. (From Jørgensen, P.L., *Scand. Actuarial J.*, 5, 372–394, 2004. With permission.)

The present values of guarantees and premiums (paid up or future) are defined as the market values.

- Guarantee benefits = (A + B) − C: The difference between the present value of the contractually specified guaranteed benefits (A + B) and the present value of the remaining future premiums (C). This can be interpreted as the market value of the remaining promised net cash flows. The guaranteed benefits can be negative.

- Bonus potential on future premiums = max{0, C − B}: The maximum of zero and the difference between the present value of the remaining future premiums (C) and the present value of the part of all total guaranteed benefits that are related to future premiums (B).

- Bonus potential on paid premiums = max{0, D − A}: The maximum of zero and the difference between the value of the retrospective reserve (D) and the present value of the part of the guaranteed benefits that are specifically related to the free policy (A).

The present-value calculations are done either by an *estimated term structure of zero coupon interest rates* or by a *simplified market interest rate*, determined and published by the FSA.

The calculation of market value (the present value) will change in accordance with the implementation of the new accounting rules from International Accounting Standards Board (IASB) (see Section 5.2).

For a thorough discussion on the new Danish fair valuation system, see Jørgensen (2004).

As part of the supervisory review process, the life insurance and pension companies have been requested to twice a year provide the Danish FSA with reports on the effect on some key solvency variables when a number of market variables are changed (stress test). The introduction of this system was done in order to accommodate the life and pensions companies' desire to be allowed to invest a larger share of their assets in less gilt-edged securities. The maximum portfolio weight in these assets (e.g., stocks) was 40% up to 1997, 50% between 1997 and 2000, and 70% from 2001.

Non-life insurance companies are from 2004 on also requested to report stress tests. The first report was for the financial year 2003 (Finanstilsynet, 2003).

As a step in the valuation of the investment risk, as compared to the available solvency capital plus the technical provisions, the Danish FSA has introduced two simple scenario tests: a red scenario and a yellow scenario. This new risk-based system is known as the *traffic light system*. A company can be in either the red, yellow, or green light zone. Some of the weights in the stress tests are given in the table on the following page.

A critical value is defined as the solvency margin requirement minus 3% of the technical provisions for life insurance. A life insurance company will be in the red light zone if its base capital is below the critical level. We will discuss these levels later.

The companies have to complete nine stress tests reports, called RG here (we will later look at the intervention that the FSA can take):

1. RG01: Summary report — There is one column for the red light scenario and one for the yellow one. Summarized data from the other eight reports are copied to this summary; see its template below.

Scenarios[a]	Red Light Zone (Medium Negative Market Development)	Yellow Light Zone (Huge Negative Market Development)
Increase or decrease in the interest rate level (the worst) RG03	Long term: ±1.0% Medium term: ±0.85% Short term: ±0.7	Long term: ±1.43% Medium term: ±1.18% Short term: ±1.0%
Increase or decrease in the interest rate level on life provisions (the worst) RG04	±0.7% A change of discount rate of + 0.595	±1.0% A change of discount rate of + 0.8
Decrease in stock prices RG05	12%	30%
Decrease in real estate investment values RG07	8%	12%
Decrease in values of counterparties (risk weighted) RG08	8%	8%
Decrease in currencies RG06	VaR 99% in 10 days	VaR 99.5% in 10 days

[a] The RGs referred to are references to the reports discussed below.

2. RG02: Basic information — Includes basic information as total balance, the available and required solvency margins *before* the stress tests are done, and for life companies, e.g., the collective bonus potential (see Figure 6.1) before the stress tests.

3. RG03: Interest rate risk (interest-bearing receivables) — The market values on bonds, financial instruments, etc., and the effect of interest rate changes, according to the table above. The worst of increase or decrease scenarios for both the red and yellow light calculations.

4. RG04: Interest rate risk (life provisions) — As for RG03, but on the technical provision part of the balance sheet (see Figure 6.1).

5. RG05: Stock price risk — The market value and the effects of the stress test are recorded.

6. RG06: Currency risk — The eight most common currencies in the Danish market are given and value-at-risk measures at 99.0 and 99.5% levels are calculated for the sum of the expected currencies. The FSA has two reports on its Web site that describe the procedures.[13]

7. RG07: Real estate investment risk — The market value and the effects of the stress test are recorded.

8. RG08: Credit and counterparty risks — The risks depend on the instrument's maturity. The capital assessment is set to 8% of the instrument's risk-weighted value. Both unweighted and weighted values are recorded.

---

[13] www.finanstilsynet.dk/sw303.asp.

9. RG09: Subsidiary risks — The exposure of the mother company to risk scenarios in insurance subsidiaries is quoted in RG09.

The RG01 Template

Summary (million kr)	Red Light Scenario	Yellow Light Scenario
1. Worst interest rate scenario		
2. Interest rate risk (interest-bearing receivables): RG03		
3. Stock price risk: RG05		
4. Currency risk: RG06		
5. Real estate investment risk: RG07		
6. Credit and counterparty risks: RG08		
7. Subsidiary risks: RG09		
8. Total risks		
9. Interest rate risk (life provisions): RG04		
10. Interest rate risk (workers' compensation): RG04		
Life insurance and pension companies		
11. Base capital after the tests		
12. Solvency requirement after the tests		
13. 3% of the provisions after the tests		
14. Available capital after the tests		
15. Collective bonus potential (cf. Figure 6.1) after the tests		
16. Bonus potential on free policies (terminated policies)		
17. Risk adjusted solvency ratio		
18. The company is in		
Non-life and reinsurance companies (including workers' compensation)		
19. Base capital after the tests		
20. Solvency requirement		
21. Available capital		
22. The company is in		

The calculations are made for both the red and the yellow light scenarios:

Base capital after the test: (Base capital before the test) – (that part of the loss, due to the test, that is part of the base capital) – (that part of the insurer's loss not taken by the collective bonus potential and the bonus potential from free policies).

Solvency requirement after the test: (Solvency requirement before the test) + 0.04 (the change in technical provisions due to interest rate changes; from RG04).

Critical level: (Solvency requirement after the test) – 0.03 (technical provisions after the test; from RG04).

Available capital: Base capital after the test – critical level (after test).

The traffic light test is now done, by comparing the available capital to zero:

If in the red light scenario, available capital $\leq 0 \rightarrow$ RED LIGHT

If in the yellow light scenario, available capital $\leq 0 \rightarrow$ YELLOW LIGHT

Otherwise, $\rightarrow$ GREEN LIGHT

If a company is in the red light zone, the policyholders' interests are considered to be in danger and the company is requested to report monthly (or even more frequently) on solvency to the Danish FSA, and also report in what way they may issue orders to reduce the risk profile. The Danish FSA will discuss the situation with the company and after that they will decide what action will be taken. A company in the red light zone is not allowed to increase the risks according to this scenario.

The yellow light scenario is seen as an extreme scenario, and therefore, if a company is in this zone, the policyholders' interests are only considered to be potentially in danger. The supervision is nevertheless intensified and the company is required to report quarterly on solvency.

The Danish stress testing system for life insurance uses the capital requirements of credit institutions for calculating the capital requirement for the financial risks and to that adds 1% of the technical reserves and 0.3% of the sums at risk as the capital requirement. There are insurance companies that have implemented investment plans with a high match to liabilities (low asset–liability mismatch). If the minimum capital requirement, as Solvency I, is defined as a fixed percentage of liabilities, it will not reflect the immunization that is made. Thus, the Danish experience has shown that in a number of cases, the stress test led to a lower capital requirement than the capital requirement of the Solvency I directives.[14] However, companies still have to fulfill the Solvency I minimum capital requirement.

---

[14] The Danish FSA sent a letter to the European Commission on this subject on November 11, 2004. The letter was signed by Jens Pagter Kristensen.

6.4   Finland

At the beginning of 1980, the Ministry of Social Affairs and Health[15] established a research group (RG) to revise the rules concerning the equalization
reserve and to examine solvency policies in a broad sense (Pentikäinen,
1982). The work was done for the non-life business.

The stochastic character of the insurance business was usually not taken
into account in earlier insurance legislation. Due to the work of Teivo Pentikäinen and others in Finland, this feature was recognized and taken into
account when the new Insurance Company Act was introduced in Finland
in 1953. A special equalization reserve was introduced and added to the
conventional reserves and solvency margins, which usually did not take into
account the stochastic fluctuation. The equalization reserve was designed to
protect against fluctuations in the annual claims' amount and to carry forward underwriting profits or losses from year to year. This is a condition
that is most essential for sound fiscal insurance operations. The concept of
risk theory was introduced in the new Insurance Company Act of 1953
(Chapter 10, paragraph 2, Section 3; see Pentikäinen, 1982, pp. 7.1–7.3): "The
claims reserve is equivalent to the amount of incurred but outstanding claims
and other expenditures related thereto and includes an amount of equalization, calculated according to risk theory, to provide for years with a high
loss frequency."

The minimum legal solvency margin must be related to the size of the
business's own retention. The *solvency test* was limited to concern only the
stochastic fluctuations of the claim expenditure. Other risks were not considered at that time. We will see later that during the 1990s the asset/
investment risk was also taken into account.

The idea of the Finnish solvency test was that the solvency margin plus
premiums and investment income should exceed 1 year's claims and other
expenses with a probability of 99.0%. The test is a going-concern system
with an underlying assumption that a portfolio transfer is possible in case
of insolvency.

The concept of *working capital*, i.e., available solvency margin, was introduced. The rule was that the capital required was approximately proportional to the square root of the volume (B, earned premiums). The parabolic
square root curve was approximated by a broken line, defined as

$$U = \begin{array}{ll} 0.2 + 0.2B & \textit{for}\ \ B < 4\ \ \textit{millions}\ \ \textit{FIM} \\ 0.6 + 0.1B & \textit{for}\ \ B \geq 4\ \ \textit{millions}\ \ \textit{FIM} \end{array}$$

---

[15] Its insurance department was earlier the supervisory authority of the insurance industry in
Finland.

The methodology of the Finnish solvency approach is described by Pentikäinen (1982 — general aspects) and Rantala (1982 — theoretical model); see also Pentikäinen et al. (1989).

6.4.1 The Risk Theoretical Model

The theoretical model will be able to capture both the empirical data that was analyzed by the RG and experience concerning the business and its risks. The basic transition formula defining the underwriting profit is

$$\Delta U = B + I - X - C - D$$

where $\Delta U = U - U_{-1}$ is the underwriting profit or loss accumulated into the risk reserve U, B is the earned premium income (including safety loadings and administration cost loadings), I is the net investment income, X is the claims paid and outstanding claims, C is the cost of administrations, reinsurance, etc., and D is the dividends, bonuses, etc.

All calculation is based on the net retention, i.e., deducting the shares of reinsurance from all the calculations.

The financial state of the insurer is measured at the balance sheet day, and U_{-1} means the risk reserve at the beginning of the financial year.

Yield of interest: Partition I into two components $I = i_{tot}W_{-1} + i_{tot}U_{-1}$, where i_{tot} is the rate of interest and W_{-1} and U_{-1} are the underwriting reserve and the solvency margin (risk reserve) at the beginning of the year, respectively. Replacing ΔU by $U - U_{-1}$ we have the basic formula $U = (1 + i_{tot})U_{-1} + B + i_{tot}W_{-1} - X - C - D$.

Dividends: The dividends D can be included in the expenses C, and hence we have $U = (1 + i_{tot})U_{-1} + B + i_{tot}W_{-1} - X - C$.

Solvency ratio: The solvency ratio $u = U/B$ is the basic solvency variable studied by the Finnish RG. Define $f = X/B$, $c = C/B$, and $w = W/B$, and then we have

$$u = (1 + i_{tot})\frac{B_{-1}}{B}u_{-1} + 1 + i_{tot}\frac{B_{-1}}{B}w_{-1} - f - c$$

It is convenient to replace the interest rate i_{tot} and the interest factor $r_{tot} = 1 + i_{tot}$ by $i_{rtot} = i_{tot}B_{-1}/B$ *and* $r_{rtot} = r_{tot}B_{-1}/B$. The interest rate and the interest factor are modified in relation to the growth of the premium volume B. Let $B/B_{-1} \equiv r_r \equiv 1 + i_r$.

Hence, $i_{rtot} = i_{tot} / r_r$ and $r_{rtot} = r_{tot} / r_r$ (cf. Pentikäinen, 1982, pp. 3.1–3.6). Using these notations we have

$$u = r_{rtot}u_{-1} + 1 + c + i_{rtot}w_{-1} - f$$

Let λ_p be the safety loading. What is left of the earned premiums when the loading for expenses and safety loadings are deducted is the net premiums $P = (1 - c - \lambda_p)B$. By definition, $P = E(X)$, the expected mean value of the claims; thus, we can define the mean loss ratio as $f = E(X) / B$ and hence $f = (1 - c - \lambda_p)$.

The solvency ratio can now be written as

$$u = r_{rtot}u_{-1} + \lambda_p + i_{rtot}w_{-1} + \overline{f} - f$$

Aggregate safety loading: The RG observed that the ratio $w = W/B$ is fairly constant and approximately equal to 1.71 (it varies between different classes of business and insurers). An aggregated safety loading is defined as

$$\lambda = \lambda_p + i_{rtot}w_{-1} \approx \lambda_p + i_{rtot}1.71$$

Basic equation: The basic equation, which was used for the analysis of the underwriting business and simulations, is now defined as

$$u = r_{rtot}u_{-1} + \lambda + \overline{f} - f$$

and interpreted in the following way. The initial solvency amount is increased by the yield of interest, but decreased by inflation and real growth. The safety loading increases due to the ordinary loading and yield of interest added to the underwriting reserve. The fluctuation deviation of the loss ratio is moved to u.

The basic equation done stochastically makes it possible to compute the future flow of the solvency ratio u year by year. A bundle of simulations are given in Figure 6.3. The bold lines are confidence contours (99%). Long-term cycles, as has been recognized, are not incorporated in the model.

When long-term cycles are included in the model we get the other classical figure, Figure 6.4.

All the outcomes of the simulation showed that the long-term cycles have a strong effect on the solvency ratio. This is a main reason for introducing the equalization reserve.

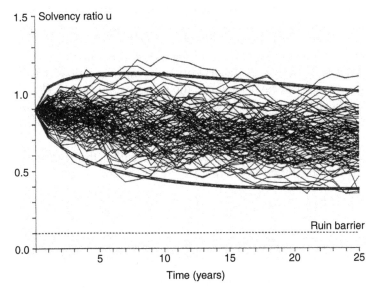

FIGURE 6.3
Simulations of the solvency ratio *u*. Long-term cycles *not* included in the model. Number of ruins: 0/100. (From Pentikäinen, T., Ed., *Solvency of Insurers and Equalization Reserves*, Vol. I, *General Aspects*, Insurance Publ. Co. Ltd., Helsinki, 1982. With permission.)

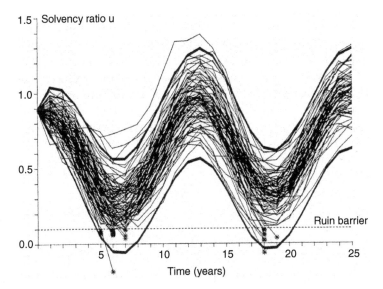

FIGURE 6.4
Simulations of the solvency ratio *u*. Long-term cycles included in the model. Number of ruins: 24/100. (From Pentikäinen, T., Ed., *Solvency of Insurers and Equalization Reserves*, Vol. I, *General Aspects*, Insurance Publ. Co. Ltd., Helsinki, 1982. With permission.)

The RG proposed to the Ministry of Social Affairs and Health that the solvency test should be preserved. The equalization reserve has been the far most important part of the Finnish non-life solvency regime. From the beginning of the 1950s, the equalization reserve had been a part of the reserve for outstanding claims, which had given biased information about the insurance industry. The RG left this question open, but the ministry decided that from the 1981 report the companies should state the amount of the equalization reserve outside the balance sheet.

The minimum solvency: U_{min} , which is often called the ruin barrier or winding-up barrier, is defined by the RG as

$$U_{min} = (0.214\overline{f} - 1.09\lambda)B + k_\alpha\sigma_X + 0.7M + U_c$$

where 0.214 and 1.09 are auxiliary constants depending on the long-term cycles, inflation, growth, and interest rate; f is the average loss ratio; λ is aggregate safety loading (standard value = 0.041); B is the gross premiums on the company's net retention; k_α is the safety factor corresponding to the confidence level proposed (1% ruin probability gives the safety factor 2.33); σ_X is the standard deviation of the annual fluctuation of the underwriting results, or of the annual aggregate loss; M is the maximum net retention per one risk unit; and U is the provisions for the risk of supercatastrophes if the company is carrying on international business.

The first term represents the maximum drop of a midline in Figure 6.4. The two last terms represent the breadth of the stochastic bundle, i.e., the confidence region. The term $0.7M$ is due to the skewness of the aggregate claims distribution. The RG proposed that every insurance company annually should check whether the actual solvency U is at least U_{min}. For small companies another rule is prescribed: $U_{min} \geq 2M$. This U_{min} will later be renamed U_{min}(insurance).

The final rule accepted by the ministry is

$$U_{min} = 0.214\sum P_j - 0.043\sum B_j + \sqrt{(7.9\sum\beta_jM_jP_j + 9.2\sum\sigma_j^2P_j^2)} + M + U_c$$

where the sums and subindexes j refer to portfolio sections and $P_j = f_jB_j$ is the net risk premiums, section j; β_j is the constants given in a table by the ministry and indicating the degree of heterogeneity of the portfolio, derived from the claims' size distributions; and σ_j is the standard deviations of the short-term variations in the risk intensities.

For the equalization reserve, the RG introduced a new concept, the *target zone*: the target zone consists of an upper limit (U_2) and a lower limit (U_1), or if the solvency ratio is used, (u_2) and (u_1). The older transfer rule should be employed. The lower limit is an early-warning limit and the upper limit is a taxation limit. The transfer to and from the equalization reserve is determined by (nearly identical to the old transfer rule)

$$\Delta U = r_{rtot} U_{-1} + (\overline{f} - f + a)B$$

where a is a control parameter, which may be set, within limits, by the insurer.

The two limits are discussed in Rantala (1982, chap. 7). They are:

The lower limit (early-warning limit):

$$U_1 = 0.221 \sum P_j - 1.13 \sum a_j B_j + \sqrt{8.4 \sum \beta_j M_j P_j + 9.9 \sum \sigma_j^2 P_j^2} + 0.7M + U_c$$

The upper limit (taxation limit):

$$U_2 = 0.75 \sum P_j + \sqrt{200 \sum \beta_j M_j P_j + 160 \sum \sigma_j^2 P_j^2} + \min(U_c, (U_c')^+)$$

where $(U_c')^+$ is the positive part of the aggregate amount of the equalization reserves of domestic reinsurance and foreign business.

The maximum of the equalization reserve is calculated as $U_{max} = 1.2 U_2(\overline{M_j})$, where $U_2(\overline{M_j})$ is given by the upper limit, with M replaced by $\overline{M_j} = 0.04(B_{tot} + V_{tot,-1} - V_{tot})$, and B is the gross premiums written and V is the total amount of unearned premiums. The upper limit is usually effective to return the equalization reserve back to the target zone when necessary. However, there are situations when this procedure is not functioning and then we have this maximum value instead.

Since the beginning of 1995 Finland has been a member of the European Union and as such has implemented the non-life and life directives. However, Finland has been keeping the equalization reserve as a complement to the ordinary solvency assessment.

The solvency test must be fulfilled simultaneously with the solvency margin requirements according to the EU directive.

The solvency system, as described above, concentrates on controlling the insurance risk. In 1994 the asset risk was taken into account in a rather rough and simple way. A discussion on a base asset model is given in Pentikäinen

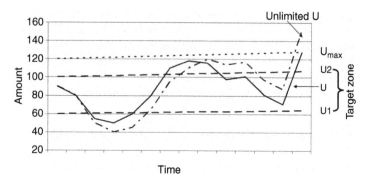

FIGURE 6.5
For sake of simplicity, limits to equalization provisions are presented as a straight line and a dashed–dotted line. In practice, they fluctuate from year to year, depending on insurance portfolio and net retention limits of the company. (From The Finnish Insurance Supervisory Authority. With permission.)

et al. (1994).[16] The solvency requirement, in a preliminary form, regarding the investment risk was introduced for statutory occupational pensions in 1997. The most important thing is that the requirement reflects the structure of the investment portfolio. Since then the system has been introduced to pension funds, pension foundations, and certain special pension institutions. After some amendments the system attained its final form in 1999 (Tuomi-koski, 2000). In the same year, the latest revision for non-life undertakings came into force.

The FIM group also proposed rules for taking account of the asset risk in life insurance. The proposed system, which is an extension of the solvency margin in the EU directive, has not yet been introduced into the insurance regulations.

The calculation is assessed in two parts:

Part 1: The fluctuation of the annual result was studied as a stochastic process of claims incurred — assuming all other variables to be deterministic. This produced the minimum solvency margin U_{min}(insurance) covering the insurance risk (see above).

Part 2: The investment return, including unrealized changes in asset values, was modeled as a separate stochastic process.[17] This produced the minimum solvency margin U_{min}(investment) covering the asset risk. This is written as

$$U_{min}(investment) = \frac{A}{100}\left(-\sum_i \alpha_i (m_i - m^*) + 2.15\sqrt{\sum_{i,j} \alpha_i \alpha_j s_i s_j r_{ij}}\right)$$

---

[16] The Finnish Insurance Modeling Group (FIM group).
[17] The model is a variant of the Wilkie model (cf. Pentikäinen et al., 1994).

where A is the total amount of assets, at market value, covering the technical provisions and the solvency requirement needed to cover insurance risk; α_i is the proportion of assets belonging to category i; m_i is the expected return on assets I; m^* is the required overall return from the insurance business on the assets; s_i is the standard deviation of the returns on assets i; and r_{ij} is the correlation between assets belonging to categories i and j. The first term in the formula represents the expected value of the investment yield exceeding the required yield, and the second term is the corresponding standard deviation. The supervisor prescribes the expected returns, standard deviations, and correlations.

The maximum probability of failure was set to 1% for both formulas. The new minimum solvency requirement was defined as

$$U_{min} = 0.80\Big[U_{min}(insurance) + \max(U_{min}(investment), Z_{nr}\Big]$$

where Z_{nr} is the maximum net retention. So if the maximum retention is larger than the risk from the investments, it is used instead. As the insurance risk and investment risk are *almost* independent, the factor 0.8 is used when combing the two risks.

There is also a target level for the solvency capital of the company. This target level is defined as $U_{min} + U_{min}(investment)$. If a company's solvency capital falls between the target level and the minimum level, it must present a plan of action, similar to the one for a new company. A good reference on the new version of the Finnish system is given in Annex 6[18] to MARKT (2002f).

According to a report[19] from the Finnish Ministry of Social Affairs and Health Insurance Department, the solvency requirement among general non-life undertakings due to asset risk is approximately 20 to 40% of premiums earned. The requirement for specialized or small companies is much higher.

Insurance undertakings may propose to the authority its own estimation of certain parameters and still remain in the solvency test framework.

6.5   The Netherlands

In mid-1999 the Dutch pensions and insurance supervisory authority, Pensioen- & Verzekeringskamer (PVK),[20] proposed a new assessment

---

[18] Annex 6: Equalization Provisions: Presentation of the Finnish System. Pertti Pulkkinen, Ministry of Social Affairs and Health Insurance Department (October 10, 2002).
[19] A report from May 2, 2000: The Finnish Solvency Test of Non-Life Insurance Companies.
[20] PVK merged with the Dutch central bank (De Nederlandsche Bank, DNB) on October 30, 2004.

framework for financial testing (PVK, 2001). In this memorandum of principles the PVK states that it wants to align its financial requirements as closely as possible to policy and control systems of the individual institutions, also acknowledging the institution's own responsibility for its business practice and risk management. One of PVK's aims has been to bring its supervision in line with international developments, e.g., on accounting standards. Preferably, there should only be one set of accounts for both annual and supervisory reporting purposes. Additionally, the new solvency requirements shall be better tuned to the institution's risk profile than before.

In October 2004, a consultation paper on the new solvency framework[21] was published (PVK, 2004). In the current Dutch regulatory framework (apart from EU requirements, i.e., Solvency I), only life insurers and pension funds have rules for regular solvency assessment. The forthcoming rules will also include non-life insurance (including health insurance). The required confidence levels for insurance undertakings will be 99.5 and 97.5% for pension funds. It is planned that the new framework will be operational from book year 2006.

The new financial assessment framework (FAF) comprises three elements:

1. Both assets and liabilities should be valued at *realistic value*.
2. Within this valuation framework each insurance undertaking and pension fund has to perform a so-called *solvency test* each year.
3. Each insurance undertaking and pension fund has to do a so-called *continuity analysis*.

We will discuss these elements below.

6.5.1   Realistic Value

Within the FAF, *realistic value* is the key valuation principle for both assets and liabilities; the key underlying issues are consistency between assets and liabilities, transparency, and comparability. The realistic value of assets is simply equal to its actual market value.

Due to a lack of a market for insurance liabilities, the realistic value of liabilities should be defined as the realistic value of the assets that would replicate the liabilities. It is made up by the expected value (best estimate) plus a realistic risk surcharge (ideally a market value margin), i.e., *the value transferred to a knowledgeable and willing third party*. The expected value, or best estimate, is the present value of the expected cash flows arising from the liability, based on underwriting principles deemed to be realistic. The risk surcharge will make the adopted level of prudence explicit. Liability cash flows should be discounted by a series of risk-free spot yields. The PVK

---

[21] PVK, "Consultatiedocument Financieel toetsingskader," October 21, 2004. (The English version, titled "Financial Assessment Framework Consultation Document," can be downloaded from the PVK's Web site, www.dnb.nl.)

intends to prescribe and publish these rates on a regular basis.[22] Smaller pension and insurance companies may be allowed to use a sole discount rate during a transition period (Annex 6 in PVK, 2004). As soon as there is a definitive International Accounting Standard (IAS)/International Financial Reporting Standard (IFRS) for the realistic valuation liabilities, the PVK will consider its applicability.

The risk surcharge can be calculated by using an *internal model* that has to be stochastic. In that case, the outcome should correspond with the 75% confidence level, as in the Australian system[23]; alternatively, a so-called *standardized approach* can be applied. This will require the portfolio to be subdivided into homogenous risk groups. For each risk group margins must be calculated for specific types of underwriting risks on the basis of certain formulas, in which a certain base must be multiplied with a percentage from a prescribed table using certain risk indicators.

For life insurers eight risk groups are proposed. Risk groups 1 and 3 have mortality risk, groups 2, 5, and 7 longevity risk, and groups 4, 6, and 8 a mix of mortality and longevity risks. Their risk surcharges consist of two components. The first component is a risk surcharge for the uncertainty related to future mortality trend uncertainty (*TSO*). The related percentage depends on the average age, the average remaining term, and the average age at expiry. The second component is a charge for possible negative stochastic variances (*NSA*). Assume that the number of insured within risk group i is n_i, then

$$TSO_i = \text{base percentage}_i \times \text{basis for risk group } i$$

and

$$NSA_i = \frac{\text{base percentage}_i}{\sqrt{n_i}} \times \text{basis for risk group } i$$

The basis for risk group 1 is the risk capital, and for the others the best-estimate expected value.

The total risk surcharge is defined by

$$\text{Risk charge} = \sqrt{\left(TSO_{total}\right)^2 + \sum_{i=1}^{8} NSA_i^2}$$

---

[22] The PVK has released a separate consultation document on this: "Notitie methode renteterm-ijnsructuur," December 3, 2004. (The English version of this document, titled "Memorandum Method Term Structure FTK," can also be downloaded from the PVK's Web site, www.dnb.nl.)
[23] Note that this confidence level is assumed to give the same prudence as the old system in Australia, i.e., with the old actuarial prudence.

where

$$TSO_{total} = \sqrt{\left(TSO_1 + TSO_3\right)^2 + \left(TSO_2 + TSO_5 + TSO_7\right)^2} + TSO_4 + TSO_6 + TSO_8$$

Similar approaches, but with a lower number of homogeneous risk groups, are proposed for funeral in-kind insurance and pension business. The charges for non-life insurance (including health insurance) will (probably) be defined by multiplying certain percentage charges from a table book with the best-estimate expected value. The percentages will be set after a field test done by PVK. For more details, see Appendix 4 of PVK (2004).

Assets may only act as cover for the liabilities of the undertaking if the company has free disposal of them, as they only can serve one purpose. If they are, for example, used as collateral for a loan, they cannot be regarded as free.

6.5.2 Solvency Test

The solvency test will basically reflect two requirements:

1. The undertaking can withstand an instantaneous event resulting in a discontinuation of business.
2. It can withstand certain adverse scenarios emerging over the next 12 months, meaning that it can transfer the existing liabilities to another institution at the end of this period.

The first requirement implies an assessment of the financial consequences in the event of immediate discontinuation of business, in particular the potential effects on the liquidation values of the assets and liabilities. For the second requirement, five main risk categories are defined: market risk, credit risk, underwriting risk, operational risk, and concentration risk. As there are no simple standard approaches for operational and concentration risks in the insurance world, the PVK will not include any risk surcharge for these risks until 2008. Three different approaches are described: a simplified method, a standardized approach, and the use of an internal model.

Smaller companies, which qualify as a result of their risk profile and operations, may perform the solvency test using a *simplified method*. When a company is given permission to use this method, the solvency test is limited to assessing the realistic value of freely disposable assets covering the foreseeable liabilities at realistic value.

Test: Realistic value of assets$_0$ > Realistic value of liabilities$_0$

To qualify for this approach, the undertaking must meet the following conditions:

1. The product range must be simple and limited, in particular:
   - The products on offer must have a simple structure.
   - They should not have embedded options, guarantees, or profit sharing.
   - For pension funds: The pension scheme is simple and unambiguous.
   - For non-life insurers: The institution only operates in a few, less dangerous branches.
2. The undertaking has a simple and risk-avoiding investment policy.
3. It has simple operations.
4. Required financial characteristics:
   - Insurers: Its solvency ratio, i.e., available divided by required solvency capital according to current EU requirements (Solvency I), should be higher than 250%.
   - Pension funds: The institution has an actual asset–liability ratio (based on realistic value) of at least 130%.

In the *standardized method*, the required risk surcharge for market, credit, and underwriting is determined. Thereafter, the required solvency margins for the individual risks are aggregated into one overall solvency margin using a prescribed formula:

Test: Realistic value of assets$_0$ > Realistic value of liabilities$_0$ + Solvency margin$_0$

6.5.2.1 Market Risks

The following risks are included in the market risk: interest rates, inflation, equity, real estate, commodities, and foreign exchange, including derivatives. Their required solvency margins are calculated as the effect of certain prescribed shocks on the surplus. The sizes of shocks, which are summarized below, are higher for insurance undertakings than for pension funds:

- Interest rate risk: Fixed percentage changes of spot yields.
- Inflation risk: Maximum of effects of best-estimate inflation × 1.5 and best-estimate inflation/1.5 (pension funds = 1.3 instead of 1.5).
- Equity risk: Mature markets, –40% (pension funds = –25%); emerging markets and private equity, –45% (pension funds = –30%); and implied volatility, –25% (pension funds = –25%). Hedge funds must be included in full in the scenarios.

- Real estate risk: –20% (pension funds = –15%).
- Commodity risk: –40% (pension funds = –30%).
- Currency risk against euro: –25% (pension funds = –20%).

All individual types of market risk are considered mutually independent apart from the interest and equity risks.

6.5.2.2   Credit Risk

The actual level of credit risk related to the assets is already expressed in the credit spread within its realistic values. The required additional solvency margin is derived by changing the observed credit spread by a fixed factor of 60% (pension funds = 40%). No margins are prescribed for credit risks related to reinsurers or intermediaries.

6.5.2.3   Underwriting Risks

Again, the solvency margin for underwriting risks will be factor based for the same homogeneous risk groups as with the calculation of the risk margins in the realistic value of the insurance liabilities. For life insurance, funeral in-kind insurance, and pension funds, the margins will be similar to the NSA factors, as described before. However, the percentages will be different.

The required solvency for non-life insurance (including health insurance) has two parts, one for claims/benefits outstanding, $OC(i)$, and one for current risks, $CR(i)$. They will be calculated as follows:

$$OC(i) = \text{Base percentage claims outstanding}_i \times F_i(n_i) \times \text{Realistic value liabilities}_i$$

$$CR(i) = \text{Base percentage current risk}_i \times G_i(n_i) \times \text{Realistic value liabilities}_i$$

The functions $F(\cdot)$ and $G(\cdot)$ will be defined later; n_i is the total number of insured in risk group i at the end of the financial year.

The total required solvency margin for underwriting risk related to group i is then calculated as the sum of $OC(i)$ and $CR(i)$, while the overall solvency margin for all underwritings is defined as

$$S_6 = \sqrt{\sum_{i=1}^{k}\sum_{j=1}^{k} \rho_{ij} x(i) x(j)}$$

where ρ_{ij} is the correlation between risk groups i and j ($\rho_{ii} = 1$).

Finally, the overall required solvency margin (the minimum capital requirement) for market risk, credit risk, and underwriting risk is determined by aggregating the individual margins as follows (for more details, see PVK, 2004, Appendix 4, Part B):

$$SM_{total} = \sqrt{S_1^2 + S_2^2 + 2\rho S_1 S_2 + S_3^2 + S_4^2 + S_5^2 + S_6^2}$$

where S_1 is the required solvency for interest rate risk (and, if applicable, inflation risk), S_2 is the required solvency for variable-yield securities, being the sum of the desired solvency for equities and real estate, S_3 is the required solvency for foreign exchange risk, S_4 is the required solvency for commodities, S_5 is the required solvency for credit risk, S_6 is the required solvency for underwriting risk, and correlation coefficient $\rho = 0.8$ (pension funds = 0.65). No other correlations are assumed.

The PVK assumes that undertakings that use internal models compute their risk surcharge from stochastic models for cash flows from liabilities and investments.

Test: Realistic value of assets$_0$ >
Realistic value of liabilities$_0$ + Solvency margin$_0$

with solvency margin$_0$ such that

Pr(Realistic value of assets$_1$ > Realistic value of liabilities$_1$)
= 0.995 (pension funds = 0.975)

The use of internal models is allowed in order to promote proper risk management. However, prior to its use, the company should demonstrate to the authority that certain requirements on the quality of the model and its internal organization have been fulfilled. The consultation document lists the following requirements on the model's assumptions, internal control procedures, and quality of data:

- *Quality of the model*: The risk factors that are modeled must be relevant to the institution. The institution has to perform a risk analysis of the relevant products/activities and must be able to show that the model properly estimates the modeled risks and that the assumptions are reasonable. Stochastic modeling of the probability distribution of the realistic value of shareholders' equity, at a time horizon of 1 year, is the basis for the internal model and solvency testing.
- *Internal organization*: The model should be integrated in the day-to-day risk management and hence be a part of the process of planning, monitoring, and managing the institution's risk profile. There must

be an independent risk management function, responsible for the internal model (its design, implementation, and maintenance), and sufficient staff with necessary skills to handle the model. There must also be a proper process for validation of the model. The model is used to evaluate risk limits, assessed by stress tests.

- *Documentation and reporting to the PVK*: The documentation of the internal model must be detailed and include information on the theoretical basis of the models and empirical evidence. It should also include an analysis of risk mitigation measures that have been taken. To the PVK, the institution must report on the desired solvency determined by the internal model, its associated probability, and the underlying probability distribution of the possible outcomes. As a reference point, the company has to compute the outcomes of the standardized method and compare them with the results of the internal model.

An institution may provide supporting evidence with respect to the requirements for the internal organization by an independent third party, such as an auditor. The approval of a model is provisional, and it will be tested regularly.

As long as the Solvency I requirements are in force, the proposed framework cannot lead to additional quantitative solvency requirements for insurers. It is expected that the solvency test of the FAF will be replaced by the future solvency capital requirement (SCR) requirement as apart of Solvency II. Thus, this test serves mainly as an additional tool for the supervisor to get a better insight into the financial position. For *pension funds* the FAF will be included in the new Dutch Pension Act, as there are currently no quantitative requirements from the EU.

6.5.3   The Continuity Analysis

The basis for the continuity analysis is the assumption that the company will continue to grow, introduce new products, change its investment policy, etc.; i.e., it will keep operating on *a going-concern basis*. Consequently, the undertaking's long-term objectives, strategy, and expectations should be taken into account. It should therefore allow the board of an undertaking, but also the PVK, to identify whether the undertaking is in a position to continue meeting its solvency requirements in the future. The continuity analysis will not result in additional capital requirements. Instead, it should be considered as an assessment tool for the supervisory review process, i.e., a part of pillar II.

The analysis charts future developments and evaluates whether long-term risks fall within the relevant risk standard. All aspects of continuity will be covered by the analysis. A spectrum of methods of analysis can be used, e.g., ALM techniques, profit testing models, embedded value techniques, etc. The horizon for the projection is 15 years for pension funds, 5 years for life insurers (including funeral insurance in kind), and 3 years for non-life insurers.

The reporting framework of the continuity analysis comprises six parts. In particular, the undertaking must describe:

1. Its business objectives, ambitions, policy, and policy instruments
2. Its best-estimate assumptions and expectations on the economic environment for the future
3. The quantified results of the assumed best-estimate (basic) scenario
4. A sensitivity analysis
5. Stress tests for the three main risks
6. An analysis of differences between previous expectations and realizations

All supervised institutions must have performed their first continuity analysis prior to 2008. For more details on the continuity analysis, see Chapter 5 of PVK (2004).

6.6 Singapore

In 2001 the Monetary Authority of Singapore (MAS) announced the adoption of a new and risk-based statutory solvency system. Draft amendments to the Insurance Act to establish the new system were released during the spring of 2003. In November the same year a consultation paper was published that described the proposals, including comments from the insurance industry (MAS, 2003). In August 2004, MAS announced the new framework for a risk-based capital (RBC) system. At the same time, MAS also issued five regulations, one notice, and one guideline.[24] The new system came into force at once.

The *valuation of assets and liabilities* should be based on market and realistic valuation, i.e., a shifting from the earlier philosophy with implicit margins and prudence in the estimates. The bases of valuation of assets and liabilities are specified in Parts IV and V of the "Insurance (Valuation and Capital) Regulations 2004" (MAS, 2004a). If the basis is not specified in these parts, then the insurer shall value the asset–liability in accordance with the accounting standards and sound actuarial principles. Such issues as overconcentration in certain types of assets are dealt with in the capital requirement rules.

---

[24] *Regulations*: Valuation and capital (MAS, 2004a), accounts and statements (MAS, 2004b), actuaries, transitional and savings provisions, and general provisions. *Notice*: On valuation of policy liabilities of life business. *Guideline*: On valuation of policy liabilities of general business.

In 2001 MAS imposed a realistic valuation requirement in the non-life sector. The non-life companies were required to determine *best estimates* (BE) of their insurance liabilities and additional *provisions for adverse deviation* (PAD), i.e., a risk margin that should reach at least 75% of sufficiency. The same type of liability valuation is now used by the life insurance sector.

The best estimate is determined by:

- Projecting future cash flows using realistic assumptions, including assumptions on expenses, mortality and morbidity rates, lapse rates, etc.

- Discounting these cash flow streams at appropriate interest rates.

The risk margin PAD is determined by using more conservative assumptions in the projection to reflect the uncertainty of the BE. This approach is similar to the entity-specific approach discussed by IASB (see Section 5.2) and should be based on sound actuarial principles.

6.6.1   Valuation of Assets (MAS, 2004a, Part IV)

The valuation of assets of an insurance fund should be made in accordance with the accounting standards if not otherwise specified in Part IV of MAS (2004a). Deposits and intangible assets, including goodwill, are not treated as assets. The valuation is briefly made in the following way:

- *Equity securities*: Its market value or its net realizable value.
- *Debt securities*: Its market value or its net realizable value.
- *Land and buildings*: Its estimated market value.
- *Loans*: Its aggregated principle amounts outstanding less any provision for doubtful debt.
- *Cash and deposits*: Its nominal amount after deducting any amount deemed uncollectible or its market value.
- *Outstanding premiums and agents' balances*: Its aggregated principle amounts outstanding less any provision for doubtful debt.
- *Deposits withheld by cedants*: Its aggregated amounts of deposits outstanding less any amount deemed uncollectible.
- *Reinsurance recoverable*: Its aggregated amounts of reinsurance recoverable outstanding less any provision for doubtful debt.

6.6.2   Valuation of Liabilities (MAS, 2004a, Part V)

The valuation of liabilities of an insurance fund should be made in accordance with the accounting standards and *sound actuarial principles*, if not otherwise specified in Part V of MAS (2004a).

6.6.2.1 General Business

The valuation of liabilities of general business (non-life) is the sum

$$\text{Premium liabilities} + \text{Claim liabilities}$$

where the *premium liabilities* = minimum [*maximum*{the unearned premium reserves, the unexpired risk reserves}], and the calculation of the unearned premium reserves and the unexpired risk reserves are given in MAS (2004a); and where the *claims' liabilities* = minimum [(best estimate) the value of the expected future payments in relation to all claims incurred prior to the valuation date, whether they have been reported to the insurer, including any expense expected to be incurred in settling those claims *plus* (risk margin) any provision for any adverse deviation from the expected experience, calculated based on the 75% level of sufficiency].

6.6.2.2 Life Business

- *Nonparticipating policy* (best estimate): The expected future payments arising from the policy, including any expense that the insurer expects to incur in administrating the policy and settling any relevant claims *plus* (risk margin) any provision made for any adverse deviation from the expected experience, *less* expected future receipts arising from the policy.
- *Investment-linked policy*: The value of the liability is the sum of the unit reserves and the nonunit reserves. The unit reserves are calculated as the value of the underlying assets backing the units and the nonunit reserves, as for nonparticipating policies.
- *Participating policy*: The value of the liability is the sum of the value of the expected future payments arising from guaranteed benefits of the policy and the value of expected payments arising from non-guaranteed benefits and any provisions made for any adverse deviation from the expected experience. The guaranteed benefits include any expense that the insurer expects to incur in administering the policies and settling the relevant claim, less expected future receipts arising from guaranteed benefit of the policy.

The *participating fund*, i.e., the life insurance fund that is comprised wholly or partly of participating policies, has been modified and divided into two parts:

- *A surplus account*: The capital provided by the shareholders kept separate from other assets of the fund (MAS, 2004a, Part VI)
- *The policy assets* (PA): The remaining assets available to meet the policy liability; the policy liability must not be less than:

- The fund minimum condition liability, representing guarantee policy benefits
- The aggregate of the policy liability of every policy of the fund

If policy assets fall short of either of the two floors, then the participating fund is deemed to be unable to support its liabilities. The shareholders must then provide capital through a deduction of the surplus account. The deductions may be recoverable when the PA no longer falls short of the two floors.

6.6.3 Total Risk Requirement

The total risk requirement (TRR), or the solvency capital requirement, is defined as the sum of three risk components:

$$TRR = C1 + C2 + C3$$

A registered insurer may use any alternative method to calculate C1, C2, or C3 requirements if the method results in a requirement that *is no less than that* determined below. The MAS may require the insurer to provide documentary evidence of that fact. In other words, using an internal model may not give a requirement less than that of a standard approach.

6.6.3.1   Component 1

Component 1 (C1) is the insurance risk as calculated in the third schedule of MAS (2004a).

General business (non-life) is

$$C1 = C1(PL) + C1(CL)$$

where C1(PL) is the premium liability risk requirement, C1(CL) is the claim liability risk requirement, and

$$C1(PL) = \sum_{i=1}^{3} \max\left\{0, p_i UR_i - PL_i\right\}$$

$$C1(CL) = \sum_{i=1}^{3} \max\left\{0, c_i C_i - C_i\right\}$$

where the sum is over three volatility groups (MAS, 2004a, Table 5):

- $i = 1$, low: personal accident, health, and fire

- $i = 2$, medium: marine and aviation cargo, motor, workmen's compensation, bonds, engineering, constructing all risk/erecting all risk, credit or political risk
- $i = 3$, high: marine and aviation (hull), professional indemnity, public liability, others

The risk factors (c_i and p_i) that apply to these groups are (MAS, 2004a, Table 4):

Volatility Category	Premium Liability, p_i	Claim Liability, c_i
Low	124%	120%
Medium	130%	125%
High	136%	130%

UR_i: The unexpired risk reserves relating to the ith volatility group

PL_i: The premium liability relating to the ith volatility group

C_i: The claim liabilities relating to the ith volatility group as defined in the valuation of liabilities

Life business is

$$C1 = C1(PL) + C1(SV)$$

where C1(PL) is the policy liability risk requirement and C1(SV) is the surrender value condition requirement.

6.6.3.1.1 Nonparticipating Fund or Investment-Linked Fund

$$C1(PL) = \max\left\{0, ML - L\right\}$$

$$C1(SV) = \max\left\{0, SV - TRR - L\right\}$$

where ML is the modified liability in respect to policies of the fund (adjustment is made according to the valuation or as set out in Table 6 of MAS (2004a), for example, change in mortality rates set out in Table 15, Table 16, etc.), L is the liability in respect to policies of the fund as determined in the valuation (unmodified), SV is the aggregate of the surrender values of the policies of the fund, and TRR is the total risk requirement of the fund.

6.6.3.1.2   *Participating Fund*

$$C1(PL) = \max\{0, MML - MCL\}$$

$$C1(SV) = \max\{0, SV - \max[TRR + MCL, L]\}$$

where *MML* is the modified minimum condition liability of the fund (adjustment is made according to the valuation or as set out in Table 6 of MAS (2004a)); *MCL* is the minimum condition liability of the fund; *SV* is the aggregate of the surrender values of the policies of the fund; *TRR* is the total risk requirement of the fund; and *L* is the liability in respect of policies of the participating fund determined in the valuation (unmodified).

6.6.3.2   Component 2

Component 2 (C2) is the market risks, credit risks, and risks arising from the mismatch, in terms of interest rate sensitivity and currency exposure, of the assets and liabilities of the insurer, calculated in the fourth schedule of MAS (2004a):

$$C2 = C2(EIR) + C2(DIR) + C2(LIR) + C2(PIR) + C2(FCR) + C2(DCR) + C2(MRR)$$

where C2(EIR) is the equity investment risk, C2(DIR) is the debt investment and duration mismatch risk, C2(LIR) is the loan investment risk, C2(PIR) is the property investment risk, C2(FCR) is the foreign currency mismatch risk, C2(DCR) is the derivative counterparty risk, and C2(MRR) is the miscellaneous risk requirement.

Details of the calculations are given in MAS (2004a), fourth schedule.

6.6.3.3   Component 3

Component 3 (C3) is the concentration risk calculated in the fifth schedule of MAS (2004a):

$$C3 = A(tot) - A(conc)$$

where A(tot) is the total asset value of the fund and A(conc) is the value of assets that do not exceed any concentration limit set out in Table 14 of MAS (2004a).

6.6.4 Financial Resources and Capital Requirements

The available solvency margin is called the *financial resources* (FR) of the insurer and is calculated as the sum of three parts: tier 1 resources are capital instruments available to absorb losses on an ongoing basis, generally consisting of the aggregate of the surplus of the insurance funds and the paid-up ordinary share capital (Singapore incorporated insurer), etc., less the reinsurance adjustments and financial resource adjustment. Tier 2 resources are capital instruments less permanent compared to tier 1 resources (can never exceed 50% of tier 1 resources). Last are provisions for nonguaranteed benefits (applicable to insurers who maintain a participating fund).

The proposed risk-based capital system consists of two requirements as defined in Part II of MAS (2004a).

- The *fund solvency requirement* (FSR) is applicable to all established insurance funds. To satisfy the FSR, the insurer shall at all times be such that the financial resources (FR) of the fund are not less than the total risk requirement of the fund:

The fund's FR ≥ the fund's TRR

- The *capital adequacy requirement* (CAR): Applicable to all insurers registered in Singapore. To satisfy the CAR, the insurer shall at all times be such that the financial resources (FR) of the insurer are not less than the highest of a minimum amount of $5 million and the sum of the aggregate of the total risk requirement of all insurance funds established and maintained by the insurer under the act (TRR(A)), and where the insurer is incorporated in Singapore, the total risk requirement arising from the assets and liabilities of the insurer that do not belong to any insurance fund established and maintained under the act (TRR(B)):

max{$5 million, TRR(A) + TRR(B)}

6.6.5 Financial Resources Warning Event

The financial resources warning event means that if the financial resources (FR) are less than max{$5 million, 120% (TRR(A) + TRR(B))}.

6.7 Sweden

In December 1999 the Swedish government decided to establish a commission with a mandate to propose changes in the current restrictions on assets covering technical provisions and changed rules for available solvency margin in insurance undertakings. The work did not start until mid-2001, and two reports were presented during 2003 (SOU, 2003a, 2003b). The chairman of the commission has published a summary in English (Ajne, 2004). The Ministry of Finance has not yet decided if it will go on with the proposals or just wait for the EU's solvency project.

The proposed changes, as briefly outlined below, are intended to strengthen policyholders' protection by increasing transparency and enhancing the insurer's incentive to manage risks. The proposals are in conformity with developments in EU's Solvency II project and in IASB's accounting project. The proposal consists of three interdependent components:

- *Realistic valuation of insurance liabilities (technical provisions)*
  - While assets are measured at market values and liabilities are prudently overestimated by implicit margins, there is no possibility of a transparent valuation of the balance sheets of undertakings. This may result in insufficient focus on risk control and disincentives for matching.
  - The technical provisions should be based on a realistic valuation of the insurance liabilities in a way symmetrical to the valuation of assets.
  - The greatest changes will take place in life insurance, as the current method is to use conservative assumptions.
  - Figure 6.6 illustrates the proposed new solvency system in comparison with the existing.
- *Amended asset restrictions and valuation of assets covering the technical provisions*
  - The financial risk of assets is, in the current regulation, considered from a static point of view. The quantitative limits — based on the limits specified in the EU directives — for different asset types serve to reduce awareness of the risk, as they may erroneously be understood as defining acceptable risk levels, irrespective of the actual financial risk.
  - The proposal suggests that the asset restrictions be amended and that the financial risk should be considered as a part in the solvency assessment; as stated in the report (cf. Figure 6.7):

This is achieved by deducting a risk sensitive safety margin from the market value of the assets. This will not impact on external accounting, as the safety margin will only be considered in a capital adequacy test. In general accounting, assets will be listed at market value and the safety margin disclosed as the difference between pledged assets and technical provisions.

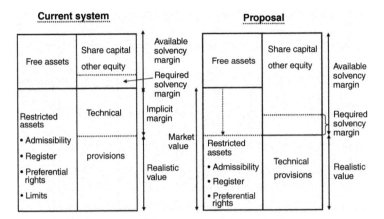

FIGURE 6.6
The proposed new solvency system in comparison with the existing one. (From Ajne, M., Proposal for a modernised solvency system for insurance undertakings, in *Swedish Society of Actuaries Centennial Book,* 2004. With permission.)

- *Assessment of risks expressed as a safety margin*
  - The safety margin will determine the amount of assets an insurer is requested to hold as collateral for its technical provisions. Today's overvaluation of the technical provisions should be replaced by a prudent valuation of this collateral, thus constituting an explicit safety margin in the assets covering the liabilities, instead of an implicit margin in the technical provisions.
  - The basis of the safety margin is the insurance and financial risks in the liabilities and corresponding assets held as collateral.

6.7.1 Technical Issues in Valuation of Liabilities

A realistic valuation involves discounting future cash flows using the market interest rate for government bonds for the relevant currency and term. The assumptions should be established prudently by statistical methods. The risk-free rate of return should be used for discounting. The proposal discusses different valuation methodologies, but calculations should be based on actuarial methods. The technical provisions are calculated using the expected value as the measure.

6.7.2 Technical Issues in Calculating the Safety Margin

The suggested design should be viewed as a standard model (illustrative example). A simple method is preferred, as it can be intuitively understood and interpreted by management and the supervisory authority. The illustrative example should be viewed as a compromise between

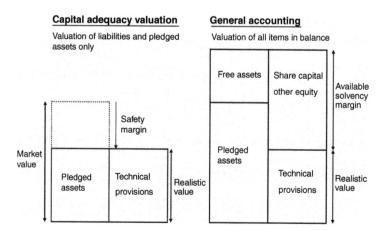

FIGURE 6.7

Illustration of the differences between a valuation for solvency and general accounting purposes. (From Ajne, M., Proposal for a modernised solvency system for insurance undertakings, in *Swedish Society of Actuaries Centennial Book*, 2004. With permission.)

prudential applicability and advanced risk measurement techniques that results in crude estimates of the risks. The standard deviation principle is proposed.

Three risk categories are considered. Each of them is divided into subrisks.

C_1: *insurance risk* — Risks that are assumed to influence the value of the insurance liabilities (i.e., the technical provisions).

C_2: *financial risk* — Risks that influence the value of the assets held as collateral.

C_3: *matching risks* — Risks that influence the value of both the insurance liabilities and the assets at the same time.

For simplicity, it is assumed that we have full independence between the three risks. This gives us the following dependence structure matrix (cf. Section 9.4).

Correlation Matrix	C_1: Insurance	C_2: Financial	C_3: Matching
C_1: Insurance	1	0	0
C_2: Financial		1	0
C_3: Matching			1

The full independence is motivated by the fact that it is possible to diversify between risks and increase transparency and disclosure, as the three parts can easily be compared.

The safety margin, C_{Tot} , is then calculated as

$$C_{Tot} = \sqrt{C_1^2 + C_2^2 + C_3^2}$$

6.7.2.1 C_1: Insurance Risk

The insurance risk is divided into a diversifiable risk (C_{11}) and a systematic risk (C_{12}). For simplicity, they are assumed to be independent.

A diversifiable insurance risk means that the relative risk decreases as the number of policies increases. The systematic insurance risk increases with increasing number of policies. A proxy of the systematic risk can be given by using the risk margin in the insurance premiums:

$$C_1^2 = C_{11}^2 + C_{12}^2$$

6.7.2.2 C_2: Financial Risk

The financial risk can also be divided into a diversifiable risk (C_{21}) and a systematic risk (C_{22}). The diversifiable financial risk will not be included, as it is limited through risk concentration limits in the asset restrictions (that are left): $C_{21} = 0$.

The systematic financial risk will be considered as a sum of a market risk (C_{221}) and a credit risk (C_{222}). They are also assumed to be independent:

$$C_2^2 = C_{221}^2 + C_{222}^2$$

6.7.2.3 C_3: Matching Risk

The interest rate risk (C_{31}) and the exchange rate risk (C_{32}) are classified as matching risks. They are also assumed to be independent:

$$C_3^2 = C_{31}^2 + C_{32}^2$$

For all the risk categories and subrisks different methods of calculation are discussed. Operational risk is not included, as it is suggested that it can be better taken care of through supervision. Liquidity risk is implicitly reflected through the market value and credit risks.

The safety margin in this illustrative example is thus

$$C_{Tot} = \sqrt{\left(C_{11}^2 + C_{12}^2\right) + \left(C_{221}^2 + C_{222}^2\right) + \left(C_{31}^2 + C_{32}^2\right)}$$

6.8   Switzerland

During the spring of 2003 the Federal Office of Private Insurance (FOPI[25]) initiated a Swiss Solvency Test (SST) project. The project was a result of a draft Insurance Supervisory Act from 2002, where the bill stated that the solvency requirement should take account of the risks in an insurance undertaking. This description is mainly based on SST (2004a), but also on different presentations made by the FOPI.

The main issues of the SST are the following:

- Assets and liabilities are market valued consistently.
- Relevant main risk categories are market, credit, and insurance risks.
- Risk is measured using the expected shortfall, or TailVar, of change of risk-bearing capital over 1 year.
- There are standard models for market, credit, and insurance risks.
- There are scenarios to take into account rare events or risks not covered by the standard models.
- Insurers must calculate two capital numbers:
  - Minimum capital (statutory solvency)
  - Target capital (market-valued solvency)
- The results of the standard models and the evaluation of the scenarios are aggregated to determine the target capital.
- In case of financial distress of an insurer, policyholders are protected by a risk margin (a market value margin (MVM)).
- The market consistent value of insurance liability is the sum of the best estimate and the risk margin, MVM.
- Internal models can be used for the calculation of target capital. The assumptions and internal models used have to be documented in an SST report and must be disclosed to the regulator.
- Reinsurance can be fully taken into account.

---

[25] Other abbreviations are BPV (Bundesamt für Privatversicherungen), OFAP (Office fédéral des assurances privées), and UFAP (Ufficio federale delle assicurazioni private).

The minimum solvency capital is based on the statutory balance sheet. It is easy to calculate but does not depend directly on the insurer's specific risk exposures.

6.8.1 Target Capital

The target capital (TC) is built on two components: a *risk margin*, or MVM, and a capital necessary for the risks emanating during a 1-year time horizon, which is denoted by $ES(\Delta C)$, or TailVar (see Section 7.4) (ES since the 1-year risk is quantified using the expected shortfall of change of risk-bearing capital[26]). The risk margin, or market value margin, is defined such that a second insurer would be compensated for the risk if taking over the assets and liabilities:

$$TC = ES(\Delta C) + RM$$

ES is defined as being the amount of risk-bearing capital that is necessary today, such that if the worst $100\alpha\%$ (e.g., $\alpha = 1\%$) of scenarios over the next year is considered, then on the average of those scenarios, the remaining risk-bearing capital will exceed the risk margin. In formal terms, ES is the minimum sum capable of compensating for $100\alpha\%$ of the worst-case expected loss. The supervisory authority will set the confidence level $(1 - \alpha)$. It may permit a higher α for certain types of insurers (e.g., for a dedicated credit insurer).

The risk margin of an insurance portfolio is defined as the hypothetical cost of regulatory capital necessary to run off all the insurance liabilities, following financial distress of the company.

In view of the regulator, it is imperative that in case of insolvency, the rightful claimants should be protected. Policyholders are seen to be best served if a third party can take over the assets and liabilities, and it will only be prepared to do this if the portfolio covers the cost of setting up the regulatory capital that would be required.

Asset allocation can be changed to optimally represent the insurance liabilities. This asset allocation is called *optimally replicating portfolio*. If an optimally replicating portfolio is achieved, target capital requirements are minimized.

The insurance company setting up the risk margin should not be penalized if, in the case of insolvency, a third party does not cover the asset portfolio to the optimally replicating portfolio as fast as possible. However, the third-party insurer (taking over the portfolio of assets and liabilities and receiving the risk margin) should not be penalized if the original insurer invested in

---

[26] The risk-bearing capital is defined as the difference between the market-consistent value of assets and the best estimate of liabilities. In terms of available solvency capital, the risk-bearing capital is the available capital plus the market value margin (risk margin).

an illiquid asset portfolio. This is allowed for in the model by assuming that future one-period risk capital requirements (ES) converge to minimal values, representing a situation where assets optimally match liabilities as fast as possible given liquidity constrains.

The speed of convergence is given by the speed with which assets could be sold off without losing significant market value.

More formally, the risk margin is defined as the cost of capital for future one-period risk capitals, i.e., expected shortfalls of risk-bearing capitals. At each time $t - 1$, the risk margin at the beginning of $t - 1$ can be decomposed

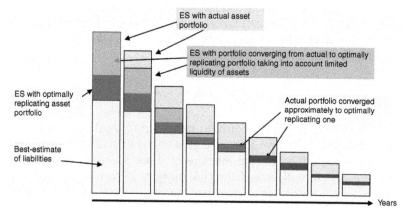

FIGURE 6.8
The run-off pattern of the insurance liabilities. It is assumed that the expected shortfall is proportional to the best estimate of liabilities. The ES with portfolio bars show the necessary 1-year regulatory capital based on the asset portfolio, which converges to the optimal replicating portfolio. (From SST, White Paper on the Swiss Solvency Test, Swiss Federal Office for Private Insurance, November 2004. With permission.)

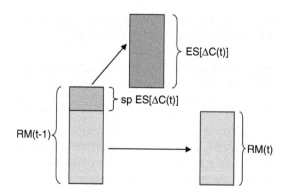

FIGURE 6.9
The necessary regulatory capital in excess of the risk margin (RM) is the expected shortfall of the change in risk-bearing capital. (From an earlier version of SST, White Paper on the Swiss Solvency Test, Swiss Federal Office for Private Insurance, November 2004. With permission.)

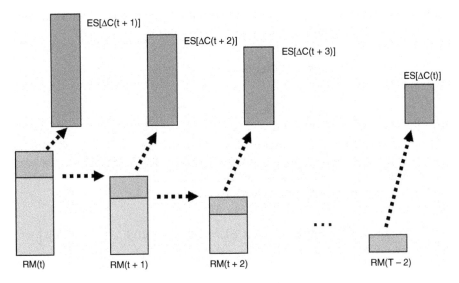

FIGURE 6.10
A company is assumed to be insolvent during $[t - 1, t]$. The company taking over the portfolio will receive the risk margin, RM(t). This is sufficient to compensate the insurer for providing the target capital for the time interval $[t, t + 1]$. (From an earlier version of SST, White Paper on the Swiss Solvency Test, Swiss Federal Office for Private Insurance, November 2004. With permission.)

in a part necessary to build up regulatory capital and a part for the risk margin necessary at the beginning of time t.

The necessary regulatory capital in excess of the risk margin is the expected shortfall (ES) of the change of risk-bearing capital:

$$RM(t-1) = spES\big(\Delta C(t)\big) + RM(t),$$

$$RM(T-1) = spES\big(\Delta C(T)\big)$$

where sp denotes the cost of capital, $RM(t)$ the risk margin at the beginning of time t, $t = 1, ..., T$, and $\Delta C(t)$ the change of risk-bearing capital during the time interval $[t - 1, t]$.

Note that all cash flows are assumed to be discounted (with the risk-free rate) to $t = 0$ and that the time interval $[T - 1, T]$ is the last period where there are still positive liabilities $L(T)$.

Consider the case when a company becomes insolvent during $[t - 1, t]$. Then a company taking over the portfolio will receive the risk margin $RM(t)$. This will be sufficient to compensate the insurer for providing the target capital for time interval $[t, t + 1]$ and to have the risk margin $RM(t + 1)$ at $t + 1$.

The recursion above leads to a simple formula for the risk margin $RM(t-1)$:

$$RM(t-1) = sp\big[ES(\Delta C(t)) + ES(\Delta C(t+1)) + ... + ES(\Delta C(T)) \big]$$

Since target capital $TC(t)$ at time t is the sum of the expected shortfall and the risk margin, we have the equivalent recursion for $TC(t-1)$:

$$TC(t-1) = ES(\Delta C(t)) + RM(t) \quad \text{for } t = 1, ..., T-1$$

$$TC(T-1) = ES(\Delta C(T))$$

It can be seen from the recursion above that target capital $TC(t-1)$ can be decomposed in a part covering the one-period risk $(ES[\Delta C(t)])$ and a part for the risk margin necessary at the end of the year.

$TC(t-1)$ can also more intuitively be written as

$$TC(t-1) = ES(\Delta C(t)) + sp\left[ES(\Delta C(t+1)) + ES(\Delta C(t+2)) + ... + ES(\Delta C(T)) \right]$$

6.8.2 Market Consistent Valuation

The assets and liabilities will be valuated at marked value (fair valuation). For this purpose, the assets are divided into three categories:

Class 1, marking-to-market valuation: Reliable market values are available (cash, government bonds, listed shares, etc.).

Class 2, mixture of marking-to-market and marking-to-model valuation: Market valuation is more difficult to determine (certain illiquid bonds, real estate, note loans, etc.).

Class 3, marking-to-model valuation: Market valuation is almost impossible to determine (private equity, some hedge funds, etc.).

The valuation of the liabilities comprises expected future obligations under insurance policies discounted by the risk-free yield curve and taking account of four principles:

Principle 1, best estimate: no explicit or implicit loadings for contingencies, losses, or other risks. Note: Loading for risks is included in the TCR and not in the valuation of the technical provisions.

Principle 2, completeness: All liabilities should be valuated, especially explicit and implied options and guarantees.

Principle 3, up-to-date information: All valuation must be made using the latest available information.

Principle 4, transparency: The models and parameters must be outlined explicitly.

6.8.3 Risks

The overall risk is broken down by main risk categories (classes) and different subrisks (risk types and risk factors). Three main risk categories are defined: insurance risk, financial risk, and operational risk. This is illustrated in Figure 7.3.

6.8.4 Standard Models

The SST consists of a number of standard models (for asset, liability, and credit risks) and a set of scenarios. Except for the credit risk model, the results of the standard models are probability distributions that describe the stochastic nature of the change of risk-bearing capital due to the modeled risk factors.

The appointed actuary also has to evaluate the scenarios and has to supplement the set with company-specific scenarios that better capture the specific risk of the company.

The results of the standard models are combined with the evaluations of the scenarios using an aggregation method. The aggregation consists — loosely speaking — of calculating the weighted mean of probability distribution given the normal situation (captured by the standard models) and special situations (described by the scenarios).

6.8.5 Asset Model

The asset model quantifies the market risks, which stem from possible changes on both the asset and liability sides due to changes in market risk factors. The asset model considers both assets and liabilities simultaneously, i.e., the A/L risk.

The model consists of around 25 risk factors, which can be divided into interest rate (approximately 10 time buckets), FX,[27] equity, and property risk factors.

All the risk factor changes are assumed to be normally distributed (with zero mean), and the joint behaviors of these risk factors are described by a covariance matrix.

Changes in risk factors lead to changes in the risk-bearing capital. For reasons of simplicity, it is assumed that the change in risk-bearing capital is a linear function of the risk factor changes. The coefficients are defined as the difference quotient (the sensitivities) for each risk factor. This means that if the share prices drop by 20%, the change in risk-bearing capital is twice the change that occurs when the share prices drop by 10%.

---

[27] Foreign exchange.

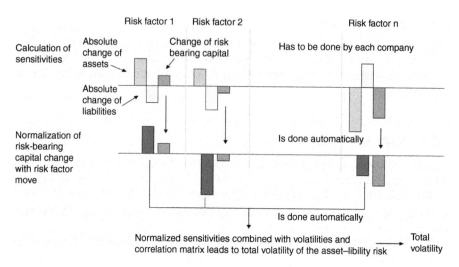

FIGURE 6.11

A depiction of the calculation of total volatility due to market risk factors. (From SST, White Paper on the Swiss Solvency Test, Swiss Federal Office for Private Insurance, November 2004. With permission.)

6.8.6 Life Insurance Model

The standard model for life insurance risks is also defined by a number of risk factors:

- Mortality
- Longevity
- Disability (BVG[28])
- Disability (non-BVG)
- Recovery rate (BVG)
- Lapse rate
- Option exercise by policyholder

The risk factor changes are assumed to be distributed by a normal distribution, analogously to the asset model. The company calculates the sensitivity of the risk-bearing capital with respect to the separate risk factors. These sensitivities are then aggregated, taking into account the volatilities of the risk factors and the correlation (specified by FOPI) between the risk factors.

---

[28] BVG: Swiss group pensions business.

6.8.7 Non-Life Insurance Model

The underlying methodology for the non-life model is similar to internal models as well as to some regulatory models, for instance, the one used in Australia or the U.K. However, in contrast to many regulatory non-life models, it is not a factor model. Instead, the appointed actuary quantifies the risk by using explicit probability distributions. This approach is more complex to implement than a factor model; however, the FOPI was of the opinion that the benefits outweigh the overhead. A distribution-based model contains enough degrees of freedom to be adapted to small as well as large insurers. Furthermore, the most common reinsurance treaties can be modeled easily and consistently. This is particularly important for small and mid-size companies, which often tend to cede a large part of their risk to reinsurers. Capturing this risk transfer is key for companies to obtain the correct capital relief.

Risk is divided between reserving risk and current year risk. Claims occurring during the current year are divided into normal claims and large claims. Normal claims are modeled by calculating for each line of business the parameter and the stochastic risk based on internal data and data supplied by the FOPI. In this way, the variance and the mean for each line of business (LOB) are obtained. Normal claims are assumed to be described by a *gamma distribution* parameterized by the first two moments and a correlation structure. Large claims are modeled as a *compound Poisson–Pareto distribution*, where the regulator supplies the Pareto parameters for each LOB. The regulator will give guidelines regarding cutoff of the different Pareto distributions.

Reserving risk is modeled using a *shifted lognormal distribution*, which is parameterized again by the first two moments, which are obtained similarly to the method used for normal claims.

The distributions for normal and large claims and reserving risks are then aggregated to arrive at the distribution for the technical result.

6.8.8 Credit Risk Model

The standard model for credit risk is the Basel II standardized approach. In order to limit the possibility for arbitrage of credit risk from the banking to the insurance sector (and the reverse), credit risk quantification follows as closely as possible the one used by the banking regulator. Therefore, a credit risk charge is calculated using an approach compatible to Basel II. This charge is then added to the target capital for insurance and market risks.

6.8.9 Scenarios

Scenarios are descriptions of possible states of the world. They are more general than simple stress tests, which consist often of stressing a single risk factor (e.g., share prices drop by 20%). Scenarios are described by stressing

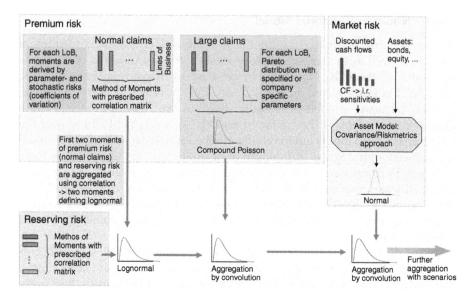

FIGURE 6.12
The calculation flow of the standard non-life model. For normal claims the gamma distribution is used. (From SST, White Paper on the Swiss Solvency Test, Swiss Federal Office for Private Insurance, November 2004. With permission.)

not one, but the whole set of risk factors. This provides a much more complete picture.

An adverse scenario is a scenario that negatively impacts the financial situation of the company.

Scenarios are an integral part of the SST. A number of adverse scenarios are prescribed. In addition, the appointed actuary should define scenarios that reflect the insurer's specific exposures.

A number of scenarios have been defined for the field test 2004:

- Industry scenario: An explosion in a chemical plant, which results in personal injuries (deaths, disablements, injuries), property damage, and business interruption.
- Pandemic event (Spanish flu epidemic of 1918 transported to 2004): Epidemic that results in personal injuries (deaths, disablements, loss of work).
- Accident scenario: (1) An accident at a company outing (bus accident), where all involved persons are insured with the insurance company. (2) A mass panic in a football stadium, resulting in many deaths, injured, and disabled.
- Hail scenario: Four hailstorms, which lead to building and motor hull damage. The definition includes storm footprints in terms of damage degrees per postal code.

- Liability for a collapsed water barrage/dam. A maximum loss and a probability for this loss are defined. Each insurance company has to estimate its own loss by taking into account the company's pool share.
- Disability scenario: Defined increase in disability rates.
- Daily allowance: Increase in rate of daily allowance.
- Default of reinsurer: The loss under this scenario is defined as the difference gross minus net of the technical result.
- Financial distress scenario: Equity values drop by 30%, downgrade to subinvestment grade (if company is rated), new business = –75%, lapse = 25%.
- Reserve scenario: 10% increase in claims provisions.
- Health insurance scenario: Antiselection.
- Terrorism.
- Historical financial risk scenarios:
  - Stock market crash, 1987
  - Nikkei crash, 1989
  - European currency crisis, 1992
  - U.S. interest rates, 1994
  - Russia/LTCM, 1998
  - Stock market crash, 2000
- Longevity: The effect of lower mortality rates on the risk capital has to be modeled.

Each scenario has an assigned probability.

6.8.10 Aggregation of Scenarios with Standard Models

In calculating the 1-year capital charge, i.e., the expected shortfall part of the target capital (TC = ES + RM), a standard density function $f_N(\bullet)$ describing the change in the risk-bearing capital (ΔC) for normal years is arrived at describing the situation of the company.

It is assumed that for most scenarios (1), there is a fixed amount shift (c_i) in the standard density, resulting in $f_i(\bullet) = f_N(\bullet - c_i)$ (see Figure 6.13).

A third assumption is that some scenarios, j, give rise to a completely new density function $f_j(\bullet)$ of the change in the risk-bearing capital.

Each scenario, i, has an assigned probability p_i of occurring.

Let $\omega = \sum_{i=1}^{n} p_i < 1$ be the sum of the n scenario probabilities.

The new hybrid stochastic scenario-based density function is given by

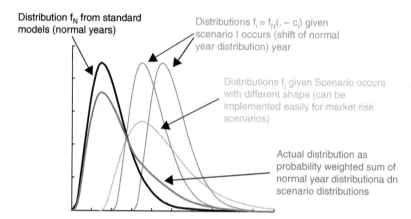

FIGURE 6.13
The aggregation of normal results with scenarios. (From SST, White Paper on the Swiss Solvency Test, Swiss Federal Office for Private Insurance, November 2004. With permission.)

$$f(\bullet) = (1 - \omega) f_N(\bullet) + \sum_{i=1}^{n} p_i f_i(\bullet) =$$

$$= (1 - \omega) f_N(\bullet) + \sum_{i=1}^{n_1} p_i f_N(\bullet - c_i) + \sum_{j=n_1+1}^{n} p_j f_j(\bullet)$$

where it is assumed that of the n scenarios there are n_1 resulting in a shift in the density function and $(n - n_1)$ scenarios resulting in completely new density functions. The expected shortfall is then calculated from this new density function.

6.8.11 Parameters

In the models used there are both specified and unspecified parameters. The SST approach has three types of parameters:

Type 1: Parameters set by the regulator and that cannot be changed (e.g., risk-free interest rate, quantile, FX rates, etc.).

Type 2: Parameters that have to be set by the companies (e.g., volatility of the hedge fund exposure, etc.).

Type 3: Parameters that are set by the regulator and that can be changed by the companies (e.g., biometric parameters, non-life parameters, etc.).

6.8.12   Internal Models

It is an aim of the supervisory authority to encourage the use of internal models. These models need to satisfy quantitative, qualitative, and organizational requirements. In particular, they must be deeply embedded into the insurer's internal processes and may not be used exclusively to calculate target capital.

6.9   U.K.

In the U.K. the traditional basis for the life insurer solvency assessment was assets measured at market value and liabilities measured by a so-called statutory method dating back to the beginning of the early 20th century. The statutory method produced a figure for the liability that was a conservative estimate of the amount needed to meet contractual obligations, but only loosely covered the potential liability for final bonuses. These calculations were generally fine for regular premium products, but less well suited to complex products with variable premiums and different guarantees or options. This drawback came into focus in the beginning of this decade, when the equity market fell and assets supporting with-profit policies fell in value, and when at the same time, the liabilities fell much more slowly as calculated by the statutory approach. Some of the deterioration in the financial position of the companies was apparent rather than real. This was a consequence of using the old system.

The capital requirements for non-life insurance undertakings operating within the U.K were widely thought of as too low and to be non-risk sensitive. The improvements in Solvency I were also seen as modest, and this has led to the adoption of an alternative approach, whereby firms in the U.K. are now expected to hold a level of capital that reflects the nature and size of risks that they have taken on. This is underpinned by a set of minimum capital requirements (MCRs) that are identical to the requirements according to the existing non-life directives.

The British Financial Services Authority (FSA) had stated in a series of publications from 1999 to 2001 that it wanted to develop a new framework for setting individual capital standards for a large range of financial firms, including insurers, that fits in a new regulatory framework (FSA, 2002a). The FSA's policy for including life and non-life insurers in its *Integrated Prudential Sourcebook* (PSB) is set out in its policy statement (FSA, 2004).

6.9.1 Pillar I: A Twin Peaks' Approach

The FSA proposed in 2002 a new risk-based minimum regulatory capital requirement for both life and non-life undertakings: the *enhanced capital requirement* (ECR). The ECR is a complement to the EU's MCR. This means that in a pillar I environment they have both the EU's MCR and the new ECR, i.e., *a twin peaks' approach* (FSA, 2002b). The concept of twin peaks is more commonly used in the context of with-profit business. Discussions on the ECRs are given for non-life insurance companies in FSA (2003a) and life insurance in FSA (2003b). After consultations, some amendments were made and the new approaches are given in FSA (2004), which include a complete package of near-final capital and provisioning rules and guidance for life and non-life insurers.

The ECR requires the companies to make more realistic assessments of their liabilities, including potential future bonuses on with-profit life policies, and to determine the need to hold capital on top of the technical provisions. Even if ECR < MCR, the MCR is binding.

The calculations of ECR are largely based on simulations and stochastic modeling of the future: dynamic financial analysis (DFA).

6.9.2 Pillar II: Individual Capital Adequacy Standards

A broader regime for assessing *individual capital adequacy standards* (ICAS) was the subject of consultation in FSA (2002a). The ICAS framework aims to reduce the likelihood of consumers suffering losses because of prudential failure. Ensuring that the amount of capital held by an undertaking is proportional to the risk profile associated with it does this. The ICAS will add more guidance on how the companies may assess adequate capital resources. The ICAS framework, which is a risk-based approach, includes two key elements:

- An *internal capital assessment* (ICA): To address risks not adequately captured in the pillar I requirements. This element would be self-assessment by the insurers: identify major risks → conduct stress tests (or use internal models) → establish probable outcome → set the ICA.

- A supervisory tool, the *individual capital guidance* (ICG): The FSA will advise or require the undertaking to hold additional capital in response to concerns about specific controls or to concerns about risks not captured by its ICA.

The ICA and ICG, which in some circumstances could be lower than the ECR, are not considered in further detail here, but we will now look closer at the approaches to non-life and life ECR assessments.

6.9.3 ECR: Non-Life Solvency Regime

According to the consultation paper CP190 (FSA, 2003a), the ECR will comprise capital charges based on asset categories and charges based on premiums written and technical provisions (depending on the class of underwriting business), i.e., $ECR = C_{assets} + C_{premiums} + C_{reserves}$.

The three charges are built up as

$$C_{assets} = \sum_{i=1}^{29} \rho_i A_i$$

where A_i is asset category i and ρ_i is an asset factor, $0 \leq \rho_i < 1$.

The asset classes and factors are given in Annex 6, PRU 3.2, in FSA (2003a), and in the policy statement, Annex 2 (FSA, 2004). For the moment, the ECR will not be a binding capital requirement and only needs to be reported privately to the FSA. The ECR calculated by the companies will be used as a starting point in the discussions between them and the FSA about their individual capital assessment (ICA). In 2006 the FSA will review whether the non-life ECR will be a hard capital requirement.

The premium and provisions charges are all based on their classes and types of business: the accounting insurance classes (e.g., motor, third-party liability motor, accident and health, property) and for each of them the type of business (i.e., direct, facultative reinsurance, proportional treaty and non-proportional treaty):

$$C_{premiums} = \sum_{i=1}^{24} p_i P_i \text{ and } C_{reserves} = \sum_{i=1}^{24} r_i R_i$$

where p and r are the premium and technical provision factors, respectively. The business classes and risk factors are given in Annex 6, PRU 7.2, in FSA (2003a).

Equalization reserves are treated as capital for the purposes of meeting the ECR: the ECR will be reduced by equalization reserves rather than capital being increased. There is no capital charge factor applied to the reserve.

Discounting of reserves is disallowed in calculating capital resources for the purpose of meeting ECR.

The FSA commissioned external actuarial work from Watson Wyatt (2003), the WW report, to analyze the volatility of relevant claims and to derive ECR factors proportional to the risks implied by this volatility. The WW report considers, for both the timescales 1 and 5 years, four different acceptable probabilities of failure (2.5, 1, 0.5, and 0.05%). In a consultation paper FSA (2003a) selected a 99.5% confidence level that an undertaking will survive for a 1-year period for its illustrations.

The approach used by Watson Wyatt (WW) can be divided into four main steps:

- *Step 1: Obtain distributions for each risk factor.* For each insurer WW has analyzed historical data to obtain distributions of possible outcomes for each asset category, reserving class, and underwriting class as specified in the ECR model from FSA.
- *Step 2: Combine the distributions for each insurer.* In this way an overall risk distribution for each insurer is obtained.
- *Step 3: Calculate the implied capital requirements.* Using the combined risk distributions for each insurer, together with the probability of failures (eight different levels), the WW then has the implied capital requirements calculated for each insurer.
- *Step 4: Select the final factors.* It is impossible to select factors that will give an ECR that is equal to the theoretical capital requirements for each insurer according to step 3. WW instead uses an optimization process over all insurers to derive the final set of risk factors to give ECRs as near as possible to the theoretical requirements in step 3.

As an example, from Watson Wyatt (2003, p. 5), the total capital requirements, expressed as percentage of net premiums, are given in the following table:

Capital Requirement Implied by ECR

1-Year Time Frame	Theoretical Requirements for All Insurers	Optimized on All Insurers	Optimized on All Nonlarge Insurers	Optimized on Small Insurers
2.5%	39	34	38	43
1.0%	49	42	49	54
0.5%	57	49	57	62
0.05%	67	58	68	73

With the selection by the FSA of a 99.5% probability over 1 year, the ECR will thus be approximately 50% of premiums to be compared with approximately 20% for the MCR, i.e., just over twice the EU requirement. This is around the same level as stock markets, and analysts generally expect non-life insurers to hold actually.

6.9.4   The Asset Risk

Watson Wyatt used its own global asset model to model the majority of assets. For each asset, it was necessary to make an assumption concerning the nature of the assets in each line in the FSA form used (e.g., line 11 — land and buildings is property), but some other lines of the FSA form are not so clear. Other important assumptions were made:

- Fixed interest securities (corporate bonds and index-linked gilts, etc.) are assumed to have an average term of 4 years.
- Equity investments are assumed to be a diversified portfolio of U.K. equities.
- Insurance dependents have been modeled as equities.
- Noninsurance dependents are modeled as property.

6.9.5 The Reserving Risk

In each financial year the distributions of the underwriting result from all prior years were modeled. The variability of the reserving results were examined by two different methods:

- Bootstrapping applied to the paid and incurred claim triangulations
- Considering the actual run-off in each financial year for each insurer

Both methods suggest the same approach: the standard deviation of the reserve run-off is dependent on the size of the account, as small accounts have greater volatility than large accounts. The formula for the standard deviation used is therefore

$$\sigma = \alpha R^{-\beta}$$

where R is the size of the reserves and α, β are parameters varying by class of business.

6.9.6 The Underwriting Risk

Separate models have been constructed for each of the classes of business being considered and also for accident year and underwriting year. The underwriting profit, net of reinsurance, for each accident year was modeled as earned premium – claims incurred in the current year – expenses incurred in the current year. Data have been taken from an FSA form.

The market underwriting cycle for all classes combined was modeled using an AR2 process:

$$UW_t = UW_{t-1} + \alpha(UW_{t-1} - \mu) + \beta(UW_{t-1} - UW_{t-2}) + \varepsilon$$

where UW_t = the underwriting result as percentage of earned premium in year t, μ = the average underwriting result, α, β are parameters, and ε is an error term.

The underwriting result of the classes was modeled using the following approach:

$$UW_{ij} = k_j UW_i - l_j$$

where UW_{ij} = the underwriting result for class i (i = 1, ..., 8) and type j (j = 1, 2, or 3: direct and facultative, proportional treaty, and nonproportional treaty, respectively), UW_i = is the underwriting result for class i (i = 1, ..., 8), and k and l are parameters.

The results of the calculations were also compared to published credit ratings. The ratio of actual capital to ECR was for 201% AAA/AA-rated companies, 142% for A-rated companies, and 95% for BBB-rated companies.

Charge Factors

Premium and Provision Charge Factors (%)

FSA Classes	Direct and Facultative		Proportional Treaty		Nonproportional Treaty	
	Premium	Provision	Premium	Provision	Premium	Provision
Accident and health	5	7.5	12	16	35	16
Motor	10	9	10	12	10	14
Aviation	32	14	33	16	61	16
Marine	22	17	22	17	38	17
Transport	12	14	12	15	16	15
Property	10	10	23	12	53	12
Third-party liability	14	14	14	14	14	14
Miscellaneous	25	14	25	14	39	14

Asset Charge Factors (%)[a]

Asset Categories (comprised)		Factor (%)
Land and buildings		7.5
Group investments	Participating interest	7.5
	Debt securities	3.5
	Shares in group undertakings	0
	Insurance dependents	
Equities		16.0
Debt securities		3.5
Loans		2.5
Deposits		0
Other financial investments		7.5
Deposit with ceding undertakings		3.5
Debtors	Policyholders	4.5
	Intermediaries	3.5
	Reinsurance	2.5
	Other	1.5
Tangible assets		7.5

[a]See Appendix 1, PRU 3.3.16R in FSA (2004) for a complete list.

6.9.7 ECR: Life Solvency Regime

The sharp falls in equity values in recent years and the regulator's desire to avoid unnecessary technical selling of equities in the light of these, coupled with other recent market developments, served to highlight deficiencies in the existing methods of assessing the capital required to back with-profit business, as outlined in the introduction to this section above.

The relative insensitivity of the valuation method of traditional liability to changes in market conditions has been addressed through a combination of a modification to the traditional approach to valuing regular premium business and adjustments to the *resilience test* (i.e., a sensitivity test; see Chapter 14). In addition, a more realistic method of assessing liabilities has been introduced, which includes a market consistent value of options and guarantees. It also allows explicitly for potential future bonuses, and for potential future actions by firms that are consistent with the fair treatment of their policyholders.

According to the consultation paper CP195 (FSA, 2003b), a company must hold *the higher of*

the calculation of mathematical reserves + the resilience test + the EU directive capital requirement

and

a realistic present value of

> expected future contractual liabilities, including options and guarantees + projected fair discretionary bonus payments plus the additional capital (RCM) required

to ensure that there would be sufficient assets to meet the realistic liabilities to policyholders after certain prescribed adverse market, credit, and persistency events.

If the sum of realistic elements + the RCM exceeds its mathematical reserves + resilience test + the EU directive capital requirement, then the undertaking must hold *top-up capital*. The higher of the two calculations is the ECR. The calculation is often referred to as the twin peaks' approach.

The rules for the ECR vary according to the business. Nonprofit business and linked business will retain the EU directive approach. In these cases, the ECR should include a risk element of 3% of the mathematical reserves, along with any additional amount needed to cover the resilience test, in order to cover investment market risk. Larger with-profit funds (> £500m) will be required to adopt the new approach to reserving and capital, i.e., the realistic basis.

With the ECR a risk capital margin (RCM) is calculated according to Rule 7.4.41R ff of FSA (2004, PRU 7.4) for U.K. life insurers having with-profit liabilities in excess of £500 million. The RCM is a part of the ECR. The FSA

reviewed the level of this proposed RCM with the assistance of a study by
Watson Wyatt (2004).

The RCM is calculated as a scenario test for each of the undertaking's with-
profit funds, looking at the combination of events, including:

- A fall in the value of equities of 20% (which may be reduced if there
  has been a recent fall during the last 3 months)
- A rise or fall in the yield on fixed-interest securities of 17.5% of the
  current yield on risk-free securities
- A widening in the credit spread on corporate bonds and reinsurance
  assets, according to a prescribed formula
- An increase or decrease by 35% in the assumed rates for termination
  of policies by voluntary surrender

The impact of the new requirements on life insurance companies is less
predictable than for non-life insurers. It will likely vary significantly between
individual undertakings, and the first published results appeared in early
2005.

6.10 U.S.

The National Association of Insurance Commissioners (NAIC) introduced a
risk-based capital (RBC) system for life and health insurers in 1992 and for
non-life insurers in 1993. The standards have, for example, been described
in the Müller report (1997). The different formulas reflect the differences in
the economic environments facing these insurers. Common risks, which
have been identified by NAIC models, include asset risk — affiliates, asset
risk — other, credit risk, underwriting risk, and business risk. From the
beginning, both life and health insurance had four main risk categories (see
below), and the non-life insurance some more categories. In 1998 the structure of the risk categories was made similar between the three insurance
types. At the end of this section we will compare the three models.

The essential steps are as follows: The existing capital, *total adjusted capital*
(TAC), is compared to the capital required to cover the risks, the *authorized
control level risk-based capital* (RBC). The TAC corresponds to our *available
solvency margin* and the RBC to our *solvency capital requirement*. The NAIC
RBC formula was designed to establish a regulatory minimum level of
capital based on risk.

The objective is to take into account all risks to which an insurance undertaking is exposed, on both the liability side and the assets side of the balance
sheet. The procedure is made in four steps:

- Fix the risk categories to which the insurance companies are exposed (also decide the dependence structure).
- Quantify each risk category (subcategories and risk factors) and equip it with a certain required capital amount.
- Combine the requirement for each risk category into a single RBC (taking into account the dependency structure).
- Compare RBC with the TAC and take appropriate supervisory measures.

We illustrate the procedure by looking at the life insurance standard.

From the start, the following four main life and health insurance risk categories were chosen:

C_1: Asset risk

C_2: Insurance risk

C_3: Interest rate risk

C_4: Business risk

The business risk was assumed to be fully correlated with the other three risks. The insurance risk was assumed to be uncorrelated with the asset risk and the interest rate risk. The latter two risks were assumed to be fully correlated. The correlation scheme between these four risks is as follows:

Correlation Matrix	C_1: Asset	C_2: Insurance	C_3: Interest Rate	C_4: Business
C_1: Asset	1	0	1	1
C_2: Insurance		1	0	1
C_3: Interest rate			1	1
C_4: Business				1

Using this *dependence structure* and the *benchmark approach*, defined in Section 9.4.2, the RBC measure is defined as (cf. Equation 9.16)

$$RBC_{L,old} = C_4 + \sqrt{C_2^2 + \left(C_1 + C_3\right)^2} \tag{6.6}$$

For each risk category basic values (subcategories) are defined and taken from the balance sheet. Each basic value is multiplied by a risk factor defined by NAIC.

The main risk factors have been updated more or less every year since 1993, and, for example, in 1996 the C_0 component was introduced. The split of C_3 and C_4, see Equation 6.7 below, was introduced in 1998. The NAIC has been cooperating with the American Academy of Actuaries (AAA) since the introduction of the RBC system. The AAA has been doing research and comparative studies. AAA did a first comparative study between the models in 1998, and a second one was published in 2002 (AAA, 2002b); cf. the comparison made at the end of this section.

As an example, AAA proposed the NAIC in 2000 (AAA, 2000) to split the C_1 category for asset risk into two separate categories: C_{1cs}, common stocks, and C_{1o}, other stocks.

The RBC model for life insurance has, at the moment, the following structure:

$$RBC_L = C_0 + C_{4a} + \sqrt{\left(C_{1o} + C_{3a}\right)^2 + (C1 \,/\, C3 - correlation) + C_{1cs}^2 + C_2^2 + C_{3b}^2 + C_{4b}^2}$$

$$(6.7)$$

where[29] C_0 is asset risk — affiliates, C_1 is asset risk — other, C_{1cs} is unaffiliated common stock and affiliated noninsurance common stock components, C_{1o} is all other asset risks, C_2 is insurance risk, C_3 is interest rate risk and health credit risk, C_{3a} is interest rate risk component, C_{3b} is health credit risk component, C_4 is business risk, C_{4a} is premium and liability component, and C_{4b} is health administrative expense component.

6.10.1 C_0 Asset Risk — Affiliates

This is the risk of default of assets for affiliated investments and it represents the RBC requirement of the downstream insurance subsidiaries owned by the insurer and applies factors to other subsidiaries. The parent is required to hold an equivalent amount of RBC to protect against defaults of affiliates.

For life insurers the risk component includes off-balance-sheet items, including noncontrolled assets, derivative instruments, guarantees for affiliates, and contingent liabilities.

6.10.2 C_{1o} Asset Risk — Other $= \sum_j \alpha_{1oj} C_{1oj}$

C_{1o} is the balance sheet amount of asset j and α_{1oj} are the risk factors defined by NAIC. The risk factors are 0 for (risk-free) government bonds and 0.20 (earlier 0.30) for risky securities. For stocks (shares) the risk factor is 0.20 (earlier 0.30) and for property (real estate) 0.10.

The risk factors are increased if there is a concentration in certain types of investments.

---

[29] The property/casualty and health formulas take a slightly different approach to the components to reflect the differences in risks associated with the different insurance types.

6.10.3 C_2 Insurance Risk $= \sum_j \alpha_{2j} C_{2j}$

The insurance risk, associated with adverse mortality and morbidity, is equivalent to underwriting risk for non-life business. A loss experience is deemed to be adverse if we have excess claims over the expected claims. To quantify this, NAIC made simulations to examine disproportional bad loss expectations in portfolios of different sizes and levels of relevant risk sums (mainly the difference between the sums insured and existing technical provisions). The conclusion was that companies with large portfolios showed lower risk sums than companies with smaller portfolios. The capital at risk was taken as basic value to quantify the insurance risk.

C_{21}: Risk capital up to $500 million $\qquad \alpha_{21} = 0.00150$

C_{22}: Risk capital, $501 to $4500 million $\qquad \alpha_{22} = 0.00100$

C_{23}: Risk capital, $4501 to $25,000 million $\qquad \alpha_{23} = 0.00075$

C_{24}: Risk capital over $25,000 million $\qquad \alpha_{24} = 0.00060$

Non-life companies calculate an underwriting risk for reserves and an underwriting risk for premiums.

6.10.4 C_3 Interest Rate Risk $= \sum_j \alpha_{3j} C_{3j}$

The interest rate risk is associated with losses due to changes in interest rate levels where the liability cash flows are not matched to asset cash flows. Three subcategories are defined. In this context products with guarantees in case of surrender are considered to be very risky. The interest rate risk has therefore been broken down into three subcategories depending on the possibility of surrender.

C_{31}: Low-risk category $\qquad \alpha_{31} = 0.0075$ or 0.0050

Life insurance contracts without right to surrender, with right to surrender at market values, and contracts with a period left to run of 1 year at the maximum.

C_{32}: Medium-risk category $\qquad \alpha_{32} = 0.0150$ or 0.0100

Life insurance contracts with surrender values in the amount of the technical provisions, from which at least 5% are deducted as a surrender charge.

C_{33}: High-risk category $\qquad \alpha_{33} = 0.0300$ or 0.0200

Life insurance contracts that guarantee surrender values almost at the same level as the technical provisions, i.e., without surrender charge.

The higher of the risk factors are applied if the insurer cannot prove that the assets are well matched with the liabilities.

This risk category does not apply to non-life or health business.

6.10.5 C_4 Business Risk $= \sum_j \alpha_{4j} C_{4j}$

The business risk category includes all those risks that are difficult to quantify in a general way and not included in the other three categories, i.e.,
overexpansion, poor management, poor business or economic conditions,
etc. The premium income is serving as the basic value with a risk factor of
0.02 in life insurance and 0.005 in health insurance.

The risk-based capital (RBC) generates the regulatory minimum amount
of capital that a company is required to maintain to avoid regulatory action.
There are five levels of action that a company can trigger under the RBC
formula. The level of the total adjusted capital (TAC) in percent of RBC is
used for different levels of intervention. The base level is the authorized
control level (ACL): the TAC is above 70% of the RBC.

If TAC $> 2 \times$ RBC, then there is no need for the supervisory authority to
intervene. If the TAC falls below the RBC, then there are four tiers of interventions.

Intervention Level	In % RBC	Measures
No intervention	≥ 200%	—
CAL: company action level	150–200%	The company must submit a plan to restore adequate capital resources within a period of 45 days
RAL: regulatory action level	100–150%	The company has to comply with a catalog of measures to the supervisory authority
ACL: authorized control level	70–100%	Right of the supervisory authority to take over the management of the company
MCL: mandatory control level	<70%	Obligation of the supervisory authority to take over the control of the company

There are discussions going on for developing the RBC system to an
enterprise risk management (ERM) system. There is also an ongoing project,
the C3 Phase 2 project, where the American Academy of Actuaries and NAIC
are working together. The C3 Phase 2 project targets variable annuities
guarantees and will use stochastic scenarios in the modeling.

6.10.6   Comparison of the Three RBC Models

The life RBC model is described above. We copy the life model below and
compare it with the non-life model (P&C RBC) and the health RBC model.
The comparison is taken from AAA (2002b).

Life:

$$RBC_L = C_0 + C_{4a} + \sqrt{\left(C_{1o} + C_{3a}\right)^2 + C_{1cs}^2 + C_2^2 + C_{3b}^2 + C_{4b}^2}$$

Non-life:

$$RBC_{NL} = R_0 + \sqrt{R_1^2 + R_2^2 + R_3^2 + R_4^2 + R_5^2}$$

R_0: Insurance affiliate investment and (nonderivative) off-balance-sheet risk

R_1: Investment asset risk — fixed income investments

R_2: Investment asset risk — equity investments

R_3: Credit risk (nonreinsurance plus half reinsurance credit risk)

R_4: Loss reserve risk, one half reinsurance credit risk, growth risk

R_5: Premium risk, growth risk

Health:

$$RBC_H = H_0 + \sqrt{H_1^2 + H_2^2 + H_3^2 + H_4^2}$$

H_0: Insurance affiliate investment and (nonderivative) off-balance-sheet risk

H_1: Investment asset risk

H_2: Insurance risk

H_3: Credit risk (health provider, reinsurance, miscellaneous receivables)

H_4: Business risk (health administrative expense risk, guaranty fund assessment risk, excessive growth)

6.11 Some Other Systems

In Sections 6.1 to 6.10 we have looked at different solvency systems that are in use or will be in use during the coming years. They are not, of course, the only systems that have been introduced or discussed. In this chapter we will briefly mention two other systems. The first is a system that has been proposed by the German insurance industry, and the second is the Norwegian complement to EU's solvency rules. A third system, in many ways similar to the U.S. RBC system, is used in Japan and is described in FSAJ (2003).

6.11.1   Germany

In the mid-1990s the European insurance supervisors, the Insurance Conference (now CEIOPS), had, on a request from the European Commission's Insurance Committee, set up a working group to look into solvency issues (see Chapter 4). Dr. Helmut Müller, from the German supervisory authority, led the group, and they presented their report in 1997 (Müller report, 1997). The main outcome from this report was a split of the commission's solvency work into two parts: Solvency I and Solvency II. It also started a discussion within various countries about the existing rules. This discussion was urged on the economical development around 2000.

As a result, the German Insurance Association (GDV)[30] decided to develop a risk-based solvency approach for supervision of German insurance undertakings, the GDV model. The GDV models, one for life and one for non-life companies, are respectively presented in GDV (2002a, 2002b). The two models have been used in practice, but have not been the official model of the German supervisory authority, and since some shortcomings were found, GDV set up a new working group in 2004 to develop a new GDV model. This new work is much concerned with correlation and diversification effects and will include more risk categories and will modulate some risk categories in another way than the GDV models of 2002.

6.11.1.1   GDV Life Model

The GDV life model, presented in GDV (2002a), had two levels of implementation. The first level was a simple system for the life undertakings, and the second focused on special conditions of individual life insurance companies by internal models.

The GDV's proposal for the first level was to adopt the existing Standard & Poor's model (S&P U.K. life model) to serve for supervision. S&P's model, which in construction is similar to the first U.S. RBC approach for life companies, compares the available solvency capital, total adjusted capital (TAC), with the solvency capital requirement as calculated in the risk-based capital (RBC). In the GDV model a *risk capital* (RC) was proposed to be calculated from the company's annual report. The RC corresponds to S&P's TAC.

The total RBC was proposed to be modeled as

$$RBC_{Life} = C_4 + \sqrt{\left(C_1 + C_3\right)^2 + C_2^2} - (E_1 + E_2)$$

where $C_1 - C_4$ are the risk charges from four risk categories (see below and compare Section 6.10) and E_1 and E_2 are deduction components from the first two risk categories (expected earnings).

---

[30] Gesamtverband der Deutschen Versicherungswirtschaft e.V., "Die Deutschen Versicherer."

C_1: *Investment risk* — This risk is assumed to capture different risks associated with investment, such as credit risk, volatility risk, currency risk, and concentration risk.

C_2: *Pricing risk* — The pricing risk is assumed to capture volatility, change and accumulation, and trend in connection with the pricing of insurance products. It includes expense and cancellation risks and the biometric risks of various insurance types.

C_3: *Interest rate risk* — This risk is assumed to cover the risk that guaranteed interest rates could not be earned.

C_4: *General business risk* — This risk should capture risks not covered by $C_1 - C_3$ and that are difficult to quantify. It was set to 1.5% of gross premium income.

Details on the GDV life model are given in GDV (2002a).

6.11.1.2   GDV Non-Life Model

As for the life model, the GDV non-life model was based on the S&P RBC model (GDV, 2002b), but modified to suit the German market.

To capture the RBC requirement, six risk categories are used: investment risk (C_1), credit risk (C_2), premium risk (C_3), loss reserve risk (C_4), life insurance and reinsurance risk (C_5), and noninsurance risk (C_6). From the available risk capital (RC) a deduction of asset risk charges is made ($A = RC - C_1 - C_2$), and this is compared to the risk charges of the four last risk categories ($B = C_3 + C_4 + C_5 + C_6$): the adequacy ratio = A/B.

For details, see GDV (2002b).

6.11.2   Norway

In 1978 the Norwegian government appointed a commission to propose a new insurance act. As there had been some failures of non-life insurance undertakings, one central point for the new act was an improvement of the supervision of the non-life undertakings. The commission presented its report in 1983 and a new Act on Insurance Activity (AIA) came into force in 1988. The AIA and especially the appurtenant regulations have later been amended, as new developments and methodologies have been introduced.

The supervisory authority envisaged a new legislation and appointed a working party (WP) in 1982 to consider technical aspects of non-life supervision. In the WP's mandate there were two central issues: to propose rules for the determination of *technical provisions* that were to be required for non-life business and the *statistical data* needed from the companies. The WP presented its report in 1984. A presentation of the basic outline of the proposed new control system is given by Norberg and Sundt (1985).[31]

---

[31] A preliminary version of this paper was presented at the XVIIth ASTIN Colloquium in Lindau, 1983.

As Norway agreed on the EEA Agreement[32] in 1992, some minor changes to the solvency regime had to be done. This means that Norway has implemented the EU solvency rules, but retained the rather detailed regulations regarding the calculation of the technical provisions.

The three basic parts of the solvency supervision are:

- The regulation on technical provisions
- The regulation on capital adequacy and solvency margin
- The regulations on asset management

The specific Norwegian, in the supervisory regulation, is the detailed rules for the technical provisions and the Bank for International Settlements (BIS) rules (see below).

We will not give a detailed description of the mathematical background to the calculation of technical provisions, but just briefly point out some interesting elements and give reference to further reading.

Reviews of the minimum requirements for the technical provisions in Norwegian non-life undertakings are given by Kristiansen (1996) and two reports from the supervisory authority, Kredittilsynet[33] (2000a, 2000b).[34]

The authorities have implemented two sets of regulations regarding the technical provisions. The Ministry of Finance laid down the general requirements in 1991, and this was supplemented by supervisory rules laid down by the supervisory authority in 1992.

The general regulation comprises the requirements and stipulates the minimum requirements for individual components to the overall technical provisions, as well as general guidelines to risk theoretic methods to be used in the estimation procedure. The supervisory authority outlines details on both the minimum requirements and the methods to be applied.

The minimum requirements for the various components of the technical provisions for non-life business are:

- *Premium provisions*: Should be at least the greatest of the unearned premiums and the premium liability, calculated on a net basis. *Theory*: Norberg and Sundt (1985), but also Pentikäinen (1982) and Rantala (1982).

- *Provisions for outstanding claims*: Should at least cover the loss liability calculated according to methods stipulated by the supervisor. Loss

---

[32] The European Economic Area Agreement was signed in 1992. It includes the EU member states and Norway, Iceland, and Lichtenstein. The three latter countries wish to participate in the internal market, but they will not have the full responsibilities of EU membership. The agreement gives them the right to be consulted by the commission during the formulation of community legislation, but not the right to a voice in decision making, which is reserved exclusively for member states.

[33] The Financial Supervisory Authority of Norway.

[34] In this report there is also a description on the regulation on technical provisions for life insurance companies.

liability is defined as the conditional expected value of future payments, on a net basis, related to both incurred but not reported claims (IBNR) and reported but not settled claims (RBNS). *Theory*: Norberg (1986).

- *Fluctuation provisions*: Should at least cover the fluctuation liability, which, in principle, is calculated by risk theoretic methods, including the utilization of the well-known NP method (cf. Chapter 9.3). In lines of business where suitable risk theoretical methods are not available, the minimum requirement equals 15% of the greater of the earned premiums for the last accounting year and the total of the minimum requirements for premium provisions and provisions for outstanding claims. *Theory*: Hesselager and Witting (1988) and Ramlau-Hansen (1988).

- *Reinsurance provisions*: Should equal a company-specific ratio multiplied by a basis of calculation defined as the total of the unearned premiums, the loss liability, and the fluctuation liability calculated on a *gross basis* less the total of the unearned premiums, the loss liability, and the fluctuation liability calculated on a *net basis* (see below).

- *Administrative provisions*: Should equal 5% of the total of the minimum requirements for premium provisions, provisions for outstanding claims, and fluctuation provisions.

With respect to the requirements on capital adequacy and solvency margin, the AIA stipulates that each undertaking shall have:

- A capital ratio that at all times constitutes at least 8% of the assets and off-balance liabilities calculated in accordance with the Basel Accord for risk weighting (cf. Chapter 5.1)

- A capital that at all times is sufficient to cover the solvency margin estimated on the basis of the undertaking's overall business (cf. the EU solvency system)

It should be noticed that both requirements should be fulfilled at all times, but they are not additive. Accordingly, the Norwegian system can be characterized as a two-track system, as there are both the BIS rules for capital adequacy and the EU's solvency rules.

6.11.2.1 Minimum Requirements for Reinsurance Provisions[35]

The main features in stipulating the reinsurance company-specific ratio, to be multiplied with the basis of calculation, is outlined below. The method

---

[35] This requirement is not a central part of the Norwegian system, but shown as an illustration of calculating the risk charge for the reinsurance counterparty risk (cf. Chapter 10.3).

uses a very simplified approach.[36] We let Q_k denote the percentage associated with rating category k.

The average percentage for the *proportional reinsurance* ceded by the company in question is given by

$$Q_{prop} = \sum_{k=1}^{K} p_k Q_k$$

where $p_k = \dfrac{P_{prop,k}}{\sum\limits_{k=1}^{K} P_{prop,k}}$ and $P_{prop,k}$ are the overall premiums for proportional

reinsurance that the company in question has ceded to reinsurers with rating category k.

The average percentage for ceded nonproportional reinsurance (XL reinsurance, stop loss reinsurance, etc.) is given by

$$Q_{XL} = \sum_{k=1}^{K} q_k Q_k$$

where $q_k = \dfrac{P_{XL,k}}{\sum\limits_{k=1}^{K} P_{XL,k}}$ and $P_{XL,k}$ are the overall premiums for non proportional

reinsurance that the company in question has ceded to reinsurers with rating category k.

The average percentage for all ceded reinsurance is defined as

$$Q_{tot} = w Q_{prop} + (1-w) Q_{XL},$$

where

$$w = \frac{\sum\limits_{k=1}^{K} P_{prop,k}}{\sum\limits_{k=1}^{K} \left(P_{prop,k} + P_{XL,k} \right)}$$

---

[36] The Norwegian FSA is now looking for alternative approaches as part of the implementation of IFRS 4.

This calculation is made for the last (t) and next-to-last ($t - 1$) financial years. The ratio to be applied when estimating the minimum requirement for reinsurance provisions is given by

$$Q_{tot}^* = aQ_{tot}(t) + (1-a)Q_{tot}(t-1),$$

where a is fixed by Kredittilsynet at the moment $a = 0.4$. An example of ratings is given in the table below.

k	Rating Categories	Q_k
1	AAA	2%
2	AA	3%
3	A	4%
4	BBB	6%
5	BB	8%
6	B	12%
7	CCC	16%
8	No rating	20%

6.12 Summary of Different Systems

We will now summarize eight of the approaches in the earlier sections, viz., the Australian (AUS), Canadian (CAN), Danish (DNK), Dutch (NLD), Singaporean (SGP), Swiss (CHE), British (GBR) and U.S. (USA). The abbreviation for each country is the ISO 3166 country code.

In the table that follows a Y means yes, and we also use the following abbreviations:

ALM	Asset liability management risk
BE	Best estimate
ECR	Enhanced capital requirement
ES	Expected shortfall or TailVaR
EU	Means the use of EU rules on solvency
FR	Financial resources
FV	Fair value, i.e., BE + a risk margin (The risk margin is usually named market value margin. In Singapore, the risk margin is called PAD, provisions for adverse deviation.)
MCR	Minimum capital requirement
MCT	Minimum capital test
MV	Market value

Comparisons	AUS APRA	CAN OSFI	DNK	NLD PVK	SGP	CHE SST	GBR FSA	USA NAIC
1. Valuation								
Liabilities	BE	Actuarial	BE	BE	BE	BE	BE	Actuarial
Technical provisions	FV	Actuarial	MV	FV	FV	FV	FV	Actuarial
Assets	MV	MV	MV	MV	MV	MV	MV	Cost or MV
Risk margin in FV	*Risk margin*	—	—	*Risk margin*	PAD	*Risk margin*	*Risk margin*	—
2. Solvency classification								
Fixed ratio	(1973)	—	EU	EU	—	EU	EU	—
Risk factor based	Y	Y	—	Y	Y	—	Y	Y
Scenario based	—	Y	Y	Y	Y	Y	Y: MCR	Y
Principles based	Y	—	Y	Y	—	Y	Y: ECR	Y
3. Minimum and target levels								
Fixed minimum amount	$2 million	—			$5 million			
Minimum level (MCR)	Y	100% of MCT 120% TAAM	EU Y: EU	EU Y: EU	—	EU Y: EU	EU Y: EU	% of RBC
Target level (SCR)	—	150% of MCT/TAAM	—	TC	TRR	TC	ECR	RBC
4. Internal models								
Use of internal model	Y	Y	—	Y	Y	Y	Y	(ALM)
5. Time horizon								
Time horizon, years	At all times	Two or five	One	One + multi	At all times	One	One	One
6. Interventions								
	Y	Y	Traffic light zones	Y	FR Warning event	Y	Y	5 levels
7. Confidence levels								
Confidence level, %	—	—	VaR: 99.0 and 99.5	99.5 and 97.5 pension	—	ES: 99.0	99.5	—
8. Available solvency capital								
ASC	Capital base	TAAM	ASC	ASC	FR	ASC	ASC	TAC

RBC   Risk-based capital

SCR   Solvency capital requirement

TAAM   Test of adequacy of assets in Canada and margin requirement

TAC   Total adjusted capital

TC   Target capital

TRR   Total risk requirement

VaR   Value at risk

(1973)   The 1973 solvency system

Summaries and comparisons between different approaches are also found in, e.g., the KPMG report (KPMG, 2002).

Note: Most factor-based systems are *retrospective*, in the sense that they use balance sheet data. On the other hand, models that are scenario or principle based are, in their nature, *prospective*, as they are considering what can happen in the future.

Part B

Present: Modeling a Standard Approach (Chapters 7–11)

In Chapters 7 to 11 we will discuss the basis for modeling and also offer an example of a simple standard model for solvency assessment.

Chapter 7 begins with the fundamental idea behind the standard approach, a twofold idea consisting of solvency assessment on the one hand and accounting assessment on the other. The starting point of the approach is the best estimate of the company's liabilities. The technical provisions measured at fair value, which is made up of the best estimate and a risk margin, usually known as a market value margin, follow this. In a full balance sheet approach we use this fair value as the mean of an unknown, but hypothetical, skew distribution of the total risk mass on the asset side. Taking a one-sided confidence interval gives us a target capital level. The difference between this level and the mean is what we call the *solvency capital requirement* (SCR). Remember that the distribution is hypothetical, which means that at this stage the SCR is also hypothetical. Different levels of capital requirements are also discussed.

The concept of risk is introduced, and different levels of diversification and its effect on a risk measure are discussed. These levels range from unit to group level, different subrisk categories to a total risk class, and within a business unit such as lines of business or asset categories.

We briefly discuss coherent risk measures and define three common risk measures: the standard deviation principle, the value at risk, and the expected shortfall.

Best estimate and fair valuation are discussed in more detail in Chapter 8. Dependency and different conservative approaches are discussed in Chapter 9.

To illustrate the theoretical discussion we provide an example of risk structure and the effect of diversification in Chapter 10. A benchmark for the model is included at the end of the chapter.

In Chapter 11 we summarize the methodology laid out in Appendix A, which contains examples of formulae and distribution-based approaches resulting in a factor-based model that can be used with the benchmark model. These approaches are then converted from a factor-based version to a spreadsheet version.

Part B

Present Modeling a Standard Approach (Chapters 7–11)

7

The Fundamental Ideas

In Section 7.1 we will discuss a fundamental idea behind the solvency assessment based on one side of a solvency–accounting coin. A best estimate of the company's liabilities will also be assumed to equal the best estimate of the technical provision of these liabilities. The concepts of *fair value*, which is *an accounting concept*, and best estimate will be discussed at length in Chapter 8. We also assume a skew distribution for a hypothetical total risk measure based on a full balance sheet.

In Section 7.2 we will discuss two distinct levels of solvency. The lower level gives us the minimum capital requirement (MCR) and the upper level the solvency capital requirement (SCR). We also discuss different triggers between these levels.

In Section 7.3 four levels of risks are discussed, from the highest in the hierarchy on units to group level (level 3) down to the lowest (level 0), which is between risk exposures. At level 3 we introduce a specific participating risk. Risks could be classified by main categories and subrisks. Further, in Section 7.3 five main risk categories are introduced, and some of their subrisks are discussed in more detail in Chapter 10.

In Section 7.4 three different risk measures are defined. These will be discussed further, in more detail, in Chapter 9.

7.1 A Model for the Solvency Assessment

The fundamental idea behind solvency assessment can be seen as the two sides of the same coin. On one side, we have the solvency assessment and its approach, and on the other side, the accounting assessment. They are, of course, not independent of each other, but to the highest possible degree interlinked. The valuation method used for accounting should, as far as possible, be used in the solvency assessment too. It would be desirable to have only one method of accounting that could be used for the solvency calculation at the same time.

As we have discussed in Chapter 2, the concept of *solvency margin* can be seen as a buffer of free assets covering the liabilities. This buffer should consist of good-quality assets, and the relative size is dependent on the time horizon. On one hand, we can think of an immediate liquidation, i.e., a run-off approach, or on the other hand, a situation where all payments are done as the debts mature, i.e., a going-concern approach. As the supervisory authority and its concern for the policyholders define the buffer frame, it will decide the time horizon. One example would be to have *a time horizon of 1 year*, corresponding to an accounting period.

We use a best estimate of the liabilities of the insurance undertaking as the start. Note that at this moment, we consider stand-alone company solvency. We will look at the problems with groups later on (Chapter 12). From this we arrive at the technical provisions. The traditional actuarial approach in establishing these provisions is to allow different sorts of prudence and loadings to be on the "safe side." This makes it impossible to compare different business units to each other. Future accounting rules will be based on the *disclosure assessment* (pillar III in the Basel II project; see Section 5.1). Hence, no implicit loadings will be used.[1]

Within the forthcoming IASB framework (cf. Section 5.2) technical provisions will be calculated using *fair valuation*. As this is an accounting concept, we avoid it in the calculation of the technical provisions within our solvency framework: the technical provisions are calculated as best estimate, i.e., equal to the best estimate of the liabilities.

For the technical provisions, the estimation of best estimate of the liabilities should be done in a similar way for life and non-life insurance. Both the approach and the level of prudence for risk margins in the solvency capital should be similar in life and non-life insurance. This is one important way of harmonizing methodology.

Consider now Figure 7.1. In the upper part (one side of the coin) we have a description of the *solvency assessment*. To the left is the *liability side*, which we assume is estimated by the best estimate.

On the *asset side* we assume a best estimate of assets corresponding to the best estimate of the technical provisions (on the liability side). On top of this we put a *prudent* solvency margin: we should achieve both a minimum capital requirement (MCR) and a solvency capital requirement (SCR), as illustrated in Figure 7.1. Both the MCR and the SCR are risk charges and should ideally be dependent on a full balance sheet approach in assessment. To get here, we assume that there is a hypothetical (skew) distribution function measuring the risk charge. This is illustrated in Figure 7.2.

In the lower part of Figure 7.1 we have the accounting assessment, reflecting the solvency side of the coin. We will not discuss taxation as a part of the accounting side of the coin, but this is very much dependent on the jurisdiction and the environment that the assessment is applied to. The

---

[1] The only reason to introduce a risk margin on and above the best estimate of the liabilities would be to take account of uncertainty in the provisioning modeling and for a time horizon of more than 1 year (accounting prudence) (cf. the Swiss run-off model).

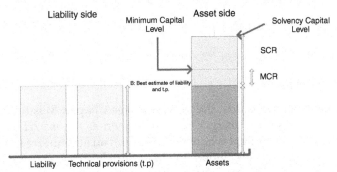
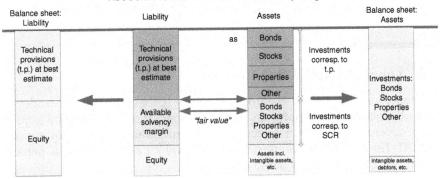

FIGURE 7.1
The upper part is a description of the solvency assessment, and below that is the corresponding accounting assessment.

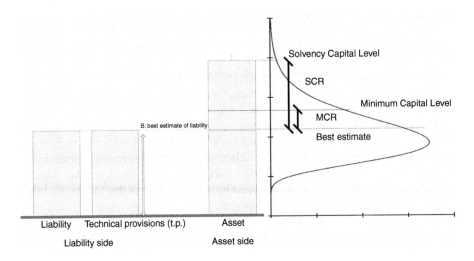

FIGURE 7.2
A probability distribution function, reflecting the uncertainty in the solvency margin, is laid out on the asset side.

accounting standards are important for transparency purposes and the disclosure; a solvency calculation should be based on common International Accounting Standards (IAS) and the full balance sheet approach, i.e., taking both sides into account.

7.1.1   Fair Value

In the accounting standards, the fair value (see Chapter 8) is formed by the best estimate plus a *risk margin*, usually known as the *market value margin* (MVM). In an environment where we have both an accounting standard and a solvency assessment regulation, the MVM can be approximated by a function of the solvency capital requirement (see Section 8.4.2).

7.2   Level of Capital Requirements

In the second phase of the EU project Solvency II, the commission introduced two distinct levels of solvency: an upper level, called solvency capital requirement (SCR), and a lower level, the minimum capital requirement (MCR). The two levels are illustrated in Figure 7.2 as emanating as percentiles from a hypothetical (skew) distribution.

The SCR can be written as the difference between the solvency capital level and the best estimate of the technical provisions.

The SCR[2] was at an earlier stage of the second phase named the *target capital requirement*. It should be seen as the target level for corrections in a going concern. This should be a *soft* level in the sense that there are no intervention measures restricting the management of the business. Of course, as we will see below, there might be measures taken by the authority to let the company, for example, submit a plan about restoring the capital level.

The MCR, or one kind of a *safety net*, should be seen as a *hard* solvency margin, with the objective of defining the level at which the management of the company is taken over by the supervisory authority. It is the absolute lowest level in the solvency system.

In this model it is supposed that between the SCR and the minimum solvency, MCR, there could be tiers reflecting various degrees of supervisory interventions; cf. the U.S. risk-based capital system in Section 6.10. Taking this approach and formulating it in our terms will give us the following table as an example of intervention measures.

In the model described above it is assumed that the MCR is calculated as the SCR, i.e., in terms of an unknown distribution function. If we had a

---

[2] Let $V_t = \mu_t$ be the best estimate of the technical provisions at time t and A_t be the assets at time t that cover both $V_t = \mu_t$ and the solvency margin, and then we can write the solvency margin in terms of the solvency capital requirement as $SCR_t = A_t - V_t$.

Intervention Level	In % of (SCR – MCR)	Measures
No intervention	≥100%	—
CAL: company action level	75–100%	The company must submit a plan to restore adequate capital level within a period of, say, 45 days
RAL: regulatory action level	50–75%	The company has to comply with a catalog of measures to the supervisory authority
ACL: authorized control level	0–50%	Right of the supervisory authority to take over the management of the company
MCL: mandatory control level	<0%	Obligation of the supervisory authority to take over control of the company

normal distribution this would give us an MCR level of the form, say, $\mu + k_{1-\alpha}\sigma$, where μ is the level of best estimate, σ the standard deviation of the distribution, and k a factor giving us the demanded level of MCR. As indicated by IAA (2004), one definition of the MCR level would be, like in Australia,[3] to set $1 - \alpha = 0.75$; i.e., in terms of the standard approach

$$\text{MCR level} = \mu + k_{0.75}\sigma$$

where $k_{0.75} \approx 0.68$.

We assume that a company always should compute the SCR by the standard approach. In Section 8.4.2 it is argued that a natural definition of MCR would be as a percentage of the SCR.

As the MCR is the last level before the company has to stop its operations in the market, as supposed here, the capital requirement level must be somewhere between the fair value of the technical provisions $V_t = \mu_t$ and the level of the solvency requirement. One way to define the MCR level[4] would be to take a percentage of μ and let, say, MCR $=x\%\mu$, $x > 100$.

In some countries courts must decide the liquidation process and then a simple and objective MCR, which may not include deliberations as the SCR to some extent includes, may be necessary.

As clarified by the U.S. FASB (see Section 8.4), the assumption of fair valuation implies that we consider a *going-concern* approach. We will use the full balance sheet approach, and hence data from the balance sheet of an entity will reflect the undertaking's going-concern approach, as balance sheet data usually do not reflect a run-off position of an undertaking.

For our purposes it is enough to have a time horizon of 1 year; i.e., the solvency assessment should be enough to let the company continue for 1 year. We could, of course, like the Dutch proposal (see Section 6.5), define the solvency so that the company could be run for 1 year and then go to run-off.

---

[3] Cf. Section 6.1. The 75% approach used in Australia reflects the old way of looking at the provisions, including prudence.
[4] Note the Danish experience (Section 6.3), where companies could get an SCR lower than the MCR.

7.3   Risks and Diversification

Risk is inherent in all actions carried out by an insurance undertaking — from designing, pricing, and marketing of its products, the underwriting procedure, the calculation of liabilities and the technical provisions, and the selection of assets backing these provisions to the overall claims and risk management. A thorough discussion of the insurer risks is given in, for example, IAA (2004).

Some of the risks face the entire economic environment and some just the insurance environment. The latter risks are difficult to protect against. For other risks there are different protections or at least actions that can be made to diminish the effects of a realization of a risky event.

The exposure to risk can be classified at three main levels:

- Risks arising at the entity level (diversifiable).
- Risks faced by the insurance industry (systematic and usually non-diversifiable). The risk affects the entire insurance market or some market segments. Some diversification and asset allocation can protect against systematic risk, as different portions of the market tend to perform differently at different times.
- Risks faced by the whole economy and whole society (systemic and nondiversifiable). The risk affects the entire financial market or the whole system, and not just some specific participants.

Everyone will die, and therefore it is not possible to take away the mortality risk from the society per se (the systemic part), but an insurance company can, by a health declaration, make a selection that diminishes the risk of dying earlier than expected according to actuarial assumptions (a systematic part).

By *diversifiable* we mean that if a risk category can be subdivided into risk classes and the risk charge of the total risk is not higher than the sum of the risk charges of each subrisk, then we have the effect of diversification (*sub-additivity*; see below). This effect can be measured as the difference between the sum of several capital charges and the total capital charge when dependency between them is taken into account. Dependency is discussed in Chapter 9. If the risk charge of risk X is $C(X)$ and Y is another risk with charge $C(Y)$, then the effect of diversification can be written as

$$C(X+Y) \leq C(X) + C(Y)$$

The diversification effect can be classified according to four main levels:

- *Level 0*: Between *risk exposures*, for example, volatility and nonvolatility, within a product line or an asset category (for example, different perils).
- *Level 1*: Between *subportfolios* within a risk category of a business unit, for example, line-of-business (LOB) or asset categories.
- *Level 2*: Between *main risk categories and subrisk classes*; for example, we can have five main risk classes as defined by IAA (2004), each consisting of one or more subrisk classes.
- *Level 3*: Between *business units* to the group or conglomerate level; for example, a financial conglomerate can have one or several insurance companies within its group.

In life insurance and at the level 0, the risk exposures can be divided into volatility, calamity, trend uncertainty, and level uncertainty.

The key components of risk, as defined by IAA (2004, p. 27), are:

- *Volatility*: The risk of random fluctuation in either or both of the frequency and severity. This risk is diversifiable as the volatility of the average claim amount declines as the number of insured risks increases (each insured risk is assumed independent of the other).
- *Uncertainty*: The risk that the model used is misspecified or that the parameters are misestimated. This risk is nondiversifiable, as it cannot reduce by increasing the number of insured. Three key elements are included (cf. IAA, 2004):
  - *Model error risk*: The model is incorrect. It cannot be reduced by changing parameters. This could be called model uncertainty.
  - *Parameter error risk*: The model may be accurate, but the estimation of the parameters may cause errors. This is also called level uncertainty.
  - *Structural error risk*: The risk structure of the parameters may change over time causing a trend uncertainty.
- *Extreme events*: Risks that are highly unlikely to occur but would result in high losses (high impact, low probability (HILP)). The HILPs are one-time shocks that cannot be extrapolated from the ordinary events. These events can be hard to model, but their impact could be measured by scenario tests (see Chapter 14).

At level 1 we could, for example, divide the asset portfolio into bonds, equities, property, cash, and other assets.

When we come to level 2 we can first look at the five main categories proposed by IAA: underwriting, credit, market, operational, and liquidity. The main risk category credit could in its part be divided into, for example, default risk, migration risk, spread risk, sovereign risk, counterparty risk, and concentration risk.

The five main categories are the underwriting risk and the four main risks discussed in Basel II (cf. Section 5.1). We follow mainly the IAA discussion (IAA, 2004, pp. 29–32).

- *Underwriting risk* (later we will use the term *insurance risk* as the concept for this main risk category). This is the risk an insurance company is faced with by underwriting the insurance contracts. They include both the perils covered by specific lines of business (LOBs) and the specific processes associated with the management and conduct of the business. There are a lot of specific risks. Some of them are the subrisks: underwriting process risk, pricing risk, claims risk, net retention risk, and the reserving risk. As another example, see the insurance risk in Figure 7.3, where the risk structure of the Swiss system is depicted.

- *Credit risk.* This is the risk of default and change in credit quality of issuers of securities in the investment portfolio, the reinsurance counterparty risk. Other risks are, for example, the direct default risk, spread risk, sovereign risk, and concentration risk.

- *Market risk.* This risk arises from the level or volatility of market prices on assets and involves different movements in the level of various financial variables. Some subrisks are, for example, the asset–liability mismatch risk, reinvestment risk, currency risk, interest rate risk, and risks in equity and property.

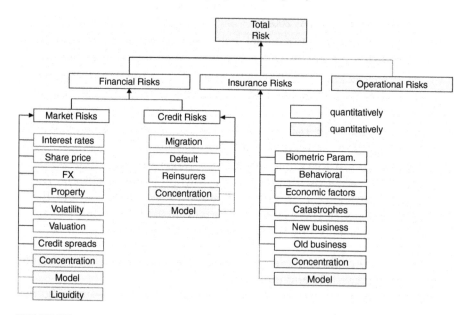

FIGURE 7.3

Three main risk categories (financial, insurance, and operational) and their subrisks as considered in the Swiss model. (From SST, White Paper on the Swiss Solvency Test, Swiss Federal Office for Private Insurance, November 2004. With permission.)

- *Operational risk.* This risk is usually seen as a residual risk not measured by the other categories. It is often defined in terms of risk of losses due to inadequate or failing processes, people, and systems such as fraud, but also risks from external events.
- *Liquidity risk.* This risk is the exposure to loss due to the fact that insufficient liquid assets are available to meet cash flow requirements.

As there has been a concentration within the financial sector of a small number of large institutions operating globally, the risk of contagion increases. "The fact that many banks adopt similar modelling techniques, assumptions, and risk management standards, and place a greater reliance on dynamic hedging might also exacerbate such market movements. Common sense suggests it would be prudent to hold liquidity cushions against such a possible outcome, particularly given the opacity of the concentrations" (Large, 2004).

Now we come to the highest level, i.e., between business units. First we can have a pure insurance group, i.e., two or more insurance companies acting as a group. If we have insurance companies, banks, and other financial units belonging to one and the same grouping, we talk about a financial conglomerate. Note that the definition of an insurance group and a financial conglomerate is the main objective of the Insurance Group Directive (IGD) (COM, 1998) and the Financial Conglomerate Directive (FCD) (COM, 2002d). Both two directives are briefly discussed in Section 12.2, although we do not stress the definition there.

The definition of a group above is from the IGD. A member of a group can be a participating undertaking, i.e., which has at least 20% or more of the voting rights or capital hold.

One definition of financial conglomerates (see Section 12.2) is any group of companies under control whose exclusive or predominant activities consist of providing significant services in at least two different financial sectors (banking, securities, and insurance).

There is also a risk of being a member of a group or a financial conglomerate. We will call this the *participating risk*. One type is the *reinsurance counterparty risk*, which we have if one of the group members is ceding parts of its business to one of the other members (the reinsurer). This type of business is usually taken care of in the insurance risk. The other type is a *credit risk.* Assume a financial conglomerate with one insurer and a bank. The bank insures its credit risk at the group's insurance company. A default in the bank will affect both the asset and the liability sides of the insurer. This means that at level 3 not only do we not have a positive effect of being a member of a group (in terms of diversification), but we also have a possible negative effect from the participating risk. Another risk type is the *affiliated investment risk* proposed by IAIS in the liquidity risk (see Section 10.5).

In Chapters 5 and 6 we briefly presented different discussions and proposals. One of the main threads in these chapters has been to show the reader that there are a lot of discussions on how to classify risks. We have, for example, the main classification made by BIS (Section 5.1) into credit, market, operational, and liquidity risks. The IAIS (Section 5.3) has its classification, and the IAA starts from the BIS classification and supplements it with the underwriting risk (Section 5.4). These main categories can then be divided into different level 2 subrisks. Different risk classifications are also made in the studies produced under the first phase of the EU Solvency II project (Section 5.5). In the same way, the different solvency systems presented in Chapter 6 all have there own classifications (more or less). Usually one and the same risk category can be found in all systems, but under different headings and references. The latter could probably depend on differences in languages and valuation of words.

One way of presenting the main risk categories and subrisks is shown in Figure 7.3. It is from the Swiss system presented in Section 6.8.

We will return to risk classification in Chapter 10, where the risks are discussed in more detail. This classification is then used in the modeling procedure in Chapter 11.

7.4   Risk Measures

In a risk-based capital environment the positive outcomes of each risk component included in the model aggregate to the total capital requirement, in one way or another. A risk measure should reflect the capital charge that a company needs to pay for its risk exposure.

We will start with some definitions.

Let the random variable $Y = \sum_{i=1}^{r} Y_i$, with expectations $E(Y) = \mu_Y$

and $E(Y_i) = \mu_{Yi}$ and variances $Var(Y) = \sigma_Y^2$ and $Var(Y_i) = \sigma_{Yi}^2$. We will drop the index of Y if the context shows what is meant.

A *standard variable* is defined as $Y_{St} = \dfrac{Y - \mu_Y}{\sigma_Y}$, with zero mean and standard

deviation 1. This variable is often approximated by the standard normal distribution.

The number of risk measures in the literature is extensive, but here we will only consider three measures. A measure should have some reasonable properties for its purpose. One such classification of measures is coherent risk measures that have been discussed extensively in the literature during the last decade.

7.4.1 Coherent Risk Measures

Artzner et al. (1999) defined four main properties that a risk measure should fulfill. A risk measure is, in principle, a mapping C from a future random outcome Y into real numbers:

$$C : Y \mapsto R$$

For any two bounded loss random variables X and Y and a risk measure $C(.)$, we define the following properties:

1. Subadditivity: $C(X + Y) \leq C(X) + C(Y)$.
2. Monotonicity: If $X \leq Y$ for all possible outcomes, then $C(X) \leq C(Y)$.
3. Positive homogeneity: For any constant $\lambda > 0, C(\lambda X) = \lambda C(X)$.
4. Translation invariance: For any constant $a > 0$, $C(X + a) = C(X) + a$.

A risk measure satisfying these four properties is said to be *coherent*.

The subadditivity property shows the effect of diversification and that "a merger doesn't create extra risk" (Artzner, 1999). The monotonicity property shows that if, for example, one loss is less than another, then the same relation holds for their risk charges.

The third property shows that if, for example, a loss is changed by a positive factor, then the risk charge is changed in the same way. This could be seen as a limiting case of the subadditivity, representing what happens when there is no diversification effect. The fourth property shows that a constant change in, for example, the loss changes the risk charge with the same amount (translation invariance).

Pros and cons of these properties may be found in the literature (e.g., Kroisandt, 2003). As an example of the latter, if we have a large number of insured units (λ) in one area with the same amount insured X, the total risk charge should be higher than the sum of each, i.e., $C(\lambda X) \geq \lambda C(X)$. Take, for example, insured buildings in an area where earthquakes are common.

7.4.2 Three Risk Measures

We will define three[5] different but, under a normality assumption, related measures. They are discussed further in Chapter 9.

7.4.2.1 Standard Deviation Principle

The traditional standard deviation principle (SDP) is a measure based on the concept of a confidence interval:

---

[5] Other risk measures are, for example, the expectation principle, variance principle, and semi-variance principle. There are also risk measures based on utility (see, for example, Kroisandt, 2003).

$$SDP_\alpha(X) = E(X) + k_{1-\alpha}\sigma_X \qquad (7.1)$$

where $k_{1-\alpha}$ is a positive scale parameter depending on α. If X is distributed as a standard normal variable, the $k_{1-\alpha}$ is clearly defined and tabulated. The measure is not coherent because the monotonicity property is violated.

7.4.2.2 Value at Risk

During the last decade or so the measure value at risk (VaR) has become a standard risk measure used to evaluate the exposure to financial risk. The VaR, which is a quantile measure, gives the amount of capital required to ensure that a company does not become insolvent, at least with a high degree of certainty.

The VaR is defined as the α-quantile x_α, which is the smallest value satisfying

$$VaR_\alpha(X) = \inf\left\{x \in R : P(X > x) \le \alpha\right\} \qquad (7.2)$$

where $0 < \alpha < 1$. It describes the highest value of the loss for more than $100(1-\alpha)\%$ of all outcomes. Typically α equals 5, 1, or 0.5%. It only measures the probability for loss, but not its magnitude.

For a normal distributed random variable $[X : N(\mu, \sigma^2)]$ this measure can be written in terms of the expected value and the standard deviation:

$$x_\alpha = \mu + k_{1-\alpha}\sigma$$

where the factor $k_{1-\alpha}$ is the $(1 - \alpha)$ quantile of the standard normal distribution. As an example, if $\alpha = 0.01$, then $k_{0.99} = 2.33$, and if $\alpha = 0.005$, then $k_{0.995} = 2.58$.

The VaR is not coherent.

7.4.2.3 Expected Shortfall or TailVaR

The expected shortfall (ES), or the tail value at risk (TailVaR), is the conditional tail expectation in the upper $\alpha\%$ of the right-hand tail of the distribution. In many situations it is not possible to compute this measure analytically.

ES is defined as

$$ES_\alpha(X) = E\left[X \mid X > VaR_\alpha(X)\right] \qquad (7.3)$$

where $0 \le \alpha < 1$. As an example, $ES_0(X) = \mu$.

This measure is coherent, and it is the *expected value of a distribution singly truncated from below*. As a measure of risk, it was introduced by Wirch (1997; see also Wirch and Hardy, 1999).

For a skewed distribution, this measure is more sensitive than VaR.

8

Valuations

In the discussion on a new solvency regime within the EU and the development of new accounting standards for insurance proposed by International Accounting Standard Board (IASB),[1] the two concepts of *best estimate* and *fair value* are used. They are used for both liabilities and assets. Here, we are going to try to explain and discuss their meanings.

The new accounting standards will probably adopt a full balance sheet approach, with the assets and liabilities being reported in the same way, probably at fair value. As this is an *accounting concept*, we avoid it in the calculation of the technical provisions within our solvency framework: the technical provisions are calculated as best estimate, i.e., equal to the best estimate of the liabilities (cf. Figure 7.1).

We will start with the accounting concept of fair value. In Section 8.1 we discuss the purposes of valuation. In Section 8.2 we discuss best estimate of insurance liabilities and technical provisions, and in Section 8.3 the concept of fair value in both an environment without solvency regulation and one with this kind of regulation. We will also give a proposal for a definition of fair value in the mixed environment with both accounting and solvency standards.

8.1 Fair Value: Introduction

The definition of fair value, according to IASB,[2] is "the amount for which an asset could be exchanged or a liability settled between knowledgeable,

---

[1] International Accounting Standard Board.

[2] IASB (2004). One of the main objectives for the new standard IFRS 4 is to define insurance contracts. A contract that is not an insurance contract will be classified as either a financial instrument (IAS 39) or a service contract (IAS 18). If the solvency calculation is only to be made on liabilities of insurance contracts, then there must be solvency rules for the financial/service contracts. As there should not be any arbitrage if an insurance company has these contracts in comparison with, for example, banks, these solvency rules should be based on existing bank rules (for example, Basel II). An open question to discuss is: Should the assets covering all liabilities be before any assets for the solvency margins?

willing parties in an arm's length transaction." For the assets there is usually a market, but for the liabilities there is no trading market. Hence, this gives rise to both conceptual and practical issues. One translation of the liability part of the definition is the one made by Clark et al. (2003): "the amount that the enterprise would have to pay a third party at the balance sheet date to take over the liability." As there is usually not a market, one may think in terms of a run-off situation, as has been proposed in the Swiss approach (see Section 6.8). The risk margin is defined such that a second insurer would be compensated for the risk of taking over the assets and liabilities. This is a hypothetical cost of regulatory capital necessary to run off all the insurance liabilities, following financial distress of the company. This situation is perhaps closer to the concept of an entity-specific valuation, as discussed in the "Draft Statement of Principles (DSOP)" by IASB (2001).

There is an interesting presentation of the fair value concept given in Jørgensen (2004).

During the last decade a lot of effort has been devoted to discussing and quantifying the concepts; e.g., IAA has been producing standard papers and in 1999 the Faculty and Institute of Actuaries established a working party (FIA WP) to consider issues arising from the initiative of the IASB.[3] The IASB's project on insurance accounting was initiated in 1997 with the objective of developing an International Accounting Standard (IAS) for insurance (cf. Section 5.2). After the reconstruction of the IASC into IASB, the IAS has been transformed into International Financial Reporting Standards (IFRS). The FIA WP presented a report in 2001 (see Hairs et al., 2002).

In the ideal situation, with assets and liabilities traded in a deep liquid market, the concept of fair value is generally taken as equal to market value. The main challenge in fair value accounting is the assessment of a fair value where the instrument is not usually traded. This applies to the majority of insurance liabilities. Thus, in this case, the (accounting) value of a liability would be a calculated value, using assumptions on future events, discount rates, mortality, etc., that *an independent marketplace participant* would make in determining *the charge* it would take to acquire the liability. As there is no liquid secondary market for insurance liabilities, this would suggest that there is no unique fair value between willing parties in the market. We will give one proposal of a valuation in Section 8.4.2.

In its DSOP, the IASB also discussed a *nonfair valuation* concept, based on a company's own assumptions and expectations (IASB, 2001). This approach is referred to as an *entity-specific value*: "the present value of the costs that the enterprise will incur in settling the liability with policyholders or other beneficiaries in accordance with its contractual terms over the life of the liability" (Abbink and Saker, 2002).

In practice, the distinction between the two definitions is not clear-cut and may not be significant. For both definitions, external and internal

---

[3] At that time IASC (International Accounting Standard Committee).

information may be used (e.g., on mortality), and on some parameters observable market prices (e.g., interest rates and asset returns). The differences between the fair value and entity-specific value are probably limited — both the market price of risk and expected value underlie both (Clark et al., 2003).

The fair value or entity-specific value of insurance contracts should reflect risk and uncertainty. The margin for risk and uncertainty is referred to as the *market value margin* (MVM), and it should reflect the market's appetite for risk based on observable market data (Clark et al., 2003).

IAA and other organizations gave comments on the IASC's insurance issues papers. It is worth noting that IAA also proposed methods of calculating the MVM and fair value[4] (IAA, 2000).

The IASB has tentatively proposed that assets and liabilities should be measured at fair value unless there is a lack of market transactions. Then an entity may use entity-specific assumptions and information when market-based information is not available without undue cost and effort (see Section 5.2 and IASB, 2004).

8.2   Purposes of Valuation

In an ideal situation all valuation, both assets and liabilities, should be made at market values, without any implicit margins that are not openly disclosed. For the assets there is a market, and hence a valuation is more or less possible. However, for the liabilities, there is no single valuation approach that can serve all possible purposes, as stated earlier.

Different assumptions and different approaches are usually used if a company is going to be valued as a going concern or in a run-off situation. It could also be valued for general purposes, e.g., its financial progress, its solvency status, or its ability to meet various obligations in extreme circumstances. The valuation could even be done to see the effect of potential future new business.

There are at least three main reasons for reporting:

- *Reporting for shareholders* and others interested in the overall financial progress of the business, who need a realistic valuation of liabilities and a consistent valuation of assets. The valuation should be done on a going-concern approach, and the degree of prudence should only be what is needed for accounting purposes.

---

[4] For example, these two reports: "Insurance Liabilities: Valuation and Capital Requirements" and "General Overview of a Possible Approach and Valuation of Risk Adjusted Cash Flows and the Setting of Discount Rates: Theory and Practice" (IAA, 2000).

- *Reporting for policyholders* and others like regulators, who need a valuation that is prudent and that shows, e.g., the solvency status of the company. It should show the company's general financial strength and its capital position both as a going concern and under run-off.

- *Reporting for tax authorities* and others, like supervisors, is needed depending on the local environment.

It would be preferable if the reporting standards could use one and only one method of valuation, or at least a family of methods showing the relationships between the valuations (Hairs et al., 2002).

The new accounting standard will offer the possibility of valuing a company on a core general-purpose valuation base. This will indeed increase transparency in the financial reporting.

_____

8.3 Best Estimate of Insurance Liability and Technical Provisions

A best estimate of an insurance liability, and hence the technical provisions, should be the actuary's best calculation of the company's future obligations depending on knowledge such as, for example, changes in the interest rate and assumed developments in longevity. As these assumptions are prognoses, there will always be deviations from the true outcome.

The liability of an insurance contract involves both an element of uncertainty in the timing and size of the cash flows and a long time horizon. The two main sources of deviations between a best-estimate assumption and a true outcome can be classified as:

1. *Uncertainty* in the best-estimate assumptions
2. *Volatility* of experience around the best-estimate assumptions

According to the FIA WP (Hairs et al., 2002, Section 6.3), best-estimate liabilities, based on a reliable statistical model, would correspond to a probability of meeting all liabilities in about 50% of cases. In these terms and for accounting, a fair value would correspond to somewhere in the range of 55 to 80%.

As the *best estimate* of the liability we use the mean value of its distribution, the expected value. This position parameter is also used as the basis from which the solvency capital requirement (SCR) is measured (see Section 7.1). The accounting concept of *fair value*, as used by IASB, is a higher value than the best estimate, as it includes a positive risk margin, the MVM.

The best estimate of insurance liabilities should reflect insurance contract features and their specific risks, which would include embedded options and guarantees.

The essential step in determining a best estimate of insurance liabilities is to measure all expected cash flows with a risk-free interest rate, including all options, profit sharing, bonuses, future benefits, and expenses. All non-financial risk drivers or nonmarket assumptions, such as mortality, morbidity, lapse, expenses, claim frequencies, and severities, have to be estimated with their risk-free counterparts. As an example, for a risk-free mortality in an n-year term life insurance we could use the population mortality, as this mortality probably is worse than the mortality in the studied insurance portfolio (due to selection effects). Of course, this does not hold for annuities, as mortality change has the opposite impact on life insurance and annuities, and therefore we need another mortality as a risk-free mortality. A discussion on best estimate and pricing the mortality risk is given in Van Broekhoven (2002). If all these parameters could be determined with certainty and the parties involved would be risk-neutral, then the expected value would equal the market value. The cash flows should be discounted. This approach is called the *direct method* (Girard, 2002). Any risk adjusting will be made in the calculation of the fair value. If it is not possible to find one single best-estimate point, we could use the management best-estimate procedure (see page 202).

8.3.1 Risk-Free Interest Rate

The theoretical interest rate is assumed to be obtained by investing in financial instruments lacking risk for default. A truly risk-free asset exists only in theory, but in practice, short-term government bonds, for example, are considered to be risk-free, as the likelihood of a government defaulting is very low (for countries with a strong economy). The concept of the risk-free interest rate is an important issue in the modern portfolio theory.[5]

As this interest rate is obtained without risk, it is implied that any additional risk that is taken by an investor should be rewarded with an interest rate higher than the risk-free rate.

8.3.2 Technical Provisions

As was stated in Chapter 7, we do not introduce any prudence in the technical provisions, as the prudence is included in the solvency margin. This means that the technical provisions are set equal to the best estimate of the liabilities.

---

[5] According to this theory rational investors will use diversification to optimize their portfolios and also how the asset should be priced given its risk relative to the market as a whole.

Solvency: Models, Assessment and Regulation

Lommele et al. (1998) discusses the concept of best estimate in accordance with the U.S. National Association of Insurance Commissioners' (NAIC) codification of statutory accounting principles. In its Statements of Statutory Accounting Principles, SSAP, No. 55, paragraphs 9 and 10 establish the use of a management's best estimate:

9. For each line of business and for all lines of business in the aggregate, management shall record its best estimate of its liabilities for unpaid claims, unpaid losses, and loss/claim adjustment expenses. Because the ultimate settlement of claims (including IBNR for death claims and accident and health claims) is subject to future events, no single loss or loss/claim adjustment expense reserve can be considered accurate with certainty. Management's analysis of the reasonableness of loss or loss/claim adjustment expense reserve estimates shall include an analysis of the amount of variability in the estimate. If, for a particular line of business, management develops its estimate considering a range of loss or loss/claim adjustment expense reserve estimates bounded by a high and a low estimate, management' s best estimate of the liability within that range shall be recorded. The high and low ends of the range shall not correspond to an absolute best and worst case scenario of ultimate settle-ments because such estimates may be the result of unlikely assumptions. Management's range shall be realistic and, therefore, shall not include the set of all possible outcomes but only those outcomes that are considered reasonable.

10. In the rare instances when, for a particular line of business, after considering the relative probability of the points within management's estimated range, it is determined that no point within management's estimate of the range is a better estimate than any other point, the midpoint within management's estimate of the range shall be accrued. It is anticipated that using the midpoint in a range will be applicable only when there is a continuous range of possible values, and no amount within that range is any more probable than any other. For purposes of this statement, it is assumed that management can quantify the high end of the range. If management determines that the high end of the range cannot be quantified, then a range does not exist, and management's best estimate shall be accrued. This guidance is not applicable when there are several point estimates, which have been determined as equally possible values, but those point estimates do not constitute a range. If there are several point estimates with equal probabilities, management should determine its best estimate of the liability.

8.4 Fair Value

We first discuss the concept of fair value where we do not have any solvency regulation regime, and after that, we give a definition of fair value in an environment where we have both accounting standards and a solvency regulation regime.

8.4.1 Fair Value in a Pure Accounting Standard Environment

We assume that the liability cash flows include a risk margin over the best-estimate assumption; this is what the market would charge for the nondiversifiable risks. This margin, in the fair value environment, is usually called the market value margin (MVM). If a policy was generating a margin larger than the MVM, it would generate a profit at point of sale. This MVM is *not* the ordinary *prudent* actuarial margin, but a margin determined by the market in recognition that the cash flows are not risk-free.

The MVM would thus reflect the premium that a marketplace participant would demand for bearing the uncertainty inherent in the cash flows. Financial risk drivers are tradable on a sufficiently liquid market. Diversifiable risks would be excluded in the MVM, but these risks would be part of the capital requirement in the solvency assessment.

The fair valuation of liabilities is not affected by the nature of, or the return of, the assets backing those liabilities. To estimate it, cash flows and MVMs can be discounted using the return on an asset portfolio, *replicating portfolio* (IAA, 2000), whose cash flows most closely replicate the liability cash flows (including MVM).

The FIA WP emphasized that the term *fair value* is "a technical term, which should not be taken as meaning that, at any given time, the value is, in absolute terms, the *financially* correct value" (Hairs et al., 2002). The WP concludes that the concept of MVM within fair value is fine in theory, but that it is questionable if it is possible to quantify it. But as the WP concludes, it is not seen as needed to estimate the MVM separately, but just to quantify the fair value.

In Clark et al. (2003) the pros and cons of different estimation methods are discussed, i.e., deterministic approaches (sensitivity analysis and bootstrapping) and a stochastic approach. Both approaches require interpretation of the market risk preferences to determine the MVM. The most obvious source of information for most books of contracts would be risk margins applied in insurance portfolio transfers and reinsurance contracts. One approach would be to see the MVM as risk margins in a run-off situation, as proposed in the Swiss system (see Section 6.8).

As this approach is new and there are no agreed-upon market value margins, the assessment as a first step will have to be done by the individual companies. As a second step, various organizations and companies such as reinsurers, consulting firms, and rating firms could be used to pool knowledge on risk margin valuation and make this information available to a wider audience. It is (Clark et al., 2003) likely that future disclosure requirements and transparency will ensure the publication of these measures.

An entity-specific valuation (see the introduction above and Section 5.2) will require companies to determine the expected value of future losses, discounted at the risk-free rate of interest and adjusted for the agreed-upon market's risk margin. Alternatively, the risk margin could be set as an adjustment of the risk-free discount rate, and as stated by Clark et al. (2003), "this approach would seem to be efficient for companies, understandable for auditors and the wider market."

The MVM can also be seen as the difference between two expected cash flow measures: expectation calculated under the risk-neutral probability distribution (*Q-measure*) minus expectation calculated under the realistic probability distribution (*P-measure*) (Girard, 2002).

Embedded value reporting will not be allowed according to IASB. In Abbink and Saker (2002, chap. 5) a comparison between embedded value and fair value methodologies is made. This is summed up in Table 8.1 (see Table 5.1 in Abbink and Saker, 2002, p. 19).

In cases where it is too complex to find a replicating portfolio for the liability cash flow it would be necessary to use stochastic methods. One of the modern techniques that could be used is the *state price deflator* method, proposed by Jarvis et al. (2001). According to Abbink and Saker (2002, p. 23), these deflators can be described as stochastic discount rates that can be used to calculate the fair value of the insurance liability by:

TABLE 8.1

Comparison between Embedded Value and Fair Value Methodologies

	Embedded Value	**Fair Value**
Methodology	Deterministic	Stochastic or deterministic
Assumptions	Expected value	Risk adjusted
Discount rate	Risk adjusted	Risk-free
Return on assets	Included in projections of future surplus	Not included in projections of cash flow unless impacts on policyholder benefits
Options	Allowance based on intrinsic value on embedded value assumptions, potentially also some allowance in discount rate	Direct allowance (market consistent stochastic modeling/price of hedging asset)
Cost of capital	Direct allowance	No allowance
Value of liabilities	Obtained indirectly as the prudential reserves less the value of in-force	Obtained directly as the discounted value of cash flows

Source: Abbink, M. and Saker, M., Getting to Grips with Fair Value, presented to the Staple Inn Actuarial Society, March 5, 2002. Available at http://www.sias.org.uk/prog.html.

- *A stochastic asset model is run, the output from which will include a deflator for each time period of each scenario;*
- *The liability cash flows are projected and are adjusted for non-financial risk;*
- *For each simulation the deflator is applied to the relevant cash flow at each point in time and these values are summed across all projection steps to obtain a deflated value; and*
- *The fair value of the liability is the mean value of the deflated cash flow.*

Consider C_t as the stochastic cash flow at time t. The present value, using the stochastic deflator D_t (a stochastic discount function), is then $Value_t = E\left[D_t C_t\right]$. This is similar in principle but more powerful than the actuarial method in multiplying the expected value of the cash flow by a deterministic discount function, v^t : $Value_t = v^t E\left[C_t\right]$ (Jarvis et al., 2001).

The U.S. Financial Accounting Standard Board (FASB)[6] has also been working on a fair value measurement project (see, e.g., CAS, 2004; Jørgensen, 2004). In 2002 FASB signed a memorandum of understanding with the IASB to "seek to reduce the existing differences between IFRS and U.S. GAAP[7] in order to accelerate progress toward the attainment of global accounting standards." The board has clarified that:

1. The fair value measurement presumes that the entity is a *going concern* without intention or need to liquidate.
2. *Willing parties* are all marketplace participants that are willing and able to transact, having the legal and financial ability to do so.

A fair value hierarchy prioritizes the market inputs that should be used for all fair value estimation:

Level 1 estimates: Obtained using quoted prices for *identical* assets or liabilities in active markets to which an entity has immediate access.

Level 2 estimates: Obtained using quoted prices for *similar* assets or liabilities in active markets, adjusted as appropriate for differences. The difference between measuring and measured assets or liabilities must be *objectively* determinable. Otherwise, the estimates will be level 3 estimates.

Level 3 estimates: Based on results of valuation techniques generally consistent with those used by marketplace participants in pricing the types of assets and liabilities being measured. FASB recognizes that level 3 estimates will vary to the extent of market inputs used, but emphasizes that the valuation techniques must be consistently applied.

---

[6] More information can be found at www.fasb.org.
[7] Generally accepted accounting principles.

For a general discussion on valuation of cash flows, see Gutterman (1999). A Fair Value Task Force of the American Academy of Actuaries (AAA, 2002a) has published a monograph on fair value accounting principles to the liabilities associated with insurance contracts. When using the hierarchy one often finds level 3 the most applicable. The monograph discusses the various levels and three principles commonly applied when computing present values and making risk adjustments:

Principle 1: If there is no risk, discount the cash flows at the risk-free rate.

Principle 2: If there is risk in the cash flows, the present value estimate should include a risk adjustment to reflect the market price of risk.

Principle 3: Include all cash flows.

One way of calculating the risk margin MVM is discussed in AAA (2002a, pp. 16–17). We start by defining C_t as the cash flow at the end of period t, MVM_t as the market value margin for period t, L_t as the liability fair value at the end of period t (or beginning of period $t + 1$), r_f as the risk-free rate, r_L as the discount rate, and r_l as the growth rate for liabilities.

If the insurer's contract involves cash flows for n periods, the calculation is made in the following steps:

1. Begin by calculating the final period MVM: $MVM_n = C_n \dfrac{r_f - r_L}{1 + r_l}$.

2. Find the fair value at the beginning of the final period:

$$L_{n-1} = \frac{\left[C_n + MVM_n \right]}{1 + r_f}.$$

3. Go to the pervious period $(n - 1)$ and compute MVM:

$$MVM_{n-1} = (L_{n-1} + C_{n-1}) \frac{r_f - r_L}{1 + r_l}.$$

4. Find the fair value at the beginning of that period:

$$L_{n-2} = \frac{\left[L_{n-1} + C_{n-1} + MVM_{n-1} \right]}{1 + r_f}.$$

5. Go back to step 3 for the next prior period $(n - 2)$ and repeat steps 3 through 5 as many times as necessary.

This cash flow adjustment is consistent with discounting at the risk-free rate and the calculations can be extended. For more details, see AAA (2002a).

8.4.2 Fair Value in a Mixed Accounting Standard and Solvency Regulation Environment

In the accounting standards the fair value is formed by the best estimate and a risk margin, the market value margin. In an environment where we have both an accounting standard and a solvency assessment regulation the fair value can be linked to the solvency capital requirement, as we will show below.

8.4.2.1 Trading between Two Companies: A Conceptual Overview

To begin with, we do not think of any accounting standard, only of an accounting date, and we just assume that company S wants to sell a portfolio of insurance contracts (its liabilities and technical provisions) to company B. As a reward, company B wants assets corresponding to the technical provisions (TP) at best estimate plus some risk margin for taking over the liabilities. At the same time, there is a solvency regulation that both companies S and B must fulfill: the solvency capital requirement (SCR) must be *fulfilled at any time*. As in Chapter 7, we assume that the SCR is calculated on top of the assets corresponding to the best estimate of the TP. The SCR would in most cases be as prescribed from the regulator, but could also be an indirect effect of requirements stated by a rating agency.

Let PV[·] be a present value operator, i.e., transforming a function to a discounted present value. Let BE(L) be the best estimate of the liabilities and BE(TP) the best estimate of the technical provisions.

We assume that PV[BE(L)] = PV[BE(TP)]; i.e., the present value of the best estimate of the liabilities equals the corresponding value of the technical provisions.

This hypothetical transaction between the two companies S (seller) and B (buyer) is divided into two parts:

- Future earnings (FE) from the portfolio; $t > 0$
- The instantaneous value of the portfolio at the time of transaction; $t = 0$

With $t = 0$ we think of either the date of transaction or a time period between the transaction and up to the next accounting date (less than 1 year).

8.4.2.1.1 Future Earning from the Portfolio

What is the need or earnings for company B in the future from these insurance contracts? First, there are the pure earnings and costs in terms of premiums and claims. Second, the company has to calculate the marginal effect on the solvency margin, $\Delta SCR_{t>0}$, due to changes in the portfolio.

The value of in-force business (VIF) is calculated as the discounted present value of the future earnings stream from the in-force portfolio less the cost of capital needed to comply with the solvency margin requirements due to changes in the portfolio:

$$VIF = PV\left[FE\,|\,t>0\right] - PV\left[c_2 \Delta SCR_{t>0}\,|\,t>0\right] \tag{8.1}$$

where FE is the future earnings and c_2 the cost of capital. $\Delta SCR_{t>0}$ is the sum of all future changes in the solvency margin due to changes in the portfolio.

8.4.2.1.2   The Instantaneous Value of the Portfolio at the Time of Transaction

What is needed for company B to take over the liabilities from company S? First, of course, there is the need to get assets corresponding to the technical provisions, including costs for handling the provisions. At the same time, the company needs to fulfill the solvency requirements.

Define the change in the solvency margin for B due to the transaction as

$$\Delta SCR_{t=0} = SCR_B(\text{after trading}) - SCR_B(\text{before trading})$$

The capital needed for B at the transaction can thus be written as (value at time, VAT)

$$VAT = PV\left[BE(TP)\,|\,t=0\right] + PV\left[c_1 \Delta SCR_{t=0}\,|\,t=0\right] \tag{8.2}$$

i.e., assets corresponding to the present value of the best estimate of the technical provisions plus the present value of the cost of capital needed to comply with the changes in the solvency margin due to the transaction. We do not define the cost of capital c_1, as it is only used for the conceptual discussion.

8.4.2.1.3   Market Value of the Liabilities

We define a market value, MV, of the transaction, and hence of the liabilities of the insurance contracts as the difference between the instantaneous value (or need) (Equation 8.2) and the future earnings (Equation 8.1):

$$MV = PV\lfloor BE(TP)\,|\,t=0\rfloor + PV\lfloor c_1 \Delta SCR_{t=0}$$
$$- PV\left[FE\,|\,t>0\right] + PV\left[c_2 \Delta SCR_{t>0}\,|\,t>0\right.$$
$$PV\lfloor BE(TP)\,|\,t=0\rfloor + PV\lfloor c_1 \Delta SCR_{t=0}$$
$$+PV\left[c_2 \Delta SCR_{t>0}\,|\,t>0\right] - PV\left[FE\,|\,t>\right.$$

$$PV\left[BE(TP)\,|\,t=0\right] + MVM - PV\left[FE\,|\,t>0\right] \tag{8.3}$$

The MVM is the market value margin defined as

$$MVM = PV\left[c_1 \Delta SCR_{t=0} \mid t = 0\right] + PV\left[c_2 \Delta SCR_{t>0} \mid t > 0\right] \qquad (8.4)$$

The first two terms in Equation 8.3 are a fair value of this transaction. Having a solvency regime with a simple standard approach in assessing the SCR by, for example, a risk factor-based approach, as the one proposed in Chapter 11, the changes in SCR in Equation 8.4 can probably be simple to calculate, especially if underlying data are openly disclosed. The change in the solvency margin depends, of course, on the company's risk and solvency environment. In this sense, this approach is more of an entity-specific approach. Both costs of capital parameters, c_1 and c_2, can be assessed according to economic theory.

The actual market value will depend on the market value of the individual parameters. The market value of cost of capital is the effect of demand and supply on the risk capital market. The market value of change in SCR depends on the solvency rules and on the solvency positions of the market participants.

In a market where the solvency rules are independent of the entity (like having SCR as a percentage of premium or technical provisions), the value of SCR is defined by these rules. In a more sophisticated solvency regime, where the solvency requirements follow the actual risk position of the risk-bearing entity, the value of SCR follows the possibility of the market participants to diversify the risk involved. In financial theory, full diversification is often possible. In reality, it is not. Insurance companies cannot freely diversify outside of insurance. They also cannot diversify by writing an infinite number of infinitely small blocks of insurance business, since each block written leads to fixed expenses. There is always a trade-off between expenses and risk diversification, and the result of this trade-off on the market will define the market cost of insurance risk.

8.4.2.2 Valuation of Insurance Liabilities

The above discussion is intended to get the idea behind the following argument in valuating the insurance liabilities.

We are going to valuate the full insurance portfolio using a time horizon of 1 year. The valuation date is $t = 0$. The fair value (FV) of the insurance liabilities is defined as the best estimate of the technical provisions plus the cost of capital c needed to comply with the solvency margin according to the solvency regulation. We assume that a standard approach is used to calculate this latter part and not any internal model; i.e., the solvency margin is transparent, as it should be possible for any third party to make this calculation from disclosed data, for example, a consultant firm. The calculation of the fair value is done *given* the value of SCR, which is assumed to be a constant:

$$FV_{t=0} = BE_{t=0}(TP) + cSCR_{t=0} \qquad (8.5)$$

where $t = 0$ indicates the date of calculating the fair value. The company's risk aversion is reflected by the fixed SCR and is inherent in the company's rating. The latter will be a part of the value c, the cost of capital parameter.[8]

The MVM of the mortality risk is discussed in Van Broekhoven (2002), where the cost of capital is mentioned as one way to calculate it. The author calculated the MVM as a "factor times the standard deviation of the underlying distribution."

8.4.2.3   Cost of Capital for Insurance Undertakings

Due to developments of, for example, ALM techniques and fair value accounting, there has been a need to find reliable methods to estimate *cost of capital*[9] for insurance undertakings. It has been recognized that the cost of capital (COC) varies across industries due to heterogeneity of the risks facing the firms. Research has shown that there is a significant industry factor for insurance. In Cummins and Phillips (2003) cost of capital models are developed to reflect the lines-of-business characteristics of undertakings in the property-liability insurance industry. It is also observed that the COC is inversely related to firms' size and that long-tail commercial lines of property-liability insurance tend to have higher COC than short-tail lines.

Different methods to estimate COC are used, for example, full-information CAPM[10] and Fama-French 3-Factor estimates (Cummins and Phillips, 2003). In the latter case,[11] the COC is 18.1% for property-liability insurance, 18.8% for life insurance, 16.9% for health insurance, and 21.1% for finance excluding insurance.

In the Swiss Solvency Test (SST) the risk margin (market value margin) is defined as the capital cost for future regulatory capital needed for the run-off of the portfolio. This is described in Figure 6.9 and Figure 6.10. In SST (2004b) the calculation of the risk margin is discussed. A review of the methodology for quantifying the risk margin within the SST is discussed by Furrer (2004). In Example 1 of op. cit. it is shown that a risk margin equals 17.3% of the target capital (our SCR).

---

[8] One simple way of determining c is to assume that the company has to borrow SCR on the loan market. Then its rating will affect the interest rate c for this loan.

[9] Cost of capital: Average of the costs to a company of its various kinds of capital (bonds, debentures, loans, shares, retained profit, and so on) (*Dictionary of International Business Terms*, Financial World Publishing, Canterbury, Kent, UK, 2001). Another definition is "the cost to borrow or invest capital."

[10] Capital Asset Pricing Model

[11] Full-information Fama-French 3-Factor estimates with sum beta adjustment, market value weighted (Cummins and Phillips, 2003, pp. 15–16).

8.4.2.4 A Relation between MVM, MCR, and SCR

From Equation 8.5 we get that the market value margin could be expressed as

$$MVM = cSCR_{t=0} < SCR_{t=0}$$

In Section 7.2 we briefly mentioned the concept of a safety net, the minimum capital requirement (MCR). We also argued that MCR could be calculated, for example, as a percentage of the expected mean, i.e., in the terms above, as a percentage of the BE(TP). In light of the above inequality, it seems plausible to define MCR as a percentage of the SCR, and that it should be greater than the MVM, i.e., $MVM < MCR = kSCR_{t=0}$.

This means that we get the following relationships between these concepts:

$$MVM = cSCR_{t=0} < MCR = kSCR_{t=0} < SCR_{t=0} \qquad (8.6)$$

i.e., $c < k$ (<1).

By the above discussion we suggest that c should be somewhere between, say, 17 and 19%. If we take half the percentage of the mandatory control level (i.e., 50% × 2RBC) of the U.S. RBC system (see Section 6.10), we have $k = 35\%$. This rough decision of the factors c and k gives us MVM = 0.18SCR < MCR = 0.35SCR < SCR. This is only given as an example.

In the Australian system a risk margin (RM), with respect to the *central estimate* of the liabilities, is required (see Section 6.1). It was thought that a 75% probability of sufficiency was enough. This means, in terms of a standard normal distribution, $\mu + 0.67\sigma = \mu + RM$, where μ is the central estimate (or best estimate) and $RM = 0.67\sigma$. If we compare this with a 99.5% confidence interval for the SCR, i.e., $\mu + 2.58\sigma = \mu + SCR$, the risk margin could be expressed as $RM = 0.26SCR$, where $SCR = 2.58\sigma$.

If we assume a skew distribution with, for simplicity, the skewness parameter $\gamma_1 = 1$, then using the NP approximation (see Section 9.3.1)

$$SCR = \left[2.58 + \left(2.58^2 - 1\right)/6\right]\sigma = 3.52\sigma \quad \text{(confidence level} = 99.5\%)$$

The risk margin would thus be

$$RM = \left[0.67 + \left(0.67^2 - 1\right)/6\right]\sigma = 0.58\sigma = 0.16 \times 3.52\sigma = 0.16SCR$$

In other words, this is in line with the MVM assumed above.

9

Dependencies, Baseline, and Benchmark Models

In Section 9.1 we will recapitulate the three risk measures defined in Section 7.4.2 and use them as a starting point. In Section 9.2 we will assume *normality*, even if we are sure that the hypothetical distribution behind the risk charge is skew, and show how these measures are defined in this simplified case. In Section 9.3 we will assume a more realistic scenario, i.e., *nonnormality*. Using the standard deviation principle (SDP) as a *baseline* (in terms of IAA (2004)) and the normal power approximation we get a simple risk charge measure that has the skewness of the hypothetical distribution inherent.

In Section 9.4 we will look at the *correlations* between risks, and to make the approach simple and robust, some level of conservative thinking can be used in setting up the structure model for the solvency capital requirement (SCR) (see Section 7.2). This SCR will be the focus in the task of finding a *simple standard model*. The different levels of conservatism are defined in terms of assumptions on the structure of the correlation matrix, irrespective of the true mathematical-statistical theory behind this structure. With this in hand, we are able to build a *factor-based model*. This is done with both the baseline approach, mentioned above, and a simplified *benchmark* approach.

In Section 9.5 we will conclude the chapter with a discussion on different types of parameters, i.e., parameters set by the EU or by the local supervisory authority and others that are set in-house by the insurance companies.

9.1 Risk Measures

A minimum level of capital and surplus (or solvency margin) is usually required by insurance companies in most countries. In a risk-based capital environment the positive outcomes of each risk component included in the model aggregate to the total capital requirement or capital charge.

If a company is willing to build its own models to calculate the target capital level, the local supervisory authority should be able to approve of its use instead of using the standard approach (SA). These models are called *internal models* (IMs). It should also be possible to gradually move from the SA toward an IM. For this reason, it is important to have a modular approach. This is discussed briefly in Chapter 14.

If a company invests resources in management and builds its own internal models, it should be possible to get a lower solvency capital than that given by the standard approach. For benchmark purposes, the company should also always compute the SCR according to the standard approach.

We recall from Section 7.1 that the SCR is defined as the difference between the solvency capital level and the best estimate of the technical provisions.

The three risk measures introduced in Section 7.4.2 are:

1. *Standard deviation principle:* $SDP_\alpha(X) = E(X) + k_{1-\alpha}\sigma_X$, where $k_{1-\alpha}$ is a positive scale parameter depending on α.
2. *Value at risk, VaR:* $VaR_\alpha(X) = \inf\{x \in \mathbb{R} : P(X > x) \leq \alpha\}$, where $0 < \alpha < 1$. VaR is defined as the $(1-\alpha)$ quantile x_α satisfying the above expression.
3. *Expected shortfall or TailVaR:* $ES_\alpha(X) = E[X \,|\, X > VaR_\alpha(X)]$, i.e., the conditional tail expectation in the upper $\alpha\%$ of the right-hand tail of the distribution, where $0 \leq \alpha < 1$. As an example, we have $ES_0(X) = \mu$.

If we let μ be the best-estimate level on the assets side (see Figure 7.2), then in terms of a risk charge the SCR can be written as

$$\text{(SCR 1) } \textit{Standard deviation principle: } C(\alpha) = SDP_\alpha(X) - \mu = k_{1-\alpha}\sigma_X \quad (9.1)$$

$$\text{(SCR 2) } \textit{Value at risk: } C(\alpha) = VaR_\alpha(X) - \mu \quad (9.2)$$

$$\text{(SCR 3) } \textit{Expected shortfall: } C(\alpha) = ES_\alpha(X) - ES_0(X) = ES_\alpha(X) - \mu \quad (9.3)$$

9.2   Assume Normality

Let the random variable $Y = \sum_{i=1}^{r} Y_i$, with expectations $E(Y) = \mu_Y$ and $E(Y_i) = \mu_{Yi}$ and variances $\text{Var}(Y) = \sigma_Y^2$ and $\text{Var}(Y_i) = \sigma^2_{Yi}$. We drop the index of Y if the context is clear what we mean.

For example, Y can be the *total* solvency capital requirement or denote one risk that is subdivided into r risk parts. The underwriting risk can be divided into different lines of business (LOBs). Hence,

$$E(Y) = \mu_Y = \sum_{i=1}^{r} \mu_i$$

$$Var(Y) = \sigma^2 = \sum_{i=1}^{r} \sigma_i^2 + \sum_{i \neq}^{r} \sum_{j}^{r} \rho_{ij} \sigma_i \sigma_j = \sum_{i=1}^{r} \sum_{j=1}^{r} \rho_{ij} \sigma_i \sigma_j \qquad (9.4)$$

where ρ_{ij} denotes the correlation between Y_i and Y_j.

We will now look at this for the normal distribution. Let X be a $N(\mu, \sigma^2)$ random variable. Let $\phi(.)$ denote the probability density function of the standard normal distribution and $\Phi(.)$ its cumulative distribution function.

9.2.1 A First-Order Approximation

The *cumulant generating function* is unique for each distribution. It can be written as a series expansion as

$$\varphi_Y(t) = \mu t + \sigma^2 \frac{t^2}{2} + \kappa_3 \frac{t^3}{3!} + \kappa_4 \frac{t^4}{4!} + \dots$$

where μ and σ^2 are the expectation and variance, respectively. κ_3, κ_4, ... are the higher cumulants of the distribution. The normal distribution has the following cumulant generating function:

$$\varphi_Y(t) = \mu t + \sigma^2 \frac{t^2}{2}$$

with all higher cumulants equal to zero. Thus, the normal distribution can be viewed as a first-order approximation of the true unknown distribution.

9.2.2 Standard Deviation Principle

From Equation 9.1 the SCR can be defined as

$$C = \mu + k_{1-\alpha}\sigma - \mu = k_{1-\alpha}\sigma \qquad (9.5)$$

i.e., the solvency capital level minus the best estimate, that is, a factor k times the standard deviation, where, e.g., $k_{0.99} = 2.33$ and $k_{0.995} = 2.58$ depending on the level of confidence (99 or 99.5%, respectively) (Table 9.1). The choice of the correct level of confidence is a political decision rather than an actuarial.

If the total risk is defined as in Equation 9.1 or 9.5, then Equation 9.4 shows that the risk can be split into a sum of components of risks and their inter-relations.

When the standard deviation is used as a risk measure and the capital requirement is a multiple of the standard deviation,[1] $C_j = k_{1-\alpha}\sigma_j$ (cf. Equation 9.5), then the capital requirement of the aggregate risk can be written as (summing over r risks)

$$C = \sqrt{\sum_{i=1}^{r}\sum_{j=1}^{r}\rho_{ij}C_iC_j} = k\sqrt{\sum_{i=1}^{r}\sum_{j=1}^{r}\rho_{ij}\sigma_i\sigma_j} = k\sqrt{\sum_{i=1}^{r}\sigma_i^2 + \sum_{i\neq}^{r}\sum_{j}^{r}\rho_{ij}\sigma_i\sigma_j} \quad (9.6a)$$

see, e.g., Appendix H in IAA (2004). The last term of Equation 9.6a can also be written as

$$C = \sqrt{\sum_{i=1}^{r}C_i^2 + \sum_{i\neq}^{r}\sum_{j}^{r}\rho_{ij}C_iC_j} \quad (9.6b)$$

If the distribution is skewed, i.e., has a heavier tail to the right, the factor $k_{1-\alpha}$ is larger than those for the normal distribution (see Section 9.3). The $k_{1-\alpha}$ might be different for different risk categories, depending on the form of the specific distribution function.

9.2.3 Value at Risk in the Case of Normality

The value at risk (VaR) equals Equation 9.5 in the case of a normal distribution; i.e., the SDP and VaR coincide:

$$C = \mu + k_{1-\alpha}\sigma - \mu = k_{1-\alpha}\sigma$$

9.2.4 Expected Shortfall in the Case of Normality

Now we will look closer at the conditional tail expectation in the case of normality. The conditional density function of X given that $x_\alpha \leq X < \infty$ is said to be a normal distribution singly truncated from below with density

$$g(y) = \begin{cases} \dfrac{1}{\sigma}\phi(\dfrac{y-\mu}{\sigma})/[1-\Phi(z_{1-\alpha})] & x_\alpha \leq y < \infty \\ 0 & \text{otherwise,} \end{cases}$$

[1] Another way of writing the capital charge would be to rewrite it as $C_j = k_{1-\alpha}\sigma_j = \mu_j k_{1-\alpha}v_j$, where μ_j is the expected value and $v_j = \sigma_j/\mu_j$ is the coefficient of variation.

where $z_{1-\alpha} = \dfrac{x_\alpha - \mu}{\sigma}$.

The mean and variance of the singly truncated normal distribution are given by (see, e.g., Johnson and Kotz, 1970)

$$\mu_{CT}(\alpha) = \mu + \sigma / R(z_{1-\alpha}) \tag{9.7a}$$

and

$$\sigma_t^2 = \sigma\left\{1 + z_{1-\alpha}R^{-1}(z_{1-\alpha}) - R^{-2}(z_{1-\alpha})\right\} \tag{9.7b}$$

where $R(z_{1-\alpha}) = R(\dfrac{x_\alpha - \mu}{\sigma})$ is *Mills ratio* (see, e.g., Kotz et al., 1985).

If we let $k_{1-\alpha}^* = 1 / R(z_{1-\alpha})$ be the distance factor, then

$$ES_\alpha(X) = \mu + k_{1-\alpha}^*\sigma \tag{9.8a}$$

and

$$\sigma_t^2 = \sigma^2\left\{1 + k_{1-\alpha}^* z_{1-\alpha} - k_{1-\alpha}^{*2}\right\} \tag{9.8b}$$

Tables of Mills ratios are usually published in most books on statistical tables. Approximations for Mills ratios are usually derived from expansions and inequality bounds. Taking the reciprocal gives the distance factor $k_{1-\alpha}^* = 1 / R(z_{1-\alpha})$ (Table 9.1).

TABLE 9.1

Distance Factors k and k^* for Different Values of α for the SDP and the Expected Shortfall in Case of Normality

$1 - \alpha$	$k_{1-\alpha}$	$k_{1-\alpha}^*$
0.990	2.33	2.67
0.995	2.58	2.90
0.999	3.09	3.37
0.9995	3.29	3.55
0.9999	3.72	3.96

In the case of *normality* we have the following risk charges or SCR:

(SCR 1) *Standard deviation principle:* $C(\alpha) = k_{1-\alpha}\sigma_X$ $\qquad$ (9.10a)

(SCR 2) *Value at risk:* $C(\alpha) = k_{1-\alpha}\sigma_X$ $\qquad$ (9.10b)

(SCR 3) *Expected shortfall:* $C(\alpha) = k_{1-\alpha}^*\sigma_X$ $\qquad$ (9.10c)

If we use Equations 9.6a and 9.6b as a *baseline*, we could rewrite the above expressions as

$$C = \sqrt{\sum_{i=1}^{r} C_i^2 + \sum_{i\neq}^{r}\sum_{j}^{r} \rho_{ij}C_iC_j} = \sqrt{\sum_{i=1}^{r}\sum_{j=1}^{r} \rho_{ij}C_iC_j} \qquad (9.11)$$

where, of course, C_i : s depends on $k_{1-\alpha}$ or $k_{1-\alpha}^*$.

From Equation 9.8a we get the risk charge or SCR measured by the expected shortfall in the case of normality as

$$C(\alpha) = k_{1-\alpha}^*\sigma \qquad (9.9)$$

For $z_{1-\alpha} > 2$ it can be seen that

$$R(z_{1-\alpha}) \approx 4\left[3z_{1-\alpha} + \sqrt{z_{1-\alpha}^2 + 8}\right]^{-1} - 2z_{1-\alpha}^{-7}$$

See Kotz et al. (1985), where also other approximations are found.

Panjer (2002) gave a discussion on the ES and its reformulation (Equation 9.8a). He also showed that if we consider the aggregate risk, say, $X = X_1 + X_2$, the allocation to risk 1 (X_1) is

$$E[X_1 \mid X > x_\alpha] = \mu_1 + k_{1-\alpha}^*\sigma_1^2[1 + \rho_{12}\frac{\sigma_2}{\sigma_1}]$$

where ρ_{12} is the correlation coefficient between X_1 and X_2.

A similar function as the expected shortfall is defined in Beard et al. (1984, p. 79) as the auxiliary function $m(x_\alpha) = E(X - x_\alpha \mid X \geq x_\alpha) = CTE(\alpha) - x_\alpha$, where the function $m(x_\alpha)/[1 + m'(x_\alpha)] = R(x_\alpha)$.

The reciprocal of Mills ratio, $k^*_{1-\alpha} = 1 / R(z_{1-\alpha})$, is the hazard rate or failure rate. In terms of insurance it has been called the rate of mortality in life insurance and mortality of claims in non-life insurance. Other names are force of mortality, intensity function, and extinction rate.

Let $S = \sum_{i=1}^{n} X_i$ and $E(S\,|\,S > x_\alpha) = \sum_{i=1}^{n} E(X_i\,|\,S > x_\alpha)$. Landsman and Valdez (2003) examined this formula in the general framework of multivariate elliptical distributions that are generalizations of multivariate distributions.

9.3 Assume Nonnormality

Our unknown and hypothetical distribution was assumed to be a normal distribution in the previous section. Here we will relax this assumption and assume that the distribution is positively skewed, but still unknown. If we have any idea of the skewness of our unknown distribution, i.e., at least a guess of the magnitude of the skewness parameter, then we could use the normal power (NP) approximation in the risk charge (Equation 9.11) to arrive at the total solvency capital requirement. The NP approximation can thus be used to estimate VaR and also expected shortfall in terms of the standard deviation principle.

The percentiles of a complicated and skew distribution can be defined in terms of the percentiles of the standard normal distribution by the NP approximation (see Figure 9.1).

9.3.1 Risk Measure Based on NP Approximation

The *normal power (NP) approximation* enables us to express the percentiles of a complicated (and skew) distribution $F(.)$ in terms of the percentiles of the standard normal distribution $\phi(.)$. Using the first terms of the inverse Cornish–Fisher expansion does this.

Assume that X has distribution function $F(x)$ with mean μ_X, variance σ^2_X, and skewness

$$\gamma_1 = \frac{E\left[(X-\mu_X)^3\right]}{\sigma^3} = \frac{E\left[(X-\mu_X)^3\right]}{E\left[(X-\mu_X)^2\right]^{3/2}}$$

and let $\Phi(y)$ be the standardized normal distribution, i.e., with $\mu_Y = 0$ and $\sigma^2_Y = 1$.

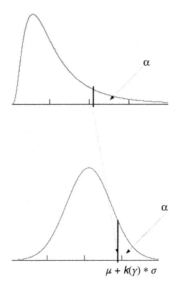

FIGURE 9.1
The percentiles of a complicated and skew distribution can be defined in terms of the percentiles of the standard normal distribution by the NP approximation. The factor k depends on the skewness in the original distribution.

We also let $u_\alpha = \dfrac{x_\alpha - \mu_x}{\sigma_x}$ and $k_{1-\alpha}$ be the $100(1-\alpha)$ percentiles of $F(.)$ and $\phi(.)$, respectively, i.e.,

$$F(u_\alpha) = \Phi(k_{1-\alpha}) = 1 - \alpha$$

Using the first two terms of the inverse Cornish–Fisher expansion we get

$$u_\alpha = k_{1-\alpha} + \frac{\gamma_1}{6}(k_{1-\alpha}^2 - 1)$$

or

$$x_\alpha = \mu_X + \sigma_X \left\{ k_{1-\alpha} + \frac{\gamma_1}{6}(k_{1-\alpha}^2 - 1) \right\} = \mu_X + \sigma_X k'_{1-\alpha}(\gamma_1)$$

where the new factor $k'_{1-\alpha}$ depends on the skewness.
For different α and $k_{1-\alpha}$ we get the following values of $k'_{1-\alpha}(\gamma_1)$ (Table 9.2).
For a skew distribution this gives us a prudent method of calculating the solvency capital requirement.

TABLE 9.2

The Distance Factors k and k' for Different Values of $1 - \alpha$ for the SDP and the Expected Shortfall in Case of Normality

$1 - \alpha$	$k_{1-\alpha}$	$k'_{1-\alpha}(\gamma_1)$
0.990	2.33	$2.33 + 0.74\,\gamma_1$
0.995	2.58	$2.58 + 0.94\,\gamma_1$
0.999	3.09	$3.09 + 1.43\,\gamma_1$
0.9995	3.29	$3.29 + 1.64\,\gamma_1$
0.9999	3.72	$3.72 + 2.14\,\gamma_1$

Our *baseline* for assessing the solvency capital requirement (SCR) is

$$C(\alpha,\gamma_1) = k'_{1-\alpha}(\gamma_1)\sigma_X \quad (9.12)$$

and hence, from Equation 9.11,

$$C(\alpha,\gamma_1) = \sqrt{\sum_{i=1}^{r} C_i^2(\alpha,\gamma_1) + \sum_{i\neq}^{r}\sum_{j}^{r} \rho_{ij} C_i(\alpha,\gamma_1) C_j(\alpha,\gamma_1)} =$$

$$\sqrt{\sum_{i=1}^{r}\sum_{j=1}^{r} \rho_{ij} C_i(\alpha,\gamma_1) C_j(\alpha,\gamma_1)} \quad (9.13)$$

In Equation 9.13 the total risk is divided by r risk classes (see the discussion in Section 7.3 and an example in Chapter 10).

9.3.2 Other Risk Measures

We have assumed that the overall (total) capital requirement is approximated by a normal distribution. This is a simple and very common assumption that can be accepted on the overall level by the approximation made in Section 9.2. On the other hand, if the distribution is moderately skewed to the right, we can use the distance factor $k^*_{1-\alpha}$, instead of $k_{1-\alpha}$, for safety.

If we know that the distribution is skewed to the right, we can apply other distance factors, e.g., based on *normal power approximation* (see above) or the *Esscher approximation* (see, e.g., Beard et al., 1984). Another approach is to use the *skew normal distribution* (for references on skew normal, multinormal, and skew elliptic distributions, see, e.g., Azzalini, 2003). If $\alpha = 0.99$, then $k = 2.57$ in the skew normal case.

9.4 Correlations between Risks: Different Levels of Conservatism

In a standard approach (SA) (or factor-based approach) it is suggested that some level of conservatism is used. As the SA model should be simple to use, one way of introducing conservatism is to make simplifying approximations on the safe side. In internal models these assumptions can be relaxed. Note that these levels of conservatism can be used on all parts of the risk measures or only at some parts. This means that they could, for example, be used on the total overall solvency capital requirement, but also only on some of its parts.

9.4.1 The Baseline Approach

The different approaches discussed below can be used on any level of diversification as defined in Section 7.3. We will define *three levels of conservatism* and illustrate this by assuming four risk categories:

C_1 : Risk category 1
C_2 : Risk category 2
C_3 : Risk category 3
C_4 : Risk category 4

The risk dependency structure is set up as the upper-right triangle of a dependency matrix.

Correlation Matrix				
	C_1:	C_2:	C_3:	C_4:
C_1:	1	ρ_{12}	ρ_{13}	ρ_{14}
C_2:		1	ρ_{23}	ρ_{24}
C_3:			1	ρ_{34}
C_4:				1

To simplify the notation we write C_i for the risk charge (category *i*), irrespective of whether it is defined by Equation 9.11 or 9.13. Hence, we get the baseline model

$$C_{tot} = \sqrt{\begin{aligned} &C_1^2 + C_2^2 + C_3^2 + C_4^2 + 2\rho_{12}C_1C_2 + 2\rho_{13}C_1C_3 + 2\rho_{14}C_1C_4 \\ &+2\rho_{23}C_2C_3 + 2\rho_{24}C_2C_4 + 2\rho_{34}C_3C_4 \end{aligned}} \qquad (9.14a)$$

If all the risk categories are pair-wise uncorrelated (fully uncorrelated), i.e., $\rho_{ij} = 0, \forall i, j$, then

$$C_{tot} = \sqrt{C_1^2 + C_2^2 + C_3^2 + C_4^2} \qquad (9.14b)$$

9.4.1.1 Level 0: No Conservative Approach

Example 9.1

We have the following dependency structure:[2]

Correlation Matrix				
	C_1:	C_2:	C_3:	C_4:
C_1:	1	0.25	0.25	1
C_2:		1	0.50	1
C_3:			1	1
C_4:				1

This gives us the following total risk measure:

$$C_{tot} = \sqrt{\begin{aligned} &C_1^2 + C_2^2 + C_3^2 + C_4^2 + 0.50C_1C_2 \\ &+0.50C_1C_3 + C_2C_3 + 2C_1C_4 + 2C_2C_4 + 2C_3C_4 \end{aligned}}$$

Due to uncertainty in estimation of the ρ *value*, one may simplify the structure by only assuming that the correlations take on the values 0, 0.2, 0.4, 0.6, 0.8, and 1.

---

[2] A linear model with one risk (C_4) fully correlated with the other risk categories (here C_1, C_2, C_3) implies that $\rho(C_1, C_2) = \rho(C_1, C_3) = \rho(C_2, C_3) = 1$. The assumption made here is an approximation for nonlinearity.

9.4.1.2   Level 1: A First Conservative Approach

Assume that

$$\rho_{ij}^* = 1 \ \text{ if } 0 < \rho_{ij} \le 1 \ (r \text{ categories})$$

and

$$\rho_{ij}^* = 0 \ \text{ if } -1 \le \rho_{ij} \le 0 \ (n - r \text{ categories})$$

Then we can write the squared variance part of it as

$$\sigma^2 = \sum_{j=1}^{n} \sigma_j^2 + \sum_{i \ne}^{n} \sum_{j}^{n} \rho_{ij}^* \sigma_i \sigma_j = \sum_{j=1}^{r} \sigma_j^2 + \sum_{i \ne}^{r} \sum_{j}^{r} \sigma_i \sigma_j$$

where the last term sums over the r risk categories that have $0 < \rho_{ij} \le 1$, i.e., $\rho_{ij}^* = 1$.

Example 9.2

We have the following dependency structure:[3]

Correlation Matrix				
	C_1:	C_2:	C_3:	C_4:
C_1:	1	0	0	1
C_2:		1	0	1
C_3:			1	1
C_4:				1

This gives us the following total risk measure:

$$C_{tot} = \sqrt{C_1^2 + C_2^2 + C_3^2 + C_4^2 + 2C_1C_4 + 2C_2C_4 + 2C_3C_4}$$

Let $C_{(123)} = C_1 + C_2 + C_3$. Then we can rewrite the above equation as

$$C_{tot} = \sqrt{\left(C_4 + C_{(123)}\right)^2 - 2C_1C_2 - 2C_1C_3 - 2C_2C_3}$$

---

[3] This is in theory incorrect, since if risk 4 is fully correlated with the other three, then, due to linear correlation, we cannot have $\rho_{12} = \rho_{13} = \rho_{23} = 0$. The structure is only used as a simplification of the correct structure. See also the comments on level 0 above.

9.4.1.3 Level 2: The Full Conservative Approach

Assume full dependencies in Equation 9.14a, i.e., $\rho_{ij} = 1, \forall(i,j)$, then the squared variance parts can be written as

$$\sigma^2 = \sum_{i=1}^{r}\sum_{j=1}^{r}\sigma_i\sigma_j = \left(\sum_{i=1}^{r}\sigma_i\right)\left(\sum_{j=1}^{r}\sigma_j\right) = \left(\sum_{i=1}^{r}\sigma_i\right)^2$$

Hence, we get

$$C = \sum_{i=1}^{r}C_i$$

Example 9.3

We have the following full dependency structure:

Correlation Matrix				
	C_1:	C_2:	C_3:	C_4:
C_1:	1	1	1	1
C_2:		1	1	1
C_3:			1	1
C_4:				1

This gives us the following total risk measure:

$$C_{tot} = C_1 + C_2 + C_3 + C_4 \tag{9.15}$$

9.4.2 A Benchmark Approach

The construction of the original U.S. NAIC risk-based capital (RBC) model for life companies (see Section 6.10) is described in the Müller report (1997, p. 55):

> Once all RBC values of the individual categories have been calculated they are combined into the total RBC. For this the individual values are, however, not simply added up but a compensation is made because not all risks will cause losses simultaneously. If it is assumed that both asset risk and interest rate risk (C1 and C3) are completely correlated and the

technical risk (C2) is not related to either of them and in addition that
the business risk (C4) is completely correlated with the other three risks
this will result in a total RBC in life insurance (RBC_{LV}) as follows:

$$RBC_{LV} = C_4 + \sqrt{C_2^2 + \left(C_1 + C_3\right)^2}$$

In general, this approach, which we will call a benchmark approach, is
based on the calculation of the risk charges in a hierarchical way:

1. Determine the individual risk charge structure on the highest level
   (level 3 or 2, according to Section 7.3; if both levels are used, this
   step will split up into two) and use an assumed dependence structure as in Section 9.4.1 to construct the benchmark model for the
   highest level.

2. In a similar way, determine the individual risk charge structure on
   the first level (level 1) and use an assumed dependence structure to
   construct the benchmark model on this level.

3. Do the same for the last level (level 0) or apply the charge model
   directly.

At each step 1 start to sort up the dependency structure in a reverse order,
as compared to the description of the NAIC RBC model above: all variables
that are fully correlated are assumed to be a single risk and the other variables another risk. The structure of this dual part is then set up. Then we
look at these risks one at the time and look at the dependency structure
within each risk. Try to split them up into dual parts and set up the risk
structure between each pair. Go on in this way until the dependency structures for all risk categories have been modeled.

Usually this type of benchmark model is used when the risk charges are
assumed to be more general, with or without any model assumption behind,
such as, for example, $\sum_{i=1}^{k} \alpha_i A_i$, where α_i are risk factors applied to some risk
measure (cf. Section 6.10). A risk charge C_i could also be the result of a stress
test, for example, the effect of, say, a 20% drop in the equities.

To illustrate this, we start with Example 9.1 of Section 9.4.1.1.

C_4 is fully correlated with the other risk categories, i.e., $\rho_{i4} = 1, \forall i$.
Let $\rho_{(123)}$ denote the total risk from the categories 1 to 3. Hence, $\rho_{(4)}$ and $\rho_{(123)}$
are fully correlated and the first step in this benchmark approach gives
us $C_{tot} = C_4 + C_{(123)}$. In the next step we look at the risk categories 1 to 3 and
get

$$C_{(123)} = \sqrt{C_1^2 + C_2^2 + C_3^2 + 0.50C_1C_2 + 0.50C_1C_3 + C_2C_3}$$

For this hypothetical example we get the following benchmark risk charge model:

$$C_{tot} = C_4 + \sqrt{C_1^2 + C_2^2 + C_3^2 + 0.50C_1C_2 + 0.50C_1C_3 + C_2C_3}$$

As another example we take Example 9.2. The benchmark model would thus be (cf. the model above)

$$C_{tot} = C_4 + \sqrt{C_1^2 + C_2^2 + C_3^2}$$

The next example is a discussion of the original U.S. NAIC RBC model for life companies, as mentioned earlier (cf. Section 6.10).

Example 9.4

Assume the following dependency structure:

Correlation Matrix				
	C_1:	C_2:	C_3:	C_4:
C_1:	1	0	1	1
C_2:		1	0	1
C_3:			1	1
C_4:				1

If we have four risk categories and the fourth (C_4) is completely correlated with the other three, which are denoted by $C_{(123)}$, then $C_{tot} = C_4 + C_{(123)}$.

If risk categories 1 and 3 are completely correlated and risk category 2 is uncorrelated with risk categories 1 and 3, we can write

$$C_{(123)}^2 = C_2^2 + \left(C_1 + C_3\right)^2$$

and hence we get the total risk charge as

$$C_{tot} = C_4 + \sqrt{C_2^2 + (C_1 + C_3)^2} \qquad (9.16)$$

The risk structure in Equation 9.16 is the one originally proposed by NAIC for life insurance companies (see Section 6.10).

The baseline approach gives

$$C_{tot} = \sqrt{C_1^2 + C_2^2 + C_3^2 + C_4^2 + 2C_1C_3 + 2C_1C_4 + 2C_2C_4 + 2C_3C_4} \qquad (9.17)$$

To compare Equations 9.16 and 9.17, we square both expressions, labeling Equation 9.16 $C_{tot}^2(BM)$ for benchmark and Equation 9.17 $C_{tot}^2(BL)$ for baseline:

$$C_{tot}^2(BM) = C_1^2 + C_2^2 + C_3^2 + C_4^2 + 2C_1C_3 + 2C_4\sqrt{C_2^2 + (C_1 + C_3)^2} \qquad (9.18)$$

Squaring Equation 9.17 gives

$$C_{tot}^2(BL) = C_1^2 + C_2^2 + C_3^2 + C_4^2 + 2C_1C_3 + 2C_4(C_2 + C_1 + C_3) \qquad (9.19)$$

Hence, the difference between the two approaches in this example is the last terms of Equations 9.18 and 9.19.

If the dependence structure is totally uncorrelated $(\rho_{ij} = 0, \forall i, j)$ or fully correlated $(\rho_{ij} = 1, \forall i, j)$, then the benchmark approach and the baseline approach in Section 9.4.1 are identical.

Even though the benchmark approach usually gives an underestimate, in comparison with the baseline approach, we will use it in Chapters 10 and 11 when we model the solvency requirement.

9.5   Parameters in a Factor-Based Model

One of the main features of a factor-based model is the possibility to change or revaluate different parameters.

  Assume that one of the four risk charges above can be transformed into the following factor-based model (this will be used in Chapter 11 for the underwriting risk):

$$C_* = \sum_{i=1}^{L} Z_{nri}\beta_i E(X_i) + \sum_{i=1}^{L} Var(LR_i)E^2(X_i) + \sum_{i \neq}^{L}\sum_{j}^{L} Cov(LR_i, LR_j)E(X_i)E(X_j)$$

(9.20)

A model like this would be dependent on at least two of three different types of parameters. We follow the Swiss supervisory authority in its proposal for a new solvency system in Switzerland (SST, 2004a; see also Section 6.8).

There are both specified and unspecified parameters. The SST approach has three types of parameters.

9.5.1 Parameter Classification

Type 1: Parameters set by the insurance regulator; they cannot be changed (e.g., risk-free interest rate, quantile, FX rates).

Type 2: Parameters that have to be set by the insurance companies (e.g., volatility of the hedge fund exposure).

Type 3: Parameters that are set by the insurance regulator and that can be changed by the companies (e.g., biometric parameters, non-life parameters).

In a standard approach, Equation 9.20 has the following types of parameters. Note that these are discussed extensively in Chapter 11 and in Appendix A:

Type 2:

$E(X_i)$: The expected claims cost for line of business i.

Type 3:

Z_{nri} : Net retention or maximum net claim for line of business i.

β_i : A parameter capturing the net claims size distribution (stable over time) for line of business i.

$Var(LR_i)$: The variance of the loss ratio for line of business i.

$Cov(LR_i, LR_j)$: The covariance between the loss ratios for lines of business i and j.

10

One Example of Risk Categories and Diversification

In Section 7.3 we briefly discussed risks and diversification and we also classified the diversification effect according to three main levels (plus a zero level of risk exposures). In this chapter we will discuss five main risk[1] categories (level 2) as defined in Figure 10.1. Most of these main risks can be subdivided into subrisks. At the stop level of these risks we have a diversification that could be made on lines of business (LOBs), asset classes, etc. (level 1) and expressed in volatility effects and nondiversifiable effects (level 0).

In Sections 10.1 to 10.5 we will discuss the five risk categories: insurance risk, market risk, credit risk, operational risk, and liquidity risk.

After the discussion on the risks we will use them and our benchmark models from Chapter 9 to set up a *standard benchmark model* in Section 10.6. This could be based on either a normal power (NP)-based approach or its counterpart under normality. The final model is given by Equation 10.4.

In Appendix A, these risks are discussed in an actuarial (mathematical/ statistical) way and formulized into factor-based formulae. In Chapter 11 we will make these formulae operational and propose measuring methods. The factor-based model is also converted into a spreadsheet approach that will simplify the understanding and calculations.

The examples discussed here, in Chapter 11, and in Appendix A are only suggestions, but hopefully they will be useful as a starting point in the development of a standard approach. In developing a solvency charging system it is important to have a close interaction between the development of theory and field tests (impact studies). One main issue is to calibrate models with realistic parameters and define the confidence level α.

Main risk categories are (Figure 10.1):

- Insurance risk
- Market risk
- Credit risk
- Operational risk
- Liquidity risk

---

[1] The risks discussed in this chapter are the same that have been discussed within a solvency sub-committee in CEA, the organization representing the European insurance industry.

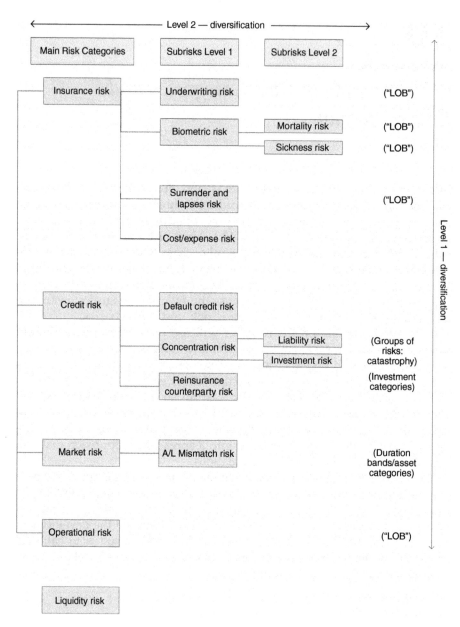

FIGURE 10.1
A visualization of risk diversification. Horizontal diversification is on level 2, and vertical diversification on level 1, according to the classification made in Section 7.3.

These risks are the same as proposed by IAA (2004) (cf. also Section 5.4), i.e., the risk specific for the insurance business (at least non-life) and the four risks proposed for banks within Basel II (see Section 5.1).

The main risks and their subrisks are discussed in Sections 10.1 to 10.5. At this stage we only discuss stand-alone business solvency. Group and conglomerate issues are outlined in Chapter 14.

10.1 Insurance Risk

"Insurance companies assume risk through the insurance contracts they underwrite" (see IAA, 2004, p. 29).

The *underwriting process risk* is addressed to the risk related to the business that will be written during the following year. We consider it *net of reinsurance*, as the reinsurance will be dealt with in the credit risk category. The underwriting process risk will thus be highly correlated with the credit risk. The actuarial approach is discussed in Appendix A.1.

This is the main non-life insurance risk, but it is also important for annuities and can be measured by the volatility of the underwriting result and by a factor measuring the expected profitability of the business. The volatility includes a separate term for the *nonsystematic underwriting risk*, i.e., the risk that is due to the variation in the frequency and severity of the claims, which can be diversified through a greater portfolio and an adequate reinsurance policy.

Another term is needed to describe the variance of *systematic underwriting risk*, i.e., the risk that is common to the whole insurance market and that cannot be reduced by pooling.

The calculations should be made by lines of business, for example, according to the insurance classes introduced by the European Commission (EC) (see Appendix B). The company's own loss experience should be used to measure the risk exposure and expected profitability.

The EC should publish volatility and correlation parameters that could be used in a standard approach, and the company would provide data on loss experience (see Chapter 11 for a hypothetical application).

Ensuring the adequacy of *claim provisions* is a pillar II issue. Hence, there should be no systematic error of the claim provision in the standard approach. However, the pure random *run-off risks* always remain, for example, in the estimation of the incurred but not reported (IBNR) and incurred but not enough reported (IBNER). We will deal with this diversifiable risk as equivalent to the diversifiable underwriting risk in the business to be written in the following year.

The capital requirement for *surrender and lapse risks* will be determined based on the assumed trend parameters and lapse ratios (given by the EC) multiplied by the guaranteed surrender value minus best-estimate liability (given by the company), but never below zero. These risks are mainly related to life insurance but will also be introduced for non-life business. Because an increase in interest rates lowers the market value of bonds, and probably also increases surrender rates and lapse rates, it is assumed that it is highly correlated to the market risk (asset–liability risk).

The *expense risk* consists of, e.g., acquisition, administrative, and claims settlement costs. Some of the costs cannot be allocated to any particular line of business and they are called overhead expenses. The risk will consist of two parts: the first is the overhead expenses part and the second is the part depending on the line of business.

For unit-linked contracts the principal risks are concentrated in expenses and lapses. The major insurance (underwriting) risk is borne by the insured. In this case, the expenses are administrative costs and the risk that these will exceed what can be earned on the policies. We will only consider the retained business and the risk will consist of weighted net technical provisions.

The *biometric risk* will consist of two parts, one part measuring volatility and trends in mortality and another measuring volatility and trends in sickness. We assume that they are uncorrelated with each other.

To sum up, the main risk category "insurance risk" will consist of the following subrisks:

- Insurance risk
  - Underwriting risk (Appendix A.1.1)
  - Biometric risks: (Appendix A.1.2)
    - Mortality risk
    - Sickness risk
  - Surrender and lapse risk (Appendix A.1.3)
  - Cost/expenses risk (Appendix A.1.4)

10.2 Market Risk

"Market risk is introduced into an insurer's operations through variations in financial markets that cause changes in asset values, products, and portfolio valuation" (IAIS, 2004b).

The market risk arises from the level or volatility of market prices on assets and involves the exposure to movements in financial variables, such as interest rates, bond prices, equities (stocks) and property (real estate) prices, and exchange rates.

We will mainly look at the *asset–liability mismatch risk* (A/L risk), which results from the volatility and uncertainty risk inherent in market values of future cash flows from the insurer's asset and liabilities. We will only consider the retained business, as risks related to reinsurance are taken into account by the credit risk.

When assessing the A/L risk, only the part of the business for which the company is bearing a risk should be taken into consideration (e.g., unit-linked business excluded). It is important to remember that a company bears the risk that the cash flows generated by the assets do not match the payouts due to the market volatility, credit default, or changes in interest rate curve.

It is essential to simultaneously consider the volatilities of assets and liabilities,when there are common background factors affecting both of them. The company can manage these risks by suitable choices of assets, by a dynamic investment policy, and by product design.

The interplay between assets and liabilities shall be reflected in the capital requirements.

The assessment of the A/L risk should be based on best estimates, and a starting point for quantifying the asset–liability risk could be a multiple of the volatility of asset return. Purchasing hedging instruments, like caps and floors, can significantly decrease the interest rate risk, and the volatility risk of equities can be reduced by the purchase of puts and calls. Those hedging strategies have a major impact when setting up an asset liability policy and are an essential tool to preventing insolvency. Using hedging instruments reduces the asset volatilities, and this should be taken into account.

The resulting asset volatilities should then be matched to corresponding volatilities in liabilities: a change in interest rates may cause simultaneous changes in asset values and in liabilities, and the company should only be required to hold sufficient capital to cover the net effect of the changes. This principle can be applied in a relatively straightforward manner for *bonds* covering technical provisions, as it is relatively easy to compare the corresponding cash flows. However, this matching principle should not be limited to bonds only. Some insurance liabilities, typically deferred annuities, have very long-term durations. The market may not offer bonds of sufficient duration to back those liabilities. So, risk managers invest more widely in *equities* and *property* to provide an adequate assets–liability policy.

The capital requirement should not be directed to *free assets* (assets exceeding technical provisions and the solvency capital requirement) of the company, since higher than necessary capital would increase the requirement, which is not reasonable. One approach is that the requirement is directed only to assets backing technical provisions, and maybe also to some extent the solvency capital requirement itself.

In a standard approach the number of asset categories should not be large, but it should be possible to enlarge the number of categories in an internal model. In our standard approach we will only classify the categories into bonds, equities, property, cash, and others.

The EC would define the parameters describing expected returns, volatilities, and correlations. The company should provide data on asset and liability values.

To simplify the calculations we will select a number of *duration bands* in a way similar to IAA (2004, p. 143). The assets corresponding to each duration band will be matched to the liabilities of similar duration.

10.2.1 Bonds

d_1: [0–1] year (median duration: $md_1 = 0.5$)

d_2: [1–2] years (median duration: $md_2 = 1.5$)

d_3: [2–5] years (median duration: $md_3 = 3.5$)

d_4: [5–8] years (median duration: $md_4 = 6.5$)

d_5: [8–12] years (median duration: $md_5 = 10.0$)

d_6: [12–16] years (median duration: $md_6 = 14.0$)

d_7: [16–24] years (median duration: $md_7 = 20.0$)

d_8: [24–] years (median duration, say: $md_8 = 28$)

10.2.2 Equities

Define a band of mean holding periods for a company's equities. This can be done by historical company data.

Duration band: Say d_4: [5–8] years (median duration: $md_4 = 6.5$).

10.2.3 Property

Define a band of mean holding periods for a company's property. This can be done by historical company data.

Duration band: Say d_5: [8–12] years (median duration: $md_5 = 10.0$).

10.2.4 Cash (and Others)

Duration band: Say d_1: [0–1] year (median duration: $md_1 = 0.5$).

Currency risk is to a large extent taken into account in the A/L risk if the calculations are made by currencies. If there is a need for a specific term for the currency mismatch risk, it could simply be defined as *coefficient times the mismatch*. A discussion of this risk is made in IAA (2004, pp. 139–140).

10.2.5 General Issues

In our theoretical approach the main risk category "market risk" will consist of the following subrisks, where the currency risk not will be included in our model:

- Market risk
  - Asset–liability risk (Appendix A.2.2)
  - Currency risk (Appendix A.2.3); not included in the model in Chapter 11

10.3 Credit Risk

"Credit risk is the risk of financial loss resulting from default or move-ment in the credit quality of issuers of securities (in the company's investment portfolio), debtors (for example, mortgagors), or counterpar-ties (for example, on reinsurance contracts, derivative contracts or depos-its given) and intermediaries, to whom the company has an exposure" (IAIS, 2004b).

The *default credit risk* is the risk that an undertaking will not receive fully or partially (or receives delayed) cash flows or assets to which it is entitled because a party with which it has a (bilateral) contract defaults in one or another way. In our model we will use the comprehensive standard model proposed in Basel II (see Section 5.1 and Appendix A.3.3). In this risk we also include downgrade/migration risk and indirect credit or spread risk. The *downgrade/migration risk* is a future risk that changes the probability of default by an obligor that will affect the present value of the contract with him. This could be measured by changed rating. By *indirect credit risk* we mean the risk due to market perception of increased risk on a micro- or macrobasis (IAIS, 2004b). The standard model of the default credit risk is the standard approach from Basel II. The reason for this (cf. also the operational risk below) is that we want to minimize the possibility for arbitrage of the risk from banking to the insurance sector, and vice versa. This approach has also been used in the Swiss system (see Section 6.8).

By *concentration risk* we mean various types of concentrations or exposures, e.g., in investments (asset concentration) and catastrophic events (concen-tration of liabilities). For example, investment in a high proportion of specific equities can be considered risky (concentration). This risk is also due to concentration in geographical areas, economic sectors, and connected par-ties.

In our model we will consider asset concentration and liability concentra-tion risks. These risks are considered to be the net of reinsurance, which will be the second risk category under the credit risk.

One alternative model would be to use some concentration measure, like the Gini coefficient, as a measure of concentration (see, e.g., Nygård and Sandström, 1981).

The *reinsurance counterparty risk* should be assessed separately. The sol-vency capital requirement for this risk could be determined by a coefficient given by the EC taking into account the new regulation system on reinsur-ance, multiplied by the technical provisions ceded to and receivables from that reinsurer. Rating factors can be used as scores.

In our approach the main risk category "credit risk" will consist of the following subrisks:

- Credit risk
  - Default credit risk (Appendix A.3.3)
  - Concentration risk (Appendix A.3.1)
  - Reinsurance counterparty risk (Appendix A.3.2)

10.4 Operational Risk

"Operational risk is the risk of loss resulting from inadequate or failed internal processes, people, systems or from external events" (IAA, 2004).

This concept has primarily emerged from the banking industry and the Basel II project. It can be seen as consisting of all risks other than insurance, market, credit, and liquidity risks. There is no unique definition of what is to be understood under operational risk, although many existing systems equate it, for instance, to a collection of exposures to:

- Failure in control and management
- Failure in IT processes
- Human errors
- Fraud
- Jurisdictional and legal risk

As such, rather than being a truly isolated risk component, operational risk is often associated and overlapping with other risk factors. The Sharma report (Section 5.5) has indicated that management shortfalls have led to many EU insurer failures.

We will model the operational risk in a manner similar to that of the standard approach of Basel II (cf. Section 5.1). The main reason for this is that we do not want any possibility for arbitrage from the banking to the insurance sector, and vice versa. This means that the operational risk formula follows as closely as possible to one of the models used by the banking sector. We will assume, as was done by NAIC in its risk-based capital system, that this risk is truly correlated to all other risks (cf. Section 6.10), meaning that the correlation is ρ (*operational risk, other risks*) = 1.

It is possible to define subrisks (see, for example, the discussion in Tripp et al., 2004).

10.5 Liquidity Risk

"The liquidity risk is the risk that an insurer, though solvent, has insufficient liquid assets to meet its obligations (such as claims payments and policy redemptions) when they fall due. The liquidity profile of an insurer is a function of both its assets and liabilities" (IAIS, 2004b).

The liquidity risk includes (IAIS, 2004b):

- *Liquidation value risk*: The risk that unexpected timing or amounts of needed cash may require the liquidation of assets in such a way that this could result in loss of realized value.
- *Capital funding risk*: The risk that the insurer will not be able to obtain sufficient outside funding, as its assets are illiquid at the same time it needs it to meet large claims.

A third risk discussed by IAIS is

- *Affiliated investment risk*: The risk that an investment in a member company of the conglomerate or group may be difficult to sell, or the affiliates may create a drain on the financial or operating resources from the insurer.

This latter risk is part of the *participating risk* in a conglomerate discussed in Section 7.3.

The liquidity risk is assumed to be a pillar II issue, i.e., belonging to the supervisory review process, and will hence not be included in our model.

10.6 Dependency

From the discussion above we can set up a dependency matrix for the four major risk categories. This table, which shows the correlations between the various risk categories, can be used to decide on what level of conservatism we are going to take.

Correlation Matrix	C_{IR}:	C_{MR}:	C_{CR}:	C_{OR}:
C_{IR}:	1	High	High	1
C_{MR}:		1	High	1
C_{CR}:			1	1
C_{OR}:				1

Note: High = high correlation; low = low correlation (used in the following table).

C_{IR}: Insurance risk (Section 10.1)

C_{MR}: Market risk (Section 10.2)

C_{CR}: Credit risk (Section 10.3)

C_{OR}: Operational risk (Section 10.4)

To simplify the suggested correlations above, we set high = 1 (and low = 0). This gives us the following matrix:

Correlation Matrix	C_{IR}:	C_{MR}:	C_{CR}:	C_{OR}:
C_{IR}:	1	1	1	1
C_{MR}:		1	1	1
C_{CR}:			1	1
C_{OR}:				1

Hence, this gives us the following risk structure (cf. Equation 9.15).

$$C_{TOT} = C_{IR} + C_{MR} + C_{CR} + C_{OR} \tag{10.1}$$

In the *insurance risk* (C_{IR}) we will include the underwriting risk (C_{ur}), biometric risk (C_{br}), surrender and lapse risk (C_{slr}), and expense risk (C_{er}). The unit-linked risk will be a part of the underwriting risk.

Correlation Matrix	C_{ur}:	C_{br}:	C_{slr}:	C_{er}:
C_{ur}:	1	High	High	High
C_{br}:		1	Low	Low
C_{slr}:			1	High
C_{er}:				1

Note:

C_{ur}: Underwriting risk

C_{br}: Biometric risk

C_{slr}: Surrender and lapse risk

C_{er}: Expenses/costs risk

We can simplify this structure in the standard benchmark approach as follows (high = 1, low = 0):

Correlation Matrix	C_{ur}:	C_{br}:	C_{slr}:	C_{er}:
C_{ur}:	1	1	1	1
C_{br}:		1	0	0
C_{slr}:			1	1
C_{er}:				1

This gives us for insurance risk

$$C_{IR} = C_{ur} + \sqrt{C_{br}^2 + \left(C_{slr} + C_{er}\right)^2} \tag{10.2}$$

In the *credit risk* (C_{CR}) we will include the default credit risk (C_{dcr}), concentration risk (C_{cor}), and the reinsurance risk (C_{rr}). We assume that the correlations between these three risks are low; i.e., we assume in the standard approach zero correlation that gives us the credit risk

$$C_{CR} = \sqrt{C_{dcr}^2 + C_{cor}^2 + C_{rr}^2} \tag{10.3}$$

The standard benchmark model is defined by combining Equations 10.1 to 10.3 above into

$$C_{TOT} = C_{ur} + \sqrt{C_{br}^2 + \left(C_{slr} + C_{er}\right)^2} + C_{MR} + \sqrt{C_{dcr}^2 + C_{cor}^2 + C_{rr}^2} + C_{OR} \tag{10.4}$$

which gives us the risk structure we will use to build up a factor-based model in Chapter 11.

11

A Proposal for a Standard Approach: From Formula to Spreadsheet

In Section 10.6 we arrived at a risk structure for the standard model, our benchmark (see Equation 10.4). The factor-based models for the different risks are developed in Appendix A, where the actuary or the more mathematically inclined can find the details. In this chapter we will first explain the theory used and the consideration done for the different risks.

In Sections 11.1 to 11.4 we will summarize the four risk factor-based parts of the model: insurance, market, credit, and operational risks.

In Section 11.5 we will present the total risk charge formula from Appendix A. This is not to scare the reader away, but just to show that from an at least seemingly complicated formula we can arrive at a simple spreadsheet solution. If you are not familiar with formulas, you can just ignore them.

In Section 11.7 we will illustrate the spreadsheet approach, from Section 11.6, by showing an example of parameters that could be used as common parameters.

It is important to have in mind that the discussion here is to show how to get to a standard model and how to handle it. It is just one example.

We have already in Equation 10.4 derived the main solvency model, our benchmark:

$$C_{TOT} = C_{ur} + \sqrt{C_{br}^2 + \left(C_{slr} + C_{er}\right)^2} + C_{MR} + \sqrt{C_{dcr}^2 + C_{cor}^2 + C_{rr}^2} + C_{OR} \quad (11.1)$$

The different parts of C_{TOT} have then been discussed in Chapter 10 and formulated in Appendix A. Combing these results into Equation 11.1, as done below, gives us the proposed standard approach for assessing the solvency capital requirement (SCR). The solvency capital level is obtained by adding the best estimate of the technical provisions (cf. Equation 9.5).

11.1 The Insurance Risk, C_{IR}

The main risk category "insurance risk" will consist of the following subrisks (cf. Section 10.1):

- Insurance risk
  - Underwriting risk (Section 11.1.1)
  - Biometric risks (Section 11.1.2)
    - Mortality risk
    - Sickness risk
  - Surrender and lapse risk (Section 11.1.3)
  - Expenses/costs risk (Section 11.1.4)

11.1.1   Underwriting Risk, C_{ur}

The model used is a mix between the classical mixed-compound Poisson model (see, e.g., Beard et al., 1984; Rantala, 2003) and the IAA proposal (IAA, 2004, p. 111). In the application below we will mainly follow the first part of the mixture, i.e., as proposed by Rantala (2003).

The model is used to describe the underwriting risk of each line of business (LOB).

The risk model has the main features that (1) the number of claims follows a Poisson distribution, (2) the individual claim sizes are independent of each other and of the number of claims and follow the same distribution,[1] and (3) the expected value of the claim number for each year is a random variable (r.v.); i.e., the Poisson parameter is of the form nq, where n is the long-term expectation of the number of claims and q is a random variable with expectation 1 and variance σ_q^2. The r.v. q is independent of the r.v.'s in numbers 1 and 2, but its values between consecutive years may be correlated. The total loss is hence a *mixed-compound Poisson* distributed variable. The idea is that the nondiversifiable risk is described by the random variable q, operating on the expected claims frequency, but not on the claims size distribution.

The division between diversifiable and nondiversifiable risks is worth preserving, as the model should be applicable to both small and large firms.

A similar reasoning can be made for the run-off risk of claim provisions. One difference would be that the systematic component might be negligible. Hence, we do not assume any systematic error of the claim provisions (CP) in the model. We assume that the diversifiable risk behaves as the corresponding pure underwriting risk.

Assume the following notation: β_i is a structure coefficient for each LOB's i. Stable coefficients, immune to inflation, are set by the EU ($0 \le \beta_i \le 1$). The

---

[1] Independent and identically distributed (i.i.d.).

variances and covariances of the loss ratios, $Var(LRi)$ and $Cov(LRi,LRj)$, are defined on the EU level.

In the standard model, the company only has to estimate the net losses, the claim provisions, and the maximum net retention of each LOB: $E(X_i)$, CP_i, and Z_{nri}, $i = 1, ..., L$.

The simplified model from Appendix A.1 is thus given by

$$\frac{1}{k_{1-a}^2}C_{ur}^2 = \sum_{i=1}^{L} Z_{nri}\beta_i E(X_i) + \sum_{i=1}^{L} Z_{nri}\beta_i CP_i + \sum_{i=1}^{L} Var(LR_i)E^2(X_i)$$

$$+ \sum_{i\neq}^{L} \sum_{j} Cov(LR_i, LR_j)E(X_i)E(X_j)$$

We also will include the unit-linked risk in the pure underwriting risk. The major insurance (underwriting) risk is borne by the insured. We assume that:

Y_1 = The net technical provisions for the financial and insurance risks for the company when there is a promised minimum guarantee

Y_2 = The net technical provisions where there are *no* financial and insurance risks for the company, but where there is a considerable business risk in the provisions or when there is a considerable administration risk

The risk charge for the unit-linked products is

$$\frac{1}{k_{1-\alpha}}C_{ulr} = \{p_1 Y_1 + p_2 Y_2\}$$

where the parameters p_i, $i = 1, 2$, are defined by the EU.

11.1.2 Biometric Risk, C_{br}

The biometric risk category is defined for both life insurance and non-life insurance (Appendix A.1.2). It could be assumed that it is uncorrelated with other risks. The risk is considered to be the net of reinsurance. The first part will take care of the *mortality risk* (M) and the second the *sickness risk* (S). The measures are divided into volatility and trend uncertainty.

Assume that $\sigma_{VOL,j}$ is the volatility in the misspecification of the probability of mortality or sickness. The parameter b_j reflects the misspecification of the trend in the probability of mortality or (remaining) sick and the terms $b_j\sigma_{VOL,j}$ could be written $\sigma_{trend,j}$. The b_j values are defined on the EU level and the $\sigma_{VOL,j}$ can be estimated by company data based on historical data.

The risk charge for the biometric risk would thus be (the sum is over LOBs)

$$\frac{1}{k_{1-\alpha}^2}C_{br} = \sum_{j=1}^{L} {}_M\sigma_{VOL,j} + 0.5\sum_{j=1}^{L} {}_Mb_j\sigma_{VOL,j} + \sum_{j=1}^{L} {}_S\sigma_{VOL,j} + 0.5\sum_{j=1}^{L} {}_Sb_j\sigma_{VOL,j}$$

If a company uses prespecified volatility measures, given by the EU, then the risk charge would be a fixed charge number. In Appendix A.1.2.3 there is a description of a simplified way of calculating the volatility in mortality proposed by the German Insurance Association (GDV).

11.1.3 Surrender and Lapse Risk, C_{slr}

Risks related to surrender and lapse are mainly defined for life insurance (see Appendix A.1.3). Since increases in interest rates lower the market value of bonds and probably also increase surrender rates, it is assumed that they are highly correlated to the asset–liability risk in Appendix A.2.2.

Let $\sigma_{VOL,j}$ be the volatility in the misspecification of the lapse ratio. Then the risk charge due to the lapse can be written as

$$\frac{1}{k_{1-\alpha}}C_{slr} = \sum_{j=1}^{L}\sigma_{VOL,j} + 0.5\sum_{j=1}^{L} l_j\sigma_{VOL,j}$$

where the trend parameters, l_j, are defined on the EU level and the volatilities $\sigma_{VOL,j}$ can be estimated from company data. A simplification in calculating the volatility is given in Appendix A.1.3:

$$\sigma_{VOL,j} \approx \hat{X}_j^+ \frac{1}{\sqrt{n_j}}\sqrt{p_j(1-p_j)}$$

where $\hat{X}_j^+$ is the positive part of an estimate of the total guaranteed surrender value *minus* the total best-estimate liability. The lapse ratios (p_j) are defined on the EU level.

11.1.4 Expenses/Costs Risk, C_{er}

The expense/cost risk consists of, e.g., acquisition, administrative, and claim settlement costs. Some of the costs cannot be allocated to any particular line of business. They are called overhead expenses. The remaining costs are attributed to the specific line of business (LOB). See Appendix A.1.4.

The part of the total capital requirement from the expense risk is defined as

$$\frac{1}{k_{1-\alpha}^2} C_{er}^2 = \sum_{i=0}^{L} c_i^2 (1+p_i)^2 E^2(X_i)$$

where the premium loading factors p_i and expense ratios c_i are defined on the EU level.

X_0 is the total claims cost, $X_0 = \sum_{i=1}^{L} X_i$, and $E(X_i)$ is the expected claims costs for LOB i, $i = 1, \ldots, L$.

11.2 Market Risk, C_{MR}

The market risk will be approximated by the asset–liability mismatch risk. This risk is a result from the volatility inherent in the market value of future cash flows of the insurer's assets and liabilities.

The discussion in Appendix A.2 is mainly based on a suggestion made by Rantala (2004a) to Comité Européen des Assurances's (CEA) solvency working group.[2] In that paper Rantala suggested *a match between bonds and interest-sensitive fair-valued[3] liabilities*. In the model a mismatch of the form

$$d_B B - d_V V$$

was assumed, where d_B, d_V are suitably defined durations of bonds and technical provisions, respectively. B is the market value of the bonds and V is the best estimate of the technical provisions, based on the current market interest rate curve. Hence, if the amount and duration of technical provisions and the bond portfolio covering these technical provisions were equal, the resulting risk charge would be zero. Another working paper (FFSA, 2003) extended the proposed model to also include a duration analysis for equities and property. We have used these approaches as the start for the modeling given in the appendix.

Here we will consider the duration of equities and property as the *holding period*, i.e., the *mean time* a company is holding its equities and property. The duration d_* will be, in accordance with IAA (2004), split up into duration bands. This means that the technical provisions must also be divided into duration bands, similar to bonds, equities, and property below.

To simplify the calculations we will select a number of *duration bands* similar to IAA (2004, p. 143). The assets corresponding to each duration band will be matched to the liabilities of similar duration[4] using a cash-flow approach.

---

[2] The originating working paper is from March 2003.
[3] In the paper by Rantala (2004a) it is assumed to be *best estimate* instead of *fair value*.
[4] This is a pragmatic solution as compared with Appendix A.2.

11.2.1   Bonds

d_1: [0–1] year (median duration: $md_1 = 0.5$)
d_2: [1–2] years (median duration: $md_2 = 1.5$)
d_3: [2–5] years (median duration: $md_3 = 3.5$)
d_4: [5–8] years (median duration: $md_4 = 6.5$)
d_5: [8–12] years (median duration: $md_5 = 10.0$)
d_6: [12–16] years (median duration: $md_6 = 14.0$)
d_7: [16–24] years (median duration: $md_7 = 20.0$)
d_8: [24–] years (median duration: say $md_8 = 28$)

11.2.2   Equities

Define a band of *mean holding periods* for a company's equities. This can be done by historical company data.

Duration band: say, d_4: [5–8] years (median duration: $md_4 = 6.5$)

11.2.3   Property

Define a band of *mean holding periods* for a company's property. This can be done by historical company data.

Duration band: say, d_5: [8–12] years (median duration: $md_5 = 10.0$)

11.2.4   Cash (and Others)

Duration band: say, d_1: [0–1] year (median duration: $md_1 = 0.5$)

For each asset category $j = 1, ..., c$, let A_j be the asset amount invested in asset j backing the liabilities. Let $\sigma(j)$ be the standard deviation of the return in asset category j and $\rho(j,k)$ be the correlation coefficient between returns of categories j and k. These are all defined on the EU level in the standard approach.

The part of the total capital requirement from the asset–liability risk is defined as

$$\frac{1}{k_{1-a}^2}C_{MR}^2 = \sum_{k=1}^{8} A_{01}^2(d_k)\sigma^2(d_k) + \sum_{j=2}^{c} A_{0j}^2\sigma^2(j) + \sum_{j\neq}^{c}\sum_{l}^{c} A_{0j}A_{0l}\rho(j,l)\sigma(j)\sigma(l)$$

where the first term is the sum ($k = 1, ..., 8$) of a mismatch part, $A_{01}(d_k)$ times the standard deviation of the return, depending on eight duration bands defined above (see also Section 10.2). The durations of bonds are split up into eight duration bands, and for equities and property, we define bands of mean holding periods based on historical data. The second term is the total variance of the portfolio, and the last term reflects the correlation between the assets in the portfolio.

The second term of the charge excludes bonds, but includes equities and property residual parts (noninterest rate return). We have assumed independence between these two residual parts, but as is seen in the data section (Section 11.7), there is a positive correlation between the residual parts of equities and property.

In the third term we have included bonds without splitting them up into duration bands.

Companies have only to list their assets in the c categories, defined by EU, and according to their durations.

11.3 Credit Risk, C_{CR}

The main risk category "credit risk" will consist of the following subrisks (cf. also Section 10.3):

- Credit risk
  - Default credit risk (Section 11.3.1)
  - Concentration risk (Section 11.3.2)
  - Reinsurance counterparty risk (Section 11.3.3)

11.3.1 Default Credit Risk, C_{dcr}

For the default credit risk we will use the standard approach used by banks according to Basel II (see Section 5.1). The main reason for doing so is that we do not want any possibility of arbitrage from the banking to the insurance sector, and vice versa.

The basis for the default credit risk will be Equation 5.5, i.e.,

$$C_{dcr}^2 = 0.08 \sum_{j,c} r_{jc} A_{jc} = \sum_{j,c} w_{jc} A_{jc}$$

where r_{jc} is the risk weight according to Table 5.2 and j the asset category and c the rating of the exposure and the risk factors $w_{jc} = 0.08 r_{jc}$. A_{jc} is the corresponding exposed asset.

11.3.2   Concentration Risk, C_{cor}

The concentration risk deals with the risk of increased exposure to both losses of investment (concentration of assets) and catastrophic events (concentration of liabilities). For example, investment in a high proportion of specific equities can be considered as concentration.

The capital charge from the concentration risk is defined as the sum of an asset concentration part and a liability concentration part

$$\frac{1}{k_{1-\alpha}^2} C_{cor}^2 = A^{*2} \{ \max(a_i^*) \bar{a}^* - \bar{a}^{*2} \} + \{ M_X \bar{X}^* - \bar{X}^{*2} \}$$

where A^* is the total assets covering the liabilities, $a_i^* = A_i^* / A^*, 0 \le a_i^* \le 1$, are the relative asset concentration category, and $\bar{a}^*$ is the mean asset concentration. M_X is the estimated maximum loss and $\bar{X}^*$ is the estimated mean loss of catastrophic events.

11.3.3   Reinsurance Counterparty Risk, C_{rr}

For each reinsurance company, $i = 1, ..., r$, the regulator (EU level) gives a score, ω_i, indicating its rating. The regulator can, for example, use company ratings from Standard & Poor's or Moody's. The scores are usually defined in the interval $0 < \omega_i < 1$.

The capital charge from the reinsurance risk is defined as

$$\frac{1}{k_{1-\alpha}^2} C_{Re}^2 = \sum_{i=1}^r (1 - \omega_i) \sigma_i^2 + \sum_{i=1}^r (1 - \omega_i) P_{i,re}^2$$

where σ_i^2 shows the volatility in the receivables from reinsurer i and P_i^2 is the squared ceded premiums from reinsurer i. The last term addresses the counterparty risk inherent in reinsurance cover.

As the score ω_i is less than 1, we see that there will always be capital requirements from the reinsurance cover. The higher the score is, the less is the risk factor, and hence the contribution to the solvency margin. An alternative approach is the one used in Norway (see Section 6.11).

11.4 Operational Risk, C_{OR}

For operational risk we will use the standard approach used by banks according to Basel II (see Section 5.1). The main reason for doing so is that we do not want any possibility for arbitrage from the banking to the insurance sector, and vice versa.

The capital requirement for the operational risk is defined as the sum over different lines of business ($i = 1, \ldots, L$)

$$C_{OR} = \sum_{i,3}^{L} \bar{B}_{i,3}\beta_i$$

where $\bar{B}_{i,3} = \dfrac{1}{3}\sum_{j=1}^{3} B_{i,j}$ is the mean of the *gross* premium income during the last 3 years and β_i is a factor, $0 \le \beta_i \le 1$, set on the EU level.

11.5 The Total Factor-Based Model

We divide with k, the value that depends on the confidence level that has been chosen, and hence if we sum up from the results from Sections 11.1 to 11.4 and insert into Equation 11.1, we arrive at

$$\tfrac{1}{k}C_{TOT}$$

$$= \{ \sum_{i=1}^{L} Z_{nri}b_i E(X_i) + \sum_{i=1}^{L} Z_{nri}\beta_i CP_i + \sum_{i=1}^{L} Var(LR_i)E^2(X_i)$$

$$+ \sum_{i\ne}^{L}\sum_{j}^{L} Cov(LR_i, LR_j)E(X_i)E(X_j) + \{p_1Y_1 + p_2Y_2\}^2\}^{1/2} \tag{11.2}$$

$$+ \left\{ \left(\sum_{j=1}^{L}{}_M\sigma_{VOL,j} + 0.5\sum_{j=1}^{L}{}_M b_j\sigma_{VOL,j} + \sum_{j=1}^{L}{}_s\sigma_{VOL,j} + 0.5\sum_{j=1}^{L}{}_s b_j\sigma_{VOL,j} \right)^2 + \right.$$

$$\left. \left(\sum_{j=1}^{L}\sigma_{VOL,j} + 0.5\sum_{j=1}^{L}l_j\sigma_{VOL,j} + \sum_{i=0}^{L}c_i^2(1+p_i)^2 E^2(X_i) \right)^2 \right\}^{1/2} + \tag{11.3}$$

$$\left\{ \sum_{k=1}^{8} A_{01}^2(d_k)\sigma^2(d_k) + \sum_{j=2}^{c} A_{0j}^2\sigma^2(j) + \sum_{j\neq}^{c}\sum_{l}^{c} A_{0j}A_{0l}\rho(j,l)\sigma(j)\sigma(l) \right\}^{1/2} + \quad (11.4)$$

$$\{0.08\sum_{j,c} r_{jc}A_{jc} + A^{*2}\{\max(a_i^*)\bar{a} - \bar{a}^{*2}\} + \{M_X\overline{X} - \overline{X}^{*2}\} +$$

$$(11.5)$$

$$\sum_{i=1}^{r}(1-\omega_i)\sigma_i^2 + \sum_{i=1}^{r}(1-\omega_i)P_{i,re}^2\}^{1/2} +$$

$$+ \sum_{i=1}^{L} \overline{B}_{i,3}\beta_i\} \qquad\qquad (11.6)$$

The first two lines (Equation 11.2) are the underwriting risk and the next two lines (Equation 11.3) are the second part of the insurance risk, combining the biometric risk, surrender and lapse risk, and expenses/costs risk.

Equation 11.4 is the market risk as defined by the A/L risk, and Equation 11.5 is the credit risk. The last term is the operational risk.

The formula above may look cumbersome, but it can be made more accessible in a spreadsheet environment. We will do that in the next chapter. But before that, we will make some assumptions about some of the division of lines of business (LOBs), etc.

The insurance classes introduced by the non-life and life directives are given in Appendix B. The number of lines of business will be $L = 16$. For the non-life business we use the following classification:

1. Accident and health (insurance classes 1 and 2)
2. Motor nonliability (class 3)
3. Property (classes 8 and 9)
4. Motor liability (class 10)
5. General liability (class 13)
6. Marine, aviation, and transport (MAT) (classes 4 to 7, 11, and 12)
7. Credit and suretyship (classes 14 and 15)
8. Legal expenses (class 17)
9. Assistance (class 18)
10. Other

For the life business we use the following classification:

11. Life insurance, annuities, and supplementary insurance (class i)
12. Unit-linked insurance (class iii)
13. Permanent health (class iv)
14. Capital redemption operations (class vi)
15. Group pension funds (class vii)
16. Other (classes ii, v, viii, and ix)

The last term in Equation 11.2 is only defined for LOB 12, so the sum will go from 1 to 11 plus 13 to 16. The run-off risk will only be used for LOBs 1 to 10.

The biometric risks will only be defined for LOBs 1 and 11 to 16.

The surrender and lapse risk are only assumed for classes 11 to 16. This part of Equation 11.3 can be simplified (see Appendix A.1.3).

11.6 A Spreadsheet Approach

From the factor-based formulas (Equations 11.2 to 11.6) it is easy to form simple spreadsheets that should be filled in by the companies.

We first look at the company data that should be given in a form like Table 11.1A and Table 11.1B. EU data (parameters) are given in Table 11.2.

TABLE 11.1A

Spreadsheet for Company Data: Insurance and Operational Risks

| LOB | Underwriting Risk | | | Insurance Risk | | | | | | Operational Risk |
	Maximum Net Retention Z	Expected Net Loss $E(X)$	Claims Provision CP	Unit Linked Y_1 Y_2	Biometric Mortality $Var(*)$-M	Biometric Sickness $Var(*)$-S	Surrender and Lapse No. of Policies n	Surrender and Lapse $X+$		Mean Gross Premium 3 years
1. Accident and health										
2. Motor, nonliability										
3. Property										
4. Motor, liability										
5. General liability										
6. MAT										
7. Credit and suretyship										
8. Legal expenses										
9. Assistance										
10. Other non-life										
11. Life insurance, annuities	0	0								
12. Unit linked	0	0								
13. Permanent health	0	0								
14. Capital redemption	0	0								
15. Group pension	0	0								
16. Other life	0	0								

TABLE 11.1B

Spreadsheet for Company Data: A/L Risk and Credit Risk

Duration Band	Median Duration dk	Bonds $B(dk)$	Equities $S(dk)$	Property $P(dk)$	Liabilities $V(dk)$	$A(dk)$ Calculated	$A(*)$ Other
1. 0–1 year	0.5						
2. 1–2 years	1.5						
3. 2–5 years	3.5						
4. 5–8 years	6.5						
5. 8–12 years	10.0						
6. 12–16 years	14.0						
7. 16–24 years	20.0						
8. 24– years	28.0						

A/L Risk

$$A(dk) = dk^*(B(dk) + S(dk) + P(dk) - V(dk))$$

Credit Risk

Reinsurer Name	Volatility Receivables $Var(*)$	Ceded Premiums P
1.		
2.		
3.		
4.		
5.		
6.		
7.		
8.		
Etc.		

Credit risk

Concentration

Total assets, covering liabilities, $A =$	
Mean asset concentration, $a =$	
Maximum estimated loss, $M =$	
Estimated maximum loss, $X=$	

TABLE 11.2A

Table for EU Parameters: Insurance and Operational Risks

EU Parameters LOB	Underwriting Risk Structure Parameter beta	Underwriting Risk Loss Ratios Var(LR_i)	Insurance Risk — Unit Linked p_1	Insurance Risk — Unit Linked p_2	Insurance Risk — Biometric Mortality b_i	Insurance Risk — Biometric Sickness b_i	Insurance Risk — Surrender and Lapse Trend Parameter l_i	Insurance Risk — Surrender and Lapse Lapse Ratio p_i	Insurance Risk — Expense Risk c_i	Insurance Risk — Expense Risk p_i	Operational Risk Weight Factor beta
1. Accident and health											
2. Motor, nonliability											
3. Property											
4. Motor, liability											
5. General liability											
6. MAT											
7. Credit and suretyship											
8. Legal expenses											
9. Assistance											
10. Other non-life											
11. Life insurance, annuities											
12. Unit linked											
13. Permanent health											
14. Capital redemption											
15. Group pension											
16. Other life											

Note: Covariances are given in Table 11.2c.

TABLE 11.2B

Table for EU Parameters: A/L and Credit Risks

A/L Risk Duration Band Interest Rate Part	Median Duration dk	(1) Bonds Var(dk)	Equities Var($d4$)	Property Var($d5$)	Weighted Var(dk)
1. 0–1 year	0.5				
2. 1–2 years	1.5				
3. 2–5 years	3.5				
4. 5–8 years	6.5				
5. 8–12 years	10.0				
6. 12–16 years	14.0				
7. 16–24 years	20.0				
8. 24– years	28.0				

Credit Risk Reinsurance	
Name	EU Scores
1.	
2.	
3.	
4.	
5.	
6.	
7.	
8.	
Etc.	

Note: Correlations are given in Table 11.2d.

TABLE 11.2C

Table for EU Parameters: Covariances between LOBs within Insurance Risk

$Cov(LR_i, LR_j)$ between Loss Ratios

LOB	1. Accident and Health	2. Motor, Nonliability	3. Property	4. Motor, Liability	5. General Liability	6. MAT	7. Credit and Suretyship	8. Legal Expenses	9. Assistance	10. Other Non-Life	11. Life Insurance, Annuities	12. Unit Linked	13. Permanent Health	14. Capital Redemption	15. Group Pension	16. Other Life
1. Accident and health													?			
2. Motor, nonliability																
3. Property																
4. Motor, liability																
5. General liability																
6. MAT																
7. Credit and suretyship																
8. Legal expenses																
9. Assistance																
10. Other non-life																
11. Life insurance, annuities																
12. Unit linked																
13. Permanent health																
14. Capital redemption																
15. Group pension																
16. Other life																

TABLE 11.2D

Table for EU Parameters: Correlations between Assets within A/L Risk

Corr(j, l) Interest Rate Dependent	Bonds	Equities	Property	Cash	Loans	Other
Bonds	1					
Equities		1				
Property			1			
Cash				1		
Loans					1	
Other						1

Corr(j, l) Other Part	Bonds	Equities	Property
Bonds	1		
Equities		1	
Property			1

Note: We have assumed independence in the book.

11.7 Parameter Estimates

Some data from Finland[5] illustrate how the commission, or CEIOPS, could start to set values on the different EU parameters. Some data have also been taken from Bateup and Reed (2001) and Djehiche and Hörfelt (2004). Note that these data, or risk-based factors, are just one example.

In arriving at the EU parameters there have to be extensive qualitative impact studies (QISs). Data from different countries have to be calibrated. If there are large differences between countries, country-specific parameters can be set by local authorities or by the EU (e.g., by CEIOPS).

There are, of course, other sources of data that could be used. As an example, there are German data published by Schmeiser (2004) and Australian data by Bateup and Reed (2001).

We illustrate the calculation by a very simple example, based on a combined model life and non-life company.

Note that we have extended the number of asset categories to include cash and loans in the illustrations above.

Illustrative data are given in Tables 11.3A – 11.3E.

---

[5] Data have been provided by Rantala (2004c).

TABLE 11.3A

Some Illustrative Data Showing the Magnitude of Some Parameters: Insurance and Operational Risk

EU Parameters LOB	Underwriting Risk Structure Parameter beta	Underwriting Risk Var(LR_i)	Unit Linked p_1	Unit Linked p_2	Insurance Risk — Biometric Mortality b_i	Insurance Risk — Biometric Sickness b_i	Insurance Risk — Surrender and Lapses Trend Parameter l_i	Insurance Risk — Surrender and Lapses Lapse Ratio p_i	Cost Risk c_i	Cost Risk p_i	Operational Risk Weight Factor beta
1. Accident and health	0.003	0.05				0.0150			0.09	0.10	0.01
2. Motor, nonliability	0.0005	0.05							0.19	0.10	0.01
3. Property	0.015	0.06							0.22	0.10	0.01
4. Motor, liability	0.0025	0.05							0.20	0.10	0.01
5. General liability	0.005	0.13							0.50	0.10	0.01
6. MAT	0.1	0.1							0.19	0.10	0.01
7. Credit and suretyship	0.6	1.2							0.12	0.10	0.01
8. Legal expenses	0.6	1.2							0.12	0.10	0.01
9. Assistance	0.6	1.2							0.12	0.10	0.01
10. Other non-life	0.015	0.9							0.15	0.10	0.01
11. Life insurance, annuities	0	0			0.0150		1	0.02	0.10	0.10	0.01
12. Unit linked		0.05	0	0.1	0.0150		1	0.02	0.05	0.10	0.01
13. Permanent health	0.003	0				0.0150	1	0.02	0.17	0.10	0.01
14. Capital redemption	0	0			0.0150		1	0.02	0.10	0.10	0.01
15. Group pension	0	0			0.0150		1	0.02	0.13	0.10	0.01
16. Other life	0	0			0.0150		1	0.02	0.10	0.10	0.01

Source: Rantala, J., Illustrations of Magnitude of Some Parameters in the Standard Approach, 19.4.04, JR, Working Paper, 2004; suggestions made from different sources.

TABLE 11.3B

Some Illustrative Data Showing the Magnitude of Some Parameters: A/L and Credit Risk

A/L Risk Duration Band Interest Rate Part	Median Duration dk	(1) Bonds SD(dk)	Equities SD($d4$)	Property SD($d5$)	Weighted Var(dk)
1. 0–1 year	0.5	0.007			0.000025
2. 1–2 years	1.5	0.012			0.000216
3. 2–5 years	3.5	0.019			0.001264
4. 5–8 years	6.5	0.025	0.17		0.061791
5. 8–12 years	10.0	0.032		0.13	0.065610
6. 12–16 years	14.0	0.037			0.019166
7. 16–24 years	20.0	0.045			0.040500
8. 24– years	28.0	0.053			0.078652

(1) Standard deviation = 0.01 → variance × duration → Var(dk) = 0.008 ×duration

Source: Rantala, J., Illustrations of Magnitude of Some Parameters in the Standard Approach, 19.4.04, JR, Working Paper, 2004c.

Credit Risk Reinsurance	
Name	EU Scores
1.	
2.	
3.	
4.	
5.	
6.	
7.	
8.	
Etc.	

Var(cash) = 0
Var(loans) = 0.04
Var(other) = 0.30

Source: Rantala, J., Illustrations of Some Parameters in the Standard Approach, 19.4.04, JR, Working Paper, 2004c.

Noninterest rate part

Var(stocks) = 0.073
Var(property) = 0.006

Source: Djehiche, B. and Hörfelt, P., Standard Approaches to Asset and Liability Risk, Research Report, Fraunhofer Chalmers Research Centre, Gothenburg, FCC, November 2004.

TABLE 11.3C

Ratings of Reinsurer

Reinsurance Recoverables Average Rating: AA	Standard & Poor's Rating $(1 - \omega)$	EU Scores ω
AAA	0.0049	0.9951
AA	0.0118	0.9882
A	0.0191	0.9809
BBB	0.0474	0.9526
BB	0.0962	0.9038
B	0.2382	0.7618
CCC	0.4972	0.5028
Not rated	0.25	0.75

TABLE 11.3D

Some Illustrative Data: Covariances between LOBs within Insurance Risk

$Cov(LR_i, LR_j)$ between Loss Ratios

LOB	1. Accident and Health	2. Motor, Nonliability	3. Property	4. Motor, Liability	5. General Liability	6. MAT	7. Credit and Suretyship	8. Legal Expenses	9. Assistance	10. Other Non-Life	11. Life Insurance, Annuities	12. Unit Linked	13. Permanent Health	14. Capital Redemption	15. Group Pension	16. Other Life
1. Accident and health	0															
2. Motor, nonliability	0	0,01														
3. Property	0,01	0,01														
4. Motor, liability	0,01	0,01														
5. General liability	0,02	0,02	0,02	0,02												
6. MAT	0	0	0	0	0											
7. Credit and suretyship	0	0	0	0	0	0										
8. Legal expenses	0	0	0	0	0	0	0									
9. Assistance	0	0	0	0	0	0	0	0								
10. Other non-life	0,05	0,05	0,06	0,05	0,09	0,08	0,26	0,26	0,26							
11. Life insurance, annuities											0					
12. Unit linked											0	0,05				
13. Permanent health											0	0	0			
14. Capital redemption											0	0	0	0		
15. Group pension											0	0	0	0	0	
16. Other life											0	0	0	0	0	0

Source: Based on Bateup, R. and Reed, L., Research and Data Analysis Relevant to the Development of Standards and Guidelines on Liability Valuation for General Insurance, The Institute of Actuaries of Australia, November 20, 2001; suggestions based on different sources.

TABLE 11.3E

Some Illustrative Data Showing the Magnitude of Some Parameters:
Correlations between Assets

Corr(*j*, *l*) Interest Rate Dependent	Bonds	Equities	Property	Cash	Loans	Other
Bonds	1	0.20	0	0	0.35	0
Equities		1	0.3	0	0	0
Property			1	0	0	0
Cash				1	0.6	0
Loans					1	0
Other						1

Source: Rantala, J., Illustrations of Magnitude of Some Parameters in the Standard Approach, 19.4.04, JR, Working Paper, 2004.

Corr(*j*, *l*) Other Part	Bonds	Equities	Property
Bonds	1		
Equities		1	0.22
Property			1

Note: We have assumed independence in the book.

Source: Djehiche, B. and Hörfelt, P., Standard Approaches to Asset and Liability Risk, Research Report, Fraunhofer Chalmers Research Centre, Gothenburg, FCC, November 2004.

11.8 An Example

The example used here is a constructed one, based on a model company that is a combined life and non-life company. The company balance sheet is given in Table 11.4 and includes four types of products (lines of business): life with profits; annuities in payout, non-life; motor; and unit linked. Different asset mixes back the liabilities of the different products. This has been done in order to reflect differences between each product. The following mixes are used:

- Life with profits: 20% equities, 10% property, and 70% bonds
- Annuities: 15% equities, 5% property, and 80% bonds
- Non-life motor: 20% equities, 5% property, and 75% bonds

For the unit-linked contracts we assume that the policyholder takes all risk.

TABLE 11.4

Balance Sheet of the Model Insurance Company, Fictitious Data.

	1000 euros		
Liabilities		**Assets**	
Group equity	1,500,000	Property	850,000
Tech provisions, life with profits	5,795,000	Equities	2,450,000
Tech provisions, annuities	305,000	Bonds	5,500,500
Tech provisions, unearned premium, gross non-life	320,000	Mortgages	0
Tech provisions, claims outstanding, gross non-life	1,280,000		
Provision for bonuses/rebates	50,000	Investments, sum	8,800,500
Tech provisions, on behalf of (obo) policyholders, gross	500,000	Investments, obo policyholders	500,000
Deposits from reinsurers	50,000	Other assets	699,500
Creditors	200,000		
Liabilities	10,000,000	Assets	10,000,000

We make the following assumptions for the model insurance company:

Gross Premiums		
Life (with profits and annuities)	488,000	8% of life provisions
Unit linked	40,000	8% of unit-linked provisions
Non-life (motor)	640,000	40% of non-life provisions
Non-life, gross claims incurred (motor)	550,000	
Reinsurance		
Life (with profits and annuities)	0	0% of gross premiums
Unit linked	0	0% of gross premiums
Non-life (motor)	44,800	7% of gross premiums

➠ One reinsurance company is used.

Assets Covering Liabilities				
	Bonds	**Equities**	**Property**	**Total**
Life: with profits	4,056,500	1,159,000	579,500	5,795,000
Life: annuities	244,000	45,750	15,250	305,000
Non-life: motor	1,200,000	320,000	80,000	1,600,000
Total:	5,500,500	1,524,750	674,750	7,700,000

The non-life bonds are assumed to have a duration of 0 to 1 year or 1 to 2 years (see Table 11.5b). The bonds corresponding to life provisions, 4,300,000, are assumed to be distributed over the six highest duration bands, i.e., duration bands 3 to 8, with the following proportions: 0.05, 0.20, 0.50, 0.15, 0.08, and 0.02. All equities and property are assumed to have a duration of 5 to 8 years and 8 to 12 years, respectively. The liabilities, 7,700,000, are assumed to be distributed over the duration bands close to the assets, i.e., given rise to low mismatch. The unit-linked provisions are not covered by the assets as the risk is taken by the insurer.

TABLE 11.5A

Spreadsheet for Model Insurance Company Data: Insurance and Operational Risks

Company Data	Insurance Risk									Operational
	Underwriting Risk			Unit Linked		Mortality	Sickness	Underwriting Risk		Unit Linked
LOB	Maximum Net Retention Z	Expected Net Loss $E(X)$	Claims Provision CP	Y_1	Y_2	$\mathrm{Var}(*)\text{-}M$	$\mathrm{Var}(*)\text{-}S$	No. of Policies n	$X+$	Mean Gross Premium 3 years
1. Accident and health	0	0	0			0	0	0	0	0
2. Motor, nonliability	100	523,776	1,280,000							640,000
3. Property	0	0	0							0
4. Motor, liability	0	0	0							0
5. General liability	0	0	0							0
6. MAT	0	0	0							0
7. Credit and suretyship	0	0	0							0
8. Legal expenses	0	0	0							0
9. Assistance	0	0	0							0
10. Other non-life	0	0	0							0
11. Life insurance, annuities	0	0		0	500,000	0.0025		10,000	6,100,000	488,000
12. Unit linked						0.0025		12,500	500,000	40,000
13. Permanent health	0	0					0.04	0	0	0
14. Capital redemption	0	0				0.0025		0	0	0
15. Group pension	0	0				0.0025		0	0	0
16. Other life	0	0				0.0025		0	0	0

TABLE 11.5B

Spreadsheet for Model Insurance Company Data: A/L Risk and Credit Risk

Duration Band	Median Duration dk	Bonds $B(dk)$	Equities $S(dk)$	Property $P(dk)$	Liabilities $V(dk)$	$A(dk)$ Calculated	$A(*)$ Other	Reinsurer Name	Volatility Receivables $Var(*)$	Ceded Premiums P
1. 0–1 year	0.5	1,080,000			1,100,000	–10,000	0	Reinsurer 1	0.001	44,800
2. 1–2 years	1.5	120,000			120,000	0		2.		
3. 2–5 years	3.5	215,000			190,000	87,500		3.		
4. 5–8 years	6.5	860,000	1,524,750		2,390,000	–34,125		4.		
5. 8–12 years	10	2,150,000		675,250	2,820,000	52,500		5.		
6. 12–16 years	14	645,000			650,000	–70,000		6.		
7. 16–24 years	20	344,000			350,000	–120,000		7.		
8. 24– years	28	86,000			80,000	168,000		8.		
								Etc.		

(Columns grouped as: A/L Risk — Bonds, Equities, Property, Liabilities, $A(dk)$, $A(*)$; Credit Risk — Reinsurer Name, Volatility Receivables, Ceded Premiums)

$A(dk) = dk*(B(dk) + S(dk) + P(dk) – V(dk))$

	Credit risk Concentration
Total assets, covering liabilities, $A =$	0
Mean asset concentration, $a =$	0
Maximum estimated loss, $M =$	1,600
Estimated maximum loss, $X =$	800

Basel II Calculation

Equities: Rating is A+ to A– with risk weights: 50%
Property: Commercial real estate with risk weight: 100%
Bonds according to the following table:

Rating	Bond Share	Bond Value	Risk Weight
AAA	25.20%	1,386,000	0%
AA	66.20%	3,641,000	0%
A	6.30%	346,500	20%
BBB	1.16%	63,800	50%
BB	0.86%	47,300	100%
B	0.09%	4950	150%
CCC	0.08%	4400	150%
Unrated	0.11%	6050	100%
	100.00%	5,500,000	

Calculation of the SCR

We assume that $1 - \alpha = 0.995$ and normality such as $k_{0.995} = 2.58$. Our benchmark, as set out in Equation 11.1, is

$$C_{TOT} = C_{ur} + \sqrt{C_{br}^2 + \left(C_{slr} + C_{cr}\right)^2} + C_{MR} + \sqrt{C_{dcr}^2 + C_{cor}^2 + C_{rr}^2} + C_{OR}$$

Insurance risk		617,651
Underwriting risk C_{ur}:	302,170	
Biometric risk C_{br}:	1	
Surrender and lapse risk C_{slr}:	33,050	
Expenses risk C_{cr}:	282,430	
Market risk		1,103,890
A/L risk C_{MR}:	1,103,890	
Credit risk		194,466
Default credit risk C_{dcr}:	179,846	
(Basel II)		
Concentration risk C_{cor}	2,064	
Reinsurance counterparty C_{rr}:	12,556	
Operational risk		11,680
Operational risk C_{or}:	11,680	
Solvency capital requirement		1,927,687

We compare this result with a rough calculation of the Solvency 0 and Solvency I requirements. As the company is a combined life and non-life company, we calculate the two different solvency requirements and sum up:

Life Solvency Requirement:

First result: 4% of technical provisions (life, excluding unit linked)

4% × 6,100,000	**244,000**

Second result: 0.3% of capital sum at risk (excluding non-life + unit linked)

0.3% × 7,900,000	**23,700**

Non-Life Solvency Requirement:

Premium index (gross premium = 640,000):

18% × 10,000	1,800
16% × 630,000	100,800
	102,600

Claims index (gross incurred claims = 550,000):

26% × 7,000	1,820
23% × 543,000	124,890
	126,710
The Solvency I requirement	**394,410**

Hence, the new risk-based factor solvency requirement is 4.89 times higher than the requirement according to the EU solvency rules. This reflects that the proposed model includes more risks, and especially the effect of the asset–liability mismatch risk.

Part C

Present and Future: EU Solvency II — Phase 2: Groups and Internal Modeling in Brief (Chapters 12–14)

In Chapters 12 to 14 we briefly discuss groups, financial conglomerates, reinsurance, the second phase of the EU's Solvency II project, and internal models and tests.

In Chapter 12 we describe the EU directives on groups, financial conglomerates, and the new reinsurance directive.

In Chapter 13 we briefly discuss the work undertaken in phase 2 of the EU's Solvency II project during the period 2003–2005.

We conclude with a brief discussion on internal models, risk management, forecasting, and stress testing procedures.

12

The European Union: Reinsurance, Insurance Groups, and Financial Conglomerates

When we talked about solvency in earlier chapters it was the stand-alone company solvency assessment that was discussed. In this chapter we will briefly look at a new proposal for solvency assessment of reinsurers and two directives on group and conglomerate solvency. After that, the proposal for a reinsurance directive was published in April 2004, which was subsequently changed as a result of consultation with the industry, for example, Comité Européen des Assurances (CEA) and direct reinsurance companies. It was first proposed that life reinsurance companies should follow the life directive. This has now been changed so that life reinsurance companies should follow the non-life directive.

12.1 Reinsurance

In April 2004 the European Commission presented a proposal for a directive on reinsurance (COM, 2004). In its explanatory memorandum it is stated that:

> A single financial market in the EU will be a key factor in promoting the competitiveness of the European economy and the lowering the capital cost to companies. An integrated market, properly regulated and prudentially sound, will deliver major benefits to consumers and other market actors, through increased security against institutional default. The vehicle to achieve this integrated market is the Financial Services Action Plan, FSAP. The Financial Services Action Plan announces a proposal for a directive on reinsurance supervision at the beginning of 2004.

> Reinsurance is a structured risk transfer between an insurance undertaking and a reinsurance company. Reinsurance fulfills the following functions for an insurance company: reduction of technical risks, permanent transfer of technical risks to the reinsurer, increase of homogeneity of insurance portfolio, reduction of volatility of technical results, substitute for capital/own funds, supply of funds for financing purposes and supply of service provision.

A reinsurer may, in a greater extent than a direct undertaking, benefit from geographical and sectoral diversification effects.

The commission also concludes that "there is no direct contractual relationship between the reinsurer and the original insured," and the policyholders normally have no priority to the assets of the reinsurer to cover their claims, i.e., not a business-to-consumer operation (B2C). But reinsurance is a business activity between professional parties. In other words, it is a business-to-business operation (B2B). "Reinsurance business is more international and has higher degrees of diversity in respect of geography and combinations of insured lines than direct insurance business."

In the memorandum it is also stated that a reinsurer faces two risks in addition to those faced by the direct insurers, namely, higher volatility in their financial results and that they may be called upon to support "ailing subsidiaries."

In the discussion on reinsurance there has been concern about potential risks, as there is no global framework for reinsurance supervision. The IAIS has initiated global work on reinsurance supervision (IAIS, 2004a). In a first step they will take charge of production of global reinsurance market statistics and preparation of market reports on an annual basis. In December 2003 the Technical Committee of IAIS approved a new standard, *Standard on Disclosures Concerning Technical Performance and Risks for Non-Life Insurers and Reinsurers*. It has not yet been adopted by the IAIS. This standard addresses the analysis of technical performance, key assumptions and sources of measurement uncertainty, as well as *sensitivity, stress testing, and scenario analysis*. But the standard does not look upon solvency assessment. In the absence of a global solvency standard for the reinsurer, the IAIS Task Force proposed to design tables that could be used to resilience test the reinsurance market.

There has not been any harmonized reinsurance supervision within the EU, which has resulted in different national rules and differences in the level of supervision of reinsurance companies.

In its early work on a reinsurance directive the Commission Services established three principles of guidance for a future regime:

- *There should be a sound and prudent regime in the interest of insurance policyholders.* Strong reinsurers will contribute to a stronger internal market and financial stability.

- *There should be an essential coordination of the member states' legislation and mutual recognition.* Main theme: Homeland license and freedom of establishment and providing services.

- *The harmonized system should lead to the abolition of systems with pledging of assets to cover outstanding claim provisions.*

The commission has chosen to propose a directive with the following feature:

- Solvency margin requirements in line with those of direct insurance, but with the possibility to increase this margin by up to 50% through the comitology (Lamfalussy) procedure for non-life insurance when this is objectively justified.

The starting point for the directive is the existing non-life and life directives. The new Solvency II project will also affect the reinsurance directive. The Commissions Services states that "insurance and reinsurance are related activities and consequently the *solvency requirements* for the two activities should be similar" (COM, 2004, p. 5).

What is said now is that for:

- *Non-life reinsurance*: The non-life directive should apply, with the possibility for increased solvency requirement. This class enhancement for certain reinsurance classes or types of reinsurance contracts (up to 50%) is seen as an essential part of the proposed directive. The proposal also provides for the subsequent adoption of the non-life directive (EEC, 1973).[1]
- *Life reinsurance*: The non-life directive should apply.

In the proposed directive Articles 35 to 41, the text on solvency margin and related issues is given (see Appendix F for the full text).

Available solvency margin

   Article 35: General rule

   Article 36: Eligible items

Required solvency margin

   Article 37: Required solvency margin for non-life reinsurance activities

   Article 38: Required solvency margin for life reinsurance activities[2]

   Article 39: Required solvency margin for a reinsurance undertaking conducting simultaneously non-life and life reinsurance

Guarantee fund

   Article 40: Amount of the guarantee fund

   Article 41: Review of the amount of the guarantee fund

The commission has on the one hand concluded that the business of a reinsurer is more like a B2B operation (see above), and on the other hand concluded that the two activities (direct insurance and reinsurance)

---

[1] Premiums, claims, provisions, and recoveries, in respect to classes 11 to 13 (cf. Appendix B), shall be increased by 50%.

[2] Shall be determined according to Article 37, i.e., the non-life directive. There is a possibility for the member states to use the life directive for business linked to investment funds or participating contracts.

should be similar when assessing the solvency. This seems to be a contradiction, and therefore direct business and reinsurance should not be treated in the same way.

One can also argue that the reinsurer has a more diversified business than a direct insurer, and thus has a lower risk and is less volatile. Life insurance includes some special problems, e.g., business that has long-term interest rate guarantees. This creates high financial risk. Normally, life reinsurance does not have such financial risks and is usually more short term, like non-life reinsurance. This would make it logical that the solvency principles for life reinsurance should follow non-life reinsurance rather than direct life insurance, as proposed in April 2004. This was changed in the final version of the reinsurance directive.

It is not always clear if a reinsurance treaty should be classified as a life or non-life treaty. It could cover both types of business in a way that makes it impossible to separate the two lines of business.

These arguments indicate *the use of the same principles for all reinsurance,* irrespective of the classification of the original business. In the light of these arguments, it was decided that the non-life directive should be the one to use.

12.2 Insurance Groups and Financial Conglomerates

The non-life and life directives (Solvency 0 and Solvency I) provide that an insurance undertaking have its own capital as a protection of its solvency (solvency margin). The *Insurance Group Directive* (IGD) (COM, 1998) and the *Financial Conglomerate Directive* (FCD) (COM, 2002d) prevent these requirements from being circumvented by groups of insurance undertakings or financial conglomerates that could otherwise allow the same regulatory capital to be used more than once to cover the solvency requirement of an insurance undertaking belonging to a group or conglomerate (*double gearing*).

These two directives ensure that a group's or conglomerate's insurance or reinsurance undertakings have a financial situation that is taken care of by an *adjusted solvency*.

By an *insurance group* we mean two or more insurance undertakings (direct insurer, reinsurer, etc.) that act as a *group*. A member of a group could be a participating undertaking (at least 20% or more of the voting rights or capital is held) (see COM, 1998, Article 1).

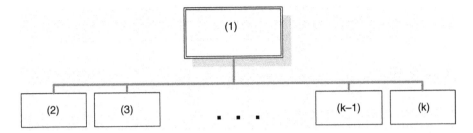

FIGURE 12.1
An organization chart of an insurance group. In terms of the IGD companies 2, ..., k are called participating undertakings.

12.2.1 Group Level Solvency

The main reason for assessing the capital adequacy of a group is to capture diversification effects. We assume a simple organization chart for the insurance group, as in Figure 12.1, and that the parent company (1) is fully dependent of the subsidiaries ($i = 2, ..., k$), which are assumed to be mutually independent with respect to solvency assessment.

The group level solvency requirement can thus be written as

$$C_{Group} = C_1 + \sqrt{\sum_{i=2}^{k} C_i^2}$$

where C_i is the solvency requirement for company i ($i = 1$ is the parent company) in accordance with the benchmark set up in Section 9.4. As we do not want double counting in the calculation of the group solvency capital, the solvency capital requirements have to be adjusted. This *adjusted solvency requirement* for insurance groups is given by Articles 9 and 10 in the IGD (COM, 1998):

Article 9

ADJUSTED SOLVENCY REQUIREMENT

1. In the case referred to in Article 2(1), Member States shall require that an adjusted solvency calculation be carried out in accordance with Annex I.

2. Any related undertaking, participating undertaking or related undertaking of a participating undertaking shall be included in the calculation referred to in paragraph 1.

3. If the calculation referred to in paragraph 1 demonstrates that the adjusted solvency is negative, the competent authorities shall take appropriate measures at the level of the insurance undertaking in question.

Article 10

REINSURANCE UNDERTAKINGS, INSURANCE HOLDING COMPANIES AND NON-MEMBER-COUNTRY INSURANCE UNDERTAKINGS

1. In the case referred to in Article 2(2), Member States shall require the method of supplementary supervision to be applied in accordance with Annex II.

2. In the case referred to in Article 2(2), the calculation shall include all related undertakings of the insurance holding company, the reinsurance undertaking or the non-member-country insurance undertaking, in the manner provided for in Annex II.

3. If, on the basis of this calculation, the competent authorities conclude that the solvency of a subsidiary insurance undertaking of the insurance holding company, the reinsurance undertaking or the non-member-country insurance undertaking is, or may be, jeopardised, they shall take appropriate measures at the level of that insurance undertaking.

Annexes I and II, which are referred to in the above articles, are given in full length in Appendix G. In the third paragraph of Annex I three methods for calculating the adjusted solvency requirements are given. The three methods are considered to be equivalent.

Method 1: Deduction and aggregation method

Method 2: Requirement deduction method (valuation by equity method)

Method 3: Accounting consolidation-based method (based on consolidated accounts)

The double use of capital, for solvency assessment purposes, must be eliminated (see Annex I of the IGD (Appendix G)). Capital that is held, directly or indirectly, by the group companies should be taken account of by a proportionate share.

If we use the following definitions:

A: The elements eligible for the solvency margin of the participating insurance undertaking

B: The proportional share of the participating insurance undertaking in the elements eligible for the solvency margin of the related insurance undertaking

C: The book value in the participating insurance undertaking of the related insurance undertaking

D: The solvency requirement of the participating insurance undertaking

E: The proportional share of the solvency requirement of the related insurance undertaking

then Method 1 equals $(A + B) - (C + D + E)$ and Method 2 equals $A - (D + E)$.

12.2.2 Financial Conglomerates

One definition of a *financial conglomerate* is "any group of companies under common control whose exclusive or predominant activities consist of providing significant services in at least two different financial sectors (banking, securities, insurance)."[3] There may be different structural features of a financial conglomerate depending on national laws and traditions. It could be possible to characterize a financial conglomerate as a securities, banking, or insurance structure.

In early 1996 the *Joint Forum on Financial Conglomerates* (Joint Forum) was established under the aegis of the Basel Committee on Banking Supervision, the International Organization of Securities Commissions (IOSCO), and the International Association of Insurance Supervisors (IAIS). The Joint Forum has published a study on the supervision of financial conglomerates (Joint Forum, 1999). "In calculating group capital, adjustment should be made to avoid double counting by deducting the amount of fund downstreamed or upstreamed from one entity to another" (Joint Forum, 1999, p. 16).[4]

A study on the allocation of solvency capital within financial conglomerates is given by Panjer (2002).

In the beginning of 1999, the Australian Prudential Regulatory Authority (APRA) issued a draft proposal for the prudential supervision of conglomerates (see APRA (1999d) and its revised version (APRA, 1999e)). A policy framework was issued in 2000 (APRA, 2000c). A conglomerate was expected to ensure that (APRA, 1999e, paragraph 62)

- Each regulated entity in the group meets the regulatory capital standard to which it is subject under solo supervision.

- The group overall has a stock of capital that is sufficient and available to meet a range of unexpected losses and adverse shocks.

---

[3] Description of financial conglomerates and their structures: http://riskinstitute.ch/136350.htm.

[4] In August 2003 the Joint Forum released two reports on risk integration and aggregation and on operational risk transfer (Joint Forum, 2003a, 2003b).

In assessing the capital strength of the group, APRA released a discussion paper on capital adequacy for conglomerates, including authorized deposit-taking institutions (ADIs) (APRA, 2001c). APRA proposed three levels for supervisory assessment, which are generalized below:

Level 1, *stand-alone entity*: The Basel Capital Accord for banks and solvency requirements for insurance companies.

Level 2, *the consolidated group*: The Basel Capital Accord for banking groups and the group solvency requirement for insurance groups (see Section 12.2.1).

Level 3, *the conglomerate group in its widest meaning, incorporating nonfinancial activities*: Could be financial conglomerates containing banking and insurance activity, or mixed conglomerates. Groups are prescribed by APRA. In the EU, this level corresponds to that described by the Financial Conglomerates Directive (see below).

A central part in EU's Financial Conglomerates Directive (COM, 2002d) is the definition and identification of an undertaking belonging to a financial conglomerate and the cooperation between supervisory authorities. A committee shall assist the commission: the *Financial Conglomerate Committee*, as defined in Article 21. Amendments to the non-life, life, and insurance group directives are given in Article 22 (see Appendix H).

In Articles 22 and 23 it is stated that the available solvency margin should be reduced by (see Appendix H):

- Participations that the insurance undertaking holds (in different insurance and reinsurance undertakings, investment firms, and credit and financial institutions)
- Items that the insurance undertaking holds with respect to the entities defined in (a), in which it holds a participation (different instruments and subordinated claims).

The second paragraph of Annex I of the directive gives the methods for the calculation of the capital adequacy requirements. The methods are considered to be equivalent.

Method 1: Accounting consolidation method

Method 2: Deduction and aggregation method

Method 3: Book value/requirement deduction method

Method 4: Combination of methods 1 to 3

A general organizational chart of a financial conglomerate is given in Figure 12.2.

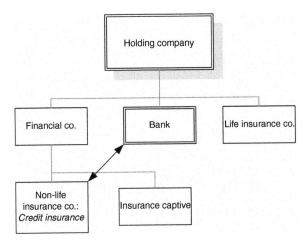

FIGURE 12.2
An organization chart of a financial conglomerate.

One problem, which we have discussed in Section 7.3, is the *participating risk*. Assume a financial conglomerate with an insurer and a bank. The bank insures its credit risk to the insurance company. A default in the bank will affect both the assets and liabilities sides of the insurer. This means that not only do we not have a positive effect of being a member of a group (in terms of diversification), but we also have a possible negative effect from the participating risk.

Assume a financial conglomerate as in Figure 12.2. Let C_{Bank} be the risk charge for the bank, according to the Basel II Accord, and $C_{Insurance}$ the corresponding risk charge for the insurance company, for example, measured as in Chapter 10. If there is no participating risk, the overall conglomerate risk charge would be

$$C_{Conglomerate} = \sqrt{C_{Bank}^2 + C_{Insurance}^2 + \sum C_i^2}$$

where the last sum is the risk charges from the other companies within the financial conglomerate (assuming independence).

On the other hand, if there is a participating risk, as discussed above, there might be full dependence between the bank and the insurance company. This would lead to the following risk charge:

$$C_{Conglomerate} = \sqrt{(C_{Bank} + C_{Insurance})^2 + \sum C_i^2}$$

13

The European Union: Solvency II — Phase II

The first phase of the Solvency II project (see Section 5.5) was a learning phase, in the sense that a series of studies was made to build the foundations for the project. During the first 2 years of the second phase the Commission Services produced some additional documents with questionnaires. At the Insurance Committee's June 2004 meeting, a road map for the development of the future work was presented, including a first wave of specific calls for advice from CEIOPS (Committee of European Insurance and Occupational Pensions Supervisors). In December the same year the second wave of specific calls for advice was released, and the third, and last, wave was released together with a new road map during spring 2005.

Before summarizing the road map we will briefly discuss the various documents and questionnaires that have preceded it.

Regarding the winding-up paper of phase I (MARKT, 2002h), the Commission Services made a questionnaire to facilitate discussion (MARKT, 2003c). The general questions presented in the paper were on the design of a prudential system, the function of the capital requirement, the harmonization of European insurance supervision, the consistency of rules between sectors (insurance and bank), and international developments (IAA (International Actuarial Association), IAIS (International Association of Insurance Supervisors), etc.).

In a paper to the Insurance Committee (MARKT, 2003d, including MARKT (2003a) as an annex), the Commission Services made recommendations on the design of a prudential supervisory system. The new system is supposed to provide the supervisory authorities with appropriate qualitative and quantitative tools to assess the overall solvency of an insurance undertaking. The Basel II three-pillar approach should be the starting point. The system should be a risk-based approach, encouraging undertakings to measure and manage their risks. There should be consistency between the financial sectors and a more efficient supervision of insurance groups and financial conglomerates. This would lead to harmonization of the supervisory methods. The Lamfalussy comitology techniques should be used to make the new solvency system effective and flexible.

13.1 Recommendations for the First Pillar

The new system was proposed to evolve two regulatory capital requirements:

- A *solvency capital requirement* (SCR) reflecting the economic capital an undertaking would need to operate with a low probability of failure. It should also reflect most risks to which a company is exposed. This would be the principal instrument for the supervisory process. The new system should allow the companies to use *internal risk models* for the calculation of the SCR.

- A *minimum capital requirement* (MCR) or *safety net* should be established to constitute a basic trigger level for ultimate supervisory action. It should be calculated in a simple and objective way, as supervisory actions at this level may need court decisions in certain jurisdictions.

The asset risk should be captured in a more explicit way in the calculation of the solvency capital requirement. If all risks an insurance company faces are considered in the risk capital calculation, it is not necessary that there are rules for assets. When insurers take all potential risks into account, they should be free to choose the way that they invest these assets.

Other main components of the first pillar are the treatment of the technical provisions, i.e., the best estimate plus a risk margin, as well as the definition of intervention levels.

13.2 Recommendations for the Second Pillar

It is essential to have quantitative tools and a strengthened supervisory review process. Several supervisory areas could be harmonized, such as principles for internal control, sound risk, and financial management, asset–liability matching, and criteria for the structure of the reinsurance program of the undertaking.

In the second pillar, risk categories not captured by the quantitative approach under pillar I will be assessed, e.g., liquidity risk and perhaps the operational risk.

There should also be minimum criteria for on-site inspections and defined intervention powers, responsibilities of the supervisory authorities, and transparency of supervisory action.

13.3 Recommendations for the Third Pillar

Transparency and disclosure requirements will be an important part of the new system. Reporting requirements should be coordinated in order to reduce administrative burden on undertakings.

13.4 General Considerations

The recommendations were welcomed by the Insurance Committee (IC) (MARKT, 2003e). The decision was that the commission should go on and prepare a more detailed paper on how the project would progress in the future. This update document was presented to the Insurance Committee in November 2003 (MARKT, 2003f). But before that, at the IC Solvency Subcommittee meeting in October 2003, a Commission Services document (MARKT, 2003b) was discussed. This document sets up a proposed structure for a framework directive, relating to the codified life directive (COM, 2002c).

In the update document, mandates for the new committee of European supervisory authorities, CEIOPS, are discussed. On November 5, 2003, the commission officially created CEIOPS. Consultations with the insurance industry and the actuarial profession (through the Groupe Consultatif (GC)) are also important issues for the coming work. CEIOPS decided to create a number of technical subgroups to perform the work related to the Solvency II project (for example, pillar I, life; pillar I, non-life; pillar II; pillar III; and group/conglomerates/cross-sector issues). These groups were fully operational from summer 2004.[1]

EIOPC (European Insurance and Occupational Pensions Committee), the successor body to the Insurance Committee, will have a key role in the development of the project; for example, strategic discussions on the overall solvency confidence level, the level of prudence in the technical provisions, and the target capital level will be important.

In February 2004 the Commission Services published a paper on the organization of the future work, pillar I work areas, and suggestions for further work on pillar II for CEIOPS (MARKT, 2004a).

---

[1] The Groupe Consultatif set up shadow working groups (WGs) to be able to cooperate with the CEIOPS WGs. These groups were operational during spring 2004. In the beginning of autumn 2004, the first concrete questions from CEIOPS WGs were delivered to GC's WGs. CEA changed its working structure in that it set up a steering group to enhance input to and dialogue with the Solvency II project key drivers. In the beginning of 2005, CEA also set up working groups to deal with similar issues. In a pilot project carried out by CEA, a comparative study between different solvency systems was made (see CEA, 2005).

It was proposed to split the work between different levels:

- *The framework directive*: The work should not differ from the usual drafting of directives; i.e., the Commission Services will draft texts and consult the EIOPC. A commission solvency working group (CSWG), now a subcommittee to the Insurance Committee, will prepare the work of EIOPC.

- *Implementing regulation: comitology*: Once the new committee architecture (Lamfalussy procedure) is in place, detailed work in these areas will be formally delegated to CEIOPS, via a mandate established by the commission (until then, the commission will give *specific calls for advice* from CEIOPS).

- *Role of the commission and the regulatory committee*: Implementing measures have the same legal value as a directive. The commission has the monopoly of regulatory initiatives at both levels. When the new EIOPC (as a regulatory committee) and CEIOPS (as a supervisory committee) are formally in place, the new structure will be:
  - The commission, after consulting EIOPC, requests advice from CEIOPS.
  - CEIOPS will determine the most appropriate way to fulfill its mandates from the commission.
  - In light of the advice given by CEIOPS, the commission will draw up its proposal of implementing measures. This proposal may be subject to an additional consultation. The European Parliament can render an opinion on this draft. The commission then presents it to the EIOPC, who votes on it.
  - Depending on the results of this vote, different procedures can be followed until final adoption.

- *Role of CEIOPS*: CEIOPS will provide technical advice on implementing measures.

- *Role of other parties — Importance of transparency*: The Commission Services considers the dialogue with the insurance industry, actuaries, and others to be essential. Transparency is an important element in the new committee architecture.

- *Role of the European Parliament*: In the extension of the new committee architecture there will be a need to involve the parliament much more closely in the technical analysis than in the past.

The document discusses general issues and provides questions to be answered by interested parties. Discussion is conducted on accounting environment issues, IAIS standards, type of harmonization, etc. Other hot topics discussed are the technical provisions in life and non-life insurance, best estimate and market value margin (MVM), level of prudence, discount rates, etc.

In calculating the solvency capital requirement (SCR), the Commission Services suggests a whole spectrum of available approaches:

- A European standard approach (SA)
- A national SA that would result from the European one with calibration of national parameters
- An internal model (IM) that would wholly or partly substitute for the SA

One way of motivating companies to develop IMs is that the resulting solvency capital level may be lower in the IM than in the SA.

Other issues that are discussed are suitable risk measures (e.g., value at risk (VaR) and TailVaR[2]) and a suitable time horizon for the SCR. Should calculations be made on a going-concern assumption or on a run-off or a winding-up basis? Classification of risk factors and the structure of the SA are other issues discussed, as well as dependencies and correlations.

One chapter deals with suggested requests on pillar II issues for CEIOPS.

A second paper with further issues for discussion was published in the beginning of April 2004 (MARKT, 2004b). Issues not discussed in COM (2004a) are dealt with in this paper.

There will be several types of measures regarding the prudent asset management in insurance companies, such as the investment risk in the SCR, and the prudent person approach, including asset liability management (ALM).

The paper also includes proposals for amended articles in the framework directive, such as Article 22 in the codified life directive (COM, 2002c) and Article 20 in the third non-life directive (EEC, 1992a):

> The assets covering the technical provisions *and the capital requirements* shall take account of the type of business carried on by an assurance undertaking in such a way as to secure the safety, yield and marketability of its investments, which the undertaking shall ensure are diversified and adequately spread. *To this end an assurance undertaking shall have an appropriate investment plan.*

Other issues discussed are the calculation of the minimum solvency margin and the guarantee fund and solvency control levels.

At the Insurance Committee's meeting on June 30, 2004, the Commissions Services made a proposal for a road map for future work and also gave a first wave of specific calls for advice from CEIOPS (MARKT, 2004c). The Insurance Committee adopted the road map and the first wave, except for

---

[2] In the second wave of specific calls for advice (MARKT, 2004d, 2004e), the TailVaR is proposed for internal models.

occupational pensions, which will be a separate subject addressed at a later date. In October, the draft proposal for a second wave of specific calls for advice from CEIOPS was published (MARKT, 2004d). Other stakeholders, such as Group Consultatif and Comité Européen des Assurances (CEA), were also asked to give comments on the requests before the Insurance Committee adopted the second wave (MARKT, 2004e).

- The commission will prepare a draft proposal for a framework directive before mid-2006. In the beginning of 2005 it was stated that this proposal would not be issued until October 2006.
- The detailed Solvency II legislation will be subsequently adopted through implementing measures under the comitology procedure with a target completion date of 2008 or 2009.
- A framework for consulting of CEIOPS and other stakeholders is drawn up. CEIOPS should be consulted by means of three waves of specific calls for advice:
  1. *The first wave of specific calls*: Pillar II issues (published in July 2004). Technical reports should be transmitted no later than June 30, 2005 or March 31, depending on the request; see below. Progress reports should be given at 4-month intervals, the first by October 31, 2004.
  2. *The second wave of specific calls*: Pillar I issues (published in December 2004). Technical reports should be transmitted no later than October 31, 2005. Progress reports should be given at 4-month intervals, the first by February 28, 2005.
  3. *The third wave of specific calls*: Pillar III issues (draft published in February 2005) (MARKT, 2005).

13.5 The First Wave of Requests (Pillar II)

Request 1: Internal control and risk management

Request 2: Supervisory review process (general)

Request 3: Supervisory review process (quantitative tools)

Request 4: Transparency of supervisory action

Request 5: Investment management rules

Request 6: Asset-liability management

13.6 The Second Wave of Requests Will Include the Following Issues (Pillar I)

Request 7: Technical provisions in life insurance

Request 8: Technical provisions in non-life insurance

Request 9: Safety measures

Request 10: Solvency capital requirement: standard formula (life and non-life)

Request 11: Solvency capital requirement: internal models (life and non-life) and their validation

Request 12: Reinsurance (and other risk mitigation techniques)

Request 13: Quantitative impact study and data-related issues

Request 14: Powers of the supervisory authorities

Request 15: Solvency control levels

Request 16: Fit and proper criteria

Request 17: Peer reviews

Request 18: Group and cross-sectoral issues

13.7 The Third Wave of Requests Will Include the Following Issues (Pillar III)

Request 19: Eligible elements to cover the capital requirements

Request 20: Independence and accountability of supervisory activities[3]

Request 21: Cooperation between supervisory authorities

Request 22: Supervisory reporting and public disclosure

Request 23: Procyclicality

Request 24: Small and medium-size enterprises

---

[3] Request 20 was removed before the third wave of requests was sent to CEIOPS. Requests 21–24 were renumbered as 20–23.

13.8 A Brief Summary

In the framework for discussion a new concept is introduced instead of the earlier target capital requirement (TCR)[4]: the solvency capital requirement (SCR). The target has now been changed to a requirement. The question whether the SCR should be a soft or hard level has to be discussed. The decision will influence the composition of trigger levels and perhaps also the confidence level.

The new solvency system should be based on a risk-oriented approach and provide the supervisors with appropriate tools and powers to assess the overall solvency of the life insurance, non-life insurance, and reinsurance undertakings.

The new solvency system should be defined in the three-pillar approach: quantitative capital requirements (pillar I), a supervisory review process (pillar II), and disclosure requirements (pillar III).

Risk management is important. The supervised institutions should be encouraged to manage their risks properly.

Issues related to insurance groups and financial conglomerates have to be addressed, including the application of internal models in a group or conglomerate.

In order to ensure convergence in financial and regulatory reporting, as well as to limit the administrative burden, supervisory reporting should be compatible with accounting rules elaborated by the International Accounting Standards Board (IASB), especially techniques and methods used to calculate technical provisions.

A general goal is to get *maximum harmonization* in the new solvency system. It should also be compatible with the approach and rules of the banking sector.

There should be a uniform level of prudence, both for technical provisions and for the SCR.

International convergence is promoted. Work done by IAIS and Groupe Consultatif (and IAA) should be considered.

One or more *quantitative impact studies* (QISs) will have to be made.

13.8.1   Pillar I Features

One cornerstone of the new system will be an increased level of harmonization for technical provisions. The SCR should reflect the risks the company faces. The SA to calculate the SCR could be based on a variety of methods, e.g., factor-based approach, probability distribution-based approach, scenarios, or a combination of these.

---

[4] The question whether the TCR should be seen as a soft or hard level has been discussed. In answer to a direct question to the Commissions Services at a meeting at the CEA in the spring of 2004, it was said to be hard. This answer reflects the change from TCR to SCR.

The MCR reflects the level of capital below which ultimate supervisory action would be triggered. It should be calculated in a simple way.

The risk factors should be based on the IAA risk classification, i.e., the Basel II risks (market risk, credit risk, and operational risk) and the underwriting risk (insurance risk).

Internal models may replace the SA to the SCR if the IMs have been validated for this purpose.

13.8.2 Pillar II Features

The supervisory review process should increase the level of harmonization of supervisory methods, tools, and processes.

13.8.3 Pillar III Features

Disclosure requirements enhance market discipline and complement requirements under Pillars II and I. They should be in line with those of IAIS and IASB.

14

Further Steps

As we learned in the previous chapter, the proposed standard approach outlined in Chapter 11 is in line with the thoughts of EU's Commission Services and the EIOPC.

Key words in the road map that was published by the Insurance Committee in July 2004 (see Chapter 13) are, for example:

- Maximum harmonization
- Standard approach (SA)
- Internal models (IMs)
- Risk classification
- IMs may replace SA
- Supervisory harmonization
- Disclosure
- Risk management

In assessing these goals, one can start with the SA, IM, and risk management and work with the following steps:

Pillar I:

1. A standard approach, using a model such as in the example in Chapter 11, should be mandatory. The coefficients and parameters should be set at a common level (for example, the EU level) for maximum harmonization.

2. As the environment is different in different parts of the EU, it should be possible (optional) for the local supervisory authority to change some or all of the common coefficients and parameters to reflect the local environment better.[1]

3. A company could change these coefficients or parameters, approved by the local authority, to take the first steps toward an internal model. This can be done coefficient by coefficient or in

---

[1] Country or market specific.

blocks, where a whole part of the SA model is changed (e.g., the SA model in calculating asset–liability (A/L) risk could be changed to a more sophisticated stochastic model).

Pillar I/Pillar II:

1. Guidance for stress testing is given at the EU level under pillar I. The companies are encouraged to carry out some simple stress testing. Some aspects may be mandated. This is one major step for the management to get a feeling of the risks the company is exposed to.

2. Under pillar II the companies are required to conduct stress tests. Depending on the situation, the authority may request a company to do some specific tests.

Figure 14.1 illustrates these steps.

In Sections 14.1 and 14.2 we will briefly discuss some steps in risk management using internal models and forecasting.

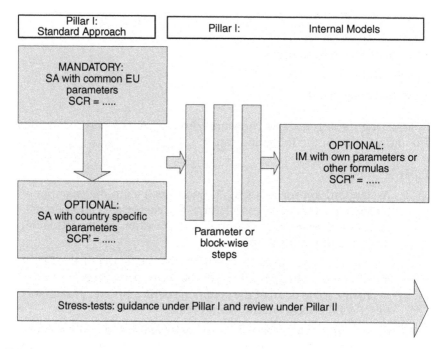

FIGURE 14.1

An illustration of the full assessment of solvency through the standard approach to internal models. Stress testing is a complement.

14.1 Internal Models and Risk Management

One way of getting knowledge about the risks to which the company is exposed is to use internal models instead of a standard approach. The main incentive to do this, besides getting knowledge, would be to get a lower solvency capital requirement (SCR) than the company would be able to get with the standard approach.

In Australia, as pointed out by Sutherland-Wong and Sherris (2004), there is no company that has adopted the advanced modeling approach (IMB method) as of the date of writing (2.5 years after its introduction). In op. cit. it is demonstrated that the IMB method produces a higher minimum capital requirement than the prescribed method (the standard approach) (see also Section 6.1). The main impact on the internal model is produced from the liability volatility assumptions. With more randomness in the model, the more volatility we get, and hence higher requirement. This is an area for further research.

In the proposed standard approach in Appendix A.2.2, the asset–liability risk is modeled by a very simple approach. Another approach, which could be used in an internal model, is one based on the theory of arbitrage pricing. A comparison between this stochastic process approach (using the VaR concept) and a simpler one (using the standard deviation concept), which is similar to the one discussed in Appendix A.2.2, is discussed in Djehiche and Hörfelt (2004). One of the examples of the two approaches given in the paper is for life insurance with annual premiums that after 10 years, or at the time of death of the insured, pay to the beneficiary a fixed amount.

The time horizon is 1 year and $\alpha = 0.01$, i.e., $k_{0.99} = 2.33$.

Proportions			Target Capital	
			SA approach	IM approach
Bonds	Stocks	Property	Similar to Appendix A.2.2	Arbitrage pricing
0.9	0.05	0.05	91	77
0.8	0.1	0.1	96	78
0.6	0.2	0.2	116	83
0.6	0.4	0	171	90
0.6	0	0.4	115	80
0.5	0.25	0.25	133	85
0.4	0.3	0.3	155	88
0.2	0.4	0.4	241	93

Use of this internal model instead of a standard approach will decrease the target capital by some 14 to 61%.

Many insurance risks can seem to be almost independent, but are heavily dependent in the extreme. One way of handling this is to use *copulas* (see, e.g., IAA, 2004). Copulas is one of the techniques that may give management insight into, and the feeling for, some of the risky parts of its business. If one recognizes a zero linear correlation between two risks, but at the same time recognizes a high tail correlation (linear correlation or other correlation methods, such as Kendall's tau), one simple way of using this knowledge on tail dependence would be to change the zero correlation for the high tail correlation in the modeling.

Another approach, connected to copulas, could be to use the *extreme value theory* (EVT), which is a statistical theory on the behavior of the tails of statistical distributions. The theory is exceptionally useful for skewed and fat-tailed distributions.

14.2 Forecasting the Future and Risk Management

Forecasting the future is one step in getting knowledge of various risks to which a company may be exposed. One way of doing this is *stochastic simulation*. It has its foundations in classical risk theory and cash flow modeling. The method strives for statistical precision such as the probability of ruin. As an example we have the Finnish case (Section 6.4), where we showed the basic transition formula and the different steps in getting a basic equation to be used for simulations. A similar approach was used in the U.K. case (Section 6.9). This latter approach, taking care of the uncertainty of the future, is usually called dynamic financial analysis (DFA).

Another approach, which does not strive for statistical precision, is *stress testing procedures*. This has been adopted as the generic term describing different techniques to assess the potential effects of exceptional events. These tests are usually more understandable to management, regulators, and nontechnical persons. One example is given by the Danish case (Section 6.3).

Discussions on stress testing are, for example, found in BIS (2000), IAIS (2003d), and the illustrative paper by MAS (2003).

A *sensitivity test* isolates the short-term impact on assets or liabilities of the insurance undertaking of a series of predefined changes in particular risk factors, such as a shock change in interest rate of ±2%, 3%, etc. One example is the *resilience test* carried out in the U.K.

A *scenario test* specifies the shocks that could affect a number of risk factors simultaneously if extreme events occur. The scenarios could be based on either *historical events* or *hypothetical events*.

A third approach, linked to scenario tests, is the *maximum loss approach* (catastrophic events' approach). It is a test for identifying the most damaging combinations of catastrophic events and their impact on the business.

Principle 10 of the IAIS principles on capital adequacy and solvency (IAIS, 2002) states that capital adequacy and solvency regimes have to be supplemented by risk management systems, and its guidance paper on stress testing (IAIS, 2003d) addresses this principle. In this guidance it is understood that *insurers would undertake stress testing as a matter of good corporate governance* and that *the supervisor may develop standard stress tests and require insurers to perform such tests*. The guidance paper also includes discussion on designing the test, the frequency, and time horizon and modeling techniques.

Part D

Appendices

Part D consists of nine appendices. The first, Appendix A, presents the basic theory of the standard approach used in Chapter 11.

Appendix B lists insurance classes used by the European Union. Appendices C and D include extracts from the first EU non-life and life directives.

Appendix E briefly discusses insurance principles and guidelines from the International Association of Insurance Supervisors (IAIS).

Appendices F through I contain excerpts of relevant EU directives: the reinsurance directive (Appendix F), the insurance group directive (Appendix G), the financial conglomerates directive (Appendix H), and the prudent person rule (Appendix I).

Appendix A

A Proposal for a Standard Approach: One Step toward Application

In Chapter 10 we discussed the main risk categories and the risk components that we will use in the sequel, and in Chapter 9 risk measures and dependency. We also arrived at a *benchmark for risk structure* (Equation 10.4), which we summarized in Chapter 11. In this appendix we will look at explicit formulae for the various risks.

The reinsurance part is taken care of by the credit risk, so all other risks will be net of reinsurance.

The four main risk categories are by dependence structure put as (Equation 10.1)

$$C_{TOT} = C_{IR} + C_{MR} + C_{CR} + C_{OR} \tag{A.1}$$

where C_{IR} is the insurance risk (discussed in Section A.1), C_{MR} is the market risk (Section A.2), C_{CR} is the credit risk (Section A.3), and C_{OR} is the operational risk (Section A.4).

A.1 Insurance Risk

As suggested in Chapter 8, the insurance risk will consist of the pure underwriting risk (C_{ur}; Section A.1.1), the biometric risk (C_{br}; Section A.1.2), surrender and lapse risk (C_{slr}; Section A.1.3), and the expense risk (C_{er}; Section A.1.2). As proposed in Chapter 10, we will use the following dependence structure (Equation 10.2):

$$C_{IR} = C_{ur} + \sqrt{C_{br}^2 + \left(C_{slr} + C_{er}\right)^2} \tag{A.2}$$

A.1.1 Underwriting Risk

The model used here is a mix between the classical mixed-compound Poisson model (see, e.g., Beard et al., 1984; Rantala, 2003) and the IAA proposal (IAA, 2004, p. 111). The main difference between the IAA proposal and the one here is that IAA assumes the gross part, i.e., including reinsurance.

In the next sections we will look at the pure underwriting risk (C_{ur}), exclusive of the random run-off risk and the unit-linked risk (C_{ulr}). Comprising the run-off risk will be done by including the claim provisions in the diversifiable part of the risk (C_{ur}) and then defining the *total underwriting risk* as

$$C_{ur} = \sqrt{C_{ur}^{"2} + C_{ulr}^2} \qquad (A.3)$$

A.1.1.1 The Capital Requirement

For each line of business (LOB), let $i = 1, 2, ..., L$, X_i be the net loss during a year, and LR_i be the loss ratio (claims/premiums).

The part of the total capital requirement from the underwriting risk is defined as

$$C_{ur}' = k\sigma_{ur}' = k\Big\{ \sum_{i=1}^{L} Z_{nri}(1+b_i)\beta_i E(X_i) + \sum_{i=1}^{L} Var(LR_i)E^2(X_i) +$$

$$\sum_{i\neq}^{L}\sum_{j}^{L} Cov(LR_i, LR_j)E(X_i)E(X_j)\Big\}^{1/2} \qquad (A.4)$$

where b_i and β_i are structure coefficients, $i = 1, 2, ..., L$, such that $b_i \geq 0$ and $0 \leq \beta_i \leq 1$, and Z_{nri} is the maximum net retention for the company and LOB i, $i = 1, 2, ..., L$.

The structure coefficients b_i influence all parts of the sums above, as they are a part of the $Var(LR_i)$ and $Cov(LR_i, LR_j)$ values. The nature of b and β is different. The first one is a random variable that rescales the claim size distribution, whereas β tries to capture the shape of the net claim size distribution (which is stable over time).

The structure coefficients and the variance of the loss ratios and the covariance between loss ratios between LOBs are to be defined on the EU level, as well as the maximum net retentions Z_{nri}. The companies only have to estimate $E(X_i)$, $i = 1, 2, ..., L$, for each LOB. For companies that have been in business for some years, it is easy to make these calculations.

For new companies, these estimates can be given by the local supervisory authority or calculated by bootstrap procedures. They could be based on business plans, e.g., with assumptions about premiums and high LR values.

For a company that wants to use internal models, a first step could be to use its own data to estimate the β values.

In Section A.1.1.3 an alternative version of Equation A.4 is given where a standard-β is used in the first term. By this, the first term will not include the maximum net retention explicitly, which will make it easier to calculate.

A.1.1.1.1 The Structure Coefficient b_i

This structure coefficient is proposed by IAA (2004, Appendix B) and is defined such that

$$b_i \geq 0, \ i = 1, 2, \ldots, L \tag{A.5}$$

See the discussion below.

A.1.1.1.2 The Structure Coefficient $\beta_i, \ 0 \leq \beta_i \leq 1$

The coefficient is approximately immune to inflation and is defined as

$$\beta_i = \frac{E(Z_i^2)}{Z_{nri}E(Z_i)}, \ i = 1, 2, \ldots L \tag{A.6}$$

It is obvious that $0 \leq \beta_i$. Now let (dropping the index i)

$$E(Z^2) = \int_0^{Z_{nr}} z^2 \, dF(z) \leq \int_0^{Z_{nr}} Z_{nr} z \, dF(z) = Z_{nr} E(Z)$$

This gives the upper bound, $\beta_i \leq 1$.

In practice, these coefficients are stable and can be recalculated by the EU as needed (Daykin et al., 1994). Z_{nr} values are the terms that bring the effect of the ceded reinsurance and are not included in the IAA formulae. The coefficients in Equation A.6 depend on how the reinsurance cuts the gross claims.

A.1.1.2 Background Theory

We will use the following conditional identities $E(Y) = E_X\{E(Y \mid X)\}$ and $\mathrm{Var}(Y) = \mathrm{Var}_X[E(Y \mid X)] + E_X[\mathrm{Var}(Y \mid X)]$ in the following discussion.

Define u_0 as the required minimum for the capital (the underwriting capital at the beginning of the year), P as the net premium income during the year, and X as the net loss during the year.

Define $Y = u_0 + P - X$ and the target capital level as

$$u_0 = E(P) - E(X) + k\sigma_{ur},$$

where

$$\sigma_{ur}^2 = Var(P - X) = Var(P) + Var(X) - 2Cov(P, X)$$

Assumptions made by both CEA and IAA include:

- Number of lines of business (LOBs): L, $i = 1, 2, \ldots, L$.
- P_i and X_i are the net premiums and net losses of LOB i.

We look at one LOB, dropping the index i:

$$X = \sum_{j=1}^{m} Z_i$$

where Z is the loss random variable in the LOB.

The number of losses, m, are assumed to be mixed Poisson distributed, $Po(nq)$, where q is a structure variable reflecting independent short period oscillations (Beard et al., 1984, p. 32). The q values also take care of correlation in claim frequencies.

The individual claim sizes (m) are independent of each other and the number of claims, and they also follow the same distribution. The total loss is hence a mixed-compound Poisson distributed variable. We also put $E(Z) = \mu_z$ and $Var(Z) = \sigma_z^2$. The variance of P is mainly due to the underwriting cycle and we will write it as

$$Var(P) = \phi E^2(X)$$

We assume that $E(q) = 1$ and $Var(q) = c$. For a Poisson distributed variable m [$Po(m)$], $E(m) = n$ and $Var(m) = n$, but for a mixed Poisson variable m we have

$$E(m) = E[E(m|q)] = E[nq] = n \tag{A.7a}$$

and

$$Var(m) = Var[E(m|q)] + E[Var(m|q)] = Var[nq] + E[nq] = n^2c + n \tag{A.7b}$$

For each line of business, define a scale parameter δ with $E(\delta) = 1$ and $Var(\delta) = b$ such that the total loss during a year is defined as

$$X = \sum_{i=1}^{L} \delta_i X_i \qquad (A.8)$$

The background for the scale parameter (see, e.g., Heckman and Meyers, 1983, where it is called a mixing parameter) is the assumption of a claim severity distribution that is known up to a scale multiplicative factor. The scale parameter could be used to model inflation. The introduction of this scale parameter is also a way of measuring the correlation in claim severities (see Wang, 1998, p. 938). Assume that the X_i values are independent and define $W_i = \delta X_i$. Then

$$Cov(W_i, W_j) = Var(\delta)E(X_i)E(X_j) = bE(X_i)E(X_j)$$

Now we have that *for each LOB* (dropping the index *j*, except the two first expressions)

$$\mu_{X_j} = E(X_j) = E[E(X \mid m)] = E[E(\sum_{i=1}^{m} Z_i \mid m)] = E[m\mu_Z] = n\mu_Z \qquad (A.9a)$$

and

$$\sigma_{X_j}^2 = Var(X_j) = Var[E(X \mid m)] + E[Var(X \mid m)] = \mu_Z^2 Var(m) + \sigma_Z^2 E(m) = \qquad (A.9b)$$
$$\mu_Z^2(n^2c + n) + n\sigma_Z^2$$

For the total business we get

$$E(X) = \sum_{j=1}^{L} E(\delta_i X_i) = \sum_{j=1}^{L} E[E(\delta_i X_i \mid \delta_i)] = \sum_{j=1}^{L} n_i \mu_{Z_i} E(\delta_i) = \sum_{j=1}^{L} n_i \mu_{Z_i} \qquad (A.10)$$

and

$$Var(X) = \sigma_X^2 = \sum_{j=1}^{L} Var(\delta_j X_j) + \sum_{i \neq}^{L} \sum_{j}^{L} Cov(\delta_i X_i, \delta_j X_j) \qquad (A.10b)$$

For each LOB we get

$$Var(\delta X) = E[Var(\delta X \mid \delta)] + Var[E(\delta X \mid \delta)] = E[\delta^2 Var(X)] + Var[\delta E(X)] \qquad (A.11)$$

and using Equations A.9a and A.9b we get

$$Var(\delta X) = E(\delta^2)Var(X) + E^2(X)Var(\delta) = (b+1)\{\mu_Z^2(n^2c + n) + n\sigma_Z^2\} + n^2\mu_Z^2 b$$
(A.12a)

Define the moments about zero as $\alpha_i = E(Z_i)$, then $E(Z) = \alpha_1$ and $Var(Z) = \alpha_2 - \alpha_1^2$ and $\alpha_2 = \sigma^2 + \alpha_1^2$. Then Equation A.12a can be written as

$$Var(\delta X) = (1+b)\{n^2\alpha_1^2 c + n\alpha_2\} + bn^2\alpha_1^2$$
(A.12b)

Define

$$\beta = \frac{\alpha_2}{Z_{nr}\alpha_1}$$
(A.13)

where Z_{nr} is the net retention (maximum net claim). This gives us, with $n\mu_z = E(X)$, the variance as

$$Var(\delta X) = (1+b)cE^2(X) + bE^2(X) + (1+b)\beta Z_{nr}E(X) = \varphi E^2(X) + (1+b)\beta Z_{nr}E(X)$$
(A.14)

where $\varphi = (1+b)c+b$.

We now get the following expression for $Var(P - X)$:

$$Var(P - X) = \phi E^2(X) + \varphi E^2(X) + (1+b)\beta Z_{nr}E(X) + \left\{-2\frac{Cov(p,x)}{E^2(X)}\right\}E^2(X)$$

If we put the term within brackets equal to v we get

$$(\phi + \varphi + v)E^2(X) + (1+b)\beta Z_{nr}E(X)$$
(A.15)

A.1.1.2.1 Estimate ($\phi + \varphi + v$)

Let $P = (1 + p)E(X)$ and $X = (1 + x)E(X)$ and the loss ratio $LR = X/P = (1 + x)/(1 + p)$. The last ratio can be approximated by $(1 + x)(1 - p) = 1 + x - p - xp$, and as xp is close to zero we get the following approximation for the loss ratio: $LR \approx 1 + x - p$. The variance of the loss ratio can now be approximated by

$$Var(LR) \approx Var(1 + x - p) = Var(x - p) = Var(\frac{X - E(X)}{E(X)} - \frac{P - E(X)}{E(X)}) = \frac{Var(X - P)}{E^2(X)}$$

Using Equation A.15 we get

$$Var(LR) = (\phi + \varphi + \upsilon) + (1+b)\beta Z_{nr} E(X) / E^2(X)$$

where the last term is insignificant for large portfolios. For each LOB we then get

$$(\phi + \varphi + \upsilon) = Var(LR)$$

The covariance of $P - X$ between LOBs can be written as

$$Cov[P_i - X_i, P_j - X_j] = E(X_i)E(X_j)Cov(p_i - x_i, p_j - x_j)$$

and letting $LR = 1 - (p - x)$ (see above), we get

$$Cov[P_i - X_i, P_j - X_j] = E(X_i)E(X_j)Cov(LR_i, LR_j)$$

Now we have got Equation A.4.

A.1.1.2.2 A Note on the Loss Ratio LR

The loss ratio could be defined, with the structure parameter δ, as

$$LR^* = \frac{\delta X}{E(\delta X)}.$$

Thus we get $E(LR^*) = 1$ and

$$Var(LR^*) = Var\left[\frac{\delta X}{E(\delta X)}\right] = \frac{Var(\delta X)}{E^2(\delta X)} = \frac{(b+1)\{\mu_Z^2(n^2 c + n) + n\sigma_Z^2\} + n^2\mu_Z^2 b}{n^2\mu_Z^2}$$

Letting the coefficient of variation, cv, be $cv = \sigma/\mu$ we get

$$Var(LR^*) = (b+1)\{\frac{1}{n} + c + \frac{1}{n}cv^2\} + b \to b + c + bc \text{ as } n \to \infty.$$

If $\beta = 1$, then the $Var(LR^*)$ tends to c as n $\to \infty$.

A.1.1.3 An Alternative to C_{ur}

The part of the total capital requirement from the underwriting risk (Equation A.4) can be written as

$$C_{ur} = k\sigma_{ur} = k\left\{\sum_{i=1}^{L}[(1+b_i)\beta_{sti} + Var(LR_i)]E^2(X_i) + \sum_{i\ne}^{L}\sum_{j}^{L}Cov(LR_i, LR_j)E(X_i)E(X_j)\right\}^{1/2}$$

(A.16)

As proposed in one of the working papers from CEA (Rantala, 2003), we could define a standard net retention as a fraction of $E(X)$ for each LOB.

$$\text{Let } Z_{st} = \zeta E(X) \text{ for each LOB} \tag{A.17}$$

and

$$\beta' = \tfrac{1}{\zeta}\beta_{st}$$

where

$$\beta_{st} = \frac{E(Z_{st}^2)}{E(X)E(Z_{st})}$$

It should be observed that $E(X_i)$ depends on the net retention Z_{sti}, $i = 1, ..., L$. Hence, the first term in Equation A.4 can be written as (dropping the sum)

$$(1+b_i)Z_{nri}\beta_iE(X_i) = (1+b_i)\zeta E(X_i)\frac{1}{\zeta}\beta_{sti}E(X_i) = (1+b_i)\beta_{sti}E^2(X_i)$$

This gives the alternative formulation of Equation A.4 given in Equation A.16.

A.1.1.4 Claim Provision (Run-Off) Risk

The claim provision risk (run-off) risk refers to the danger of having insufficient loss reserves when the insurer's license is withdrawn at a certain time point. No new business is written after this point, nor is any renewed. The claims (and other commitments) are paid from the assets.

The adequacy of the claim provisions (CP) is a pillar 2 issue. Hence, we should not deal with the systematic part, only the pure random run-off risk. We only consider the net part of the claim provisions.

A.1.1.4.1 The Capital Requirement

We use the underwriting risk of Equation A.4 and include the claim provision CP_j, $j = 1, ..., L$, for the L lines of business, in the diversifiable part of risk. This gives us

$$C_{ur}'' = k\sigma_{ur}'' = k\{\sum_{i=1}^{L} Z_{nri}(1+b_i)\beta_i E(X_i) + \sum_{i=1}^{L} Z_{nri}(1+b_i)\beta_i CP_i$$

$$+ \sum_{i=1}^{L} Var(LR_i)E^2(X_i) + \sum_{i\neq}^{L}\sum_{j}^{L} Cov(LR_i, LR_j)E(X_i)E(X_j)\}^{1/2}$$

(A.18)

A.1.1.4.2 Background Theory

As stated above, we will only deal with the diversifiable risk, i.e., the random run-off risk, which includes the estimation of incurred but not reported (IBNR). The claims provision could, e.g., be added to the risk exposure for the diversifiable part of the underwriting risk. This means that we use $E(X_j)$ + CP_j, $j = 1, ..., L$ (line of business), in Equation A.4.

This would lead to the underwriting risk in Equation A.18.

A.1.1.5 Unit-Linked Risk

For unit-linked contracts the principal risks are concentrated in expenses and lapses. The major insurance (underwriting) risk is taken by the insured unless there are guarantees of capital or return. In this case, the expenses are administration costs and the risk is that these will exceed the charges that can be made on the policies. The extent to which a contract permits the insurer to increase the charges can be limited by supervisory authorities or by other restrictions.

We only consider the retained business, as the reinsurance risks are taken account of by the credit risk. Lapses are also taken care of by a separate term.

A.1.1.5.1 The Capital Requirement, C_{ulr}

The part of the total capital requirement from the unit-linked risk is defined as

$$C_{ulr} = k\sigma_{ulr} = k\{p_1 Y_1 + p_2 Y_2\}$$

(A.19)

where the weights p_1 and p_2 are defined by EU, Y_1 is the net technical provisions for the financial and insurance risks for the company when there is a promised minimum guarantee, and Y_2 is the net technical provisions where there are *no* financial and insurance risks for the company, but where there is a considerable business risk in the provisions or when there is a considerable administration risk.

A.1.1.5.2 Background Theory

The risks that should be taken care of are:

1. The net technical provisions for the financial and insurance risks for the company when there is a promised minimum guarantee $TP(1)_{net}$
2. The net technical provisions where there are *no* financial and insurance risks for the company, but where there is a considerable business risk in the provisions or when there is a considerable administration risk $TP(2)_{net}$

The *net* liability for unit-linked contracts can be written as (numbers 1 and 2 above)

$$V_1 = TP(1) + TP(2) = w_1 TP + w_2 TP, w_1 + w_2 \le 1$$

where TP is the total technical provisions. The volatility of technical provisions other than numbers 1 and 2 are considered to equal zero.
For simplicity, we put $X = TP$. Hence,

$$Var(V_1) = Var(w_1 X + w_2 X) = (w_1 + w_2)^2 Var(X) = (w_1 + w_2)^2 \sigma_X^2$$

and

$$\sqrt{Var(V_1)} = (w_1 + w_2)\sigma_X = w_1 \sigma_X + w_2 \sigma_X$$

Let the total volatility of the technical provisions, σ_X, be split into numbers 1 and 2 above. Let

$$q_1 + q_2 = 1, w_1 = p_1 q_1 \text{ and } w_2 = p_2 q_2 = p_2(1 - q_1)$$

which gives us

$$\sigma_X = w_1 \sigma_X + w_2 \sigma_X = p_1 \sigma_{X1} + p_2 \sigma_{X2}$$

where

$$\sigma_{X1} = q_1 \sigma_X \text{ and } \sigma_{X2} = (1 - q_1)\sigma_X, \text{ and } p_1 + p_2 \le 1.$$

In our prudent environment we can write $\sigma_{Xi}^2 = E(X_i^2) - E^2(X_i) \le E(X_i^2)$, $i = 1, 2$. Hence, $\sigma_{Xi} \le \sqrt{E(X_i^2)}$. If we put $E(X_i^2) \approx Y_i^2$ we get a prudent estimate of the volatility by

$$\sigma_X \cong p_1 Y_1 + p_2 Y_2 \qquad (A.20)$$

where Y_1 is the net technical provisions in number 1 and Y_2 is the net technical provisions in number 2. This gives us Equation A.19.

A.1.2 Biometric Risk

The biometric risk category is defined for both life insurance and non-life insurance. It could be assumed that it is uncorrelated with other risks. The risk is considered net of reinsurance. The first part will take care of the *mortality risk* and the second the *sickness risk*.

A.1.2.1 The Capital Requirement

The part of the total capital requirement from the biometric risk is defined as a summation over line of business ($j = 1, ..., L$):

$$C_{br} = k\sigma_{br} = k\left\{ \sum_{j=1}^{L} {}_M\sigma_{VOL,j} + 0.5 \sum_{j=1}^{L} {}_M b_j s_{VOL,j} + \sum_{j=1}^{L} {}_S\sigma_{VOL,j} + 0.5 \sum_{j=1}^{L} {}_S b_j \sigma_{VOL,j} \right\}$$

$$(A.21)$$

where M stands for the mortality part and S for the sickness and $\sigma_{VOL,j}$ is the volatility in the misspecification of the probability of mortality or sickness.

The parameter b_j reflects the misspecification of the trend in the probability of mortality or (remaining) sick, and the terms $b_j\sigma_{VOL,j}$ could be written $\sigma_{trend,j}$. The b_j values are defined on the EU level, and $\sigma_{VOL,j}$ can be estimated by company data or given on the EU level. In the background theory (below), we only discuss mortality, but the same approach applies for sickness. For simplicity, we assume full dependence between mortality and sickness to get the simple form in Equation A.21.

A.1.2.2 Background Theory

Let Z_i, the individual claim amount, be independent and identically distributed (i.i.d.) and n the number of insurance contracts. X is the sum of the n Z_i values, i.e., $X = \sum_{i=1}^{n} Z_i$. Here we have not used any index for the line of business (LOB).

Assume that q_x is the true probability of death and q_x' is the used probability of death.

q_x' could, for example, be defined as $q_x' = (a + bt)q_x$, where t is a horizon parameter. Simplify this by assuming that $a = t = 1$, i.e.,

$$q_x' = q_x + bq_x$$

where b is a slope parameter defining the incorrectness of choosing q_x' instead of q_x.

The variance of X, $\sigma_X^2 = Var(X)$, can be simplified and approximated by (using the i.i.d. assumption above)

$$\sigma_X^2 = Var(X) = Var(\sum_{i=1}^{n} Z_i) = \sum_{i=1}^{n} Var(Z_i) \doteq nVar(Z_1) = nE(Z_1^2) - nE(Z_1)^2$$

(A.22)

Make the following approximation of $E(Z_1^2)$ and $E(Z_1)$:

$$E(Z_1^2) = Y^2 q_x' + 0^2 (1 - q_x') = Y^2 (q_x + bq_x) = Y^2 q_x + bY^2 q_x \qquad \text{(A.23a)}$$

and

$$E(Z_1) = Yq_x' + 0(1 - q_x') = Y(q_x + bq_x) = Yq_x + bYq_x \qquad \text{(A.23b)}$$

Using Equations A.23a and A.23b we get

$$n[E(Z_1^2) - E(Z_1)^2] = n[Y^2 q_x + bY^2 q_x - (Yq_x + bYq_x)^2]$$
$$= n[Y^2 q_x (1 - q_x) + bY^2 q_x (1 - q_x) - b(1 + b)Y^2 q_x^2] \qquad \text{(A.24)}$$

The variance of X can thus be written as

$$\sigma_X^2 \doteq \sigma_{Vol}^2 + b\sigma_{Voll}^2 - nb(1 + b)Y^2 q_x^2 \qquad \text{(A.25)}$$

where $\sigma_{Vol}^2 = nY^2 q_x (1 - q_x)$ describes the volatility and the other part a trend risk.

In a prudent standard environment we can neglect the last term in Equation A.25 and write

$$\sigma_X^2 \doteq \sigma_{Vol}^2 + \sigma_{trend}^2 = (1 + b)\sigma_{Vol}^2 \qquad \text{(A.26a)}$$

where

$$\sigma^2_{trend} = b\sigma^2_{Vol} \tag{A.26b}$$

For each line of business (LOB), $j = 1, ..., L$, let X_j be the sum of claim amount and $X = \sum_{j=1}^{L} X_j$. For simplicity, we assume that there is a perfect correlation between the LOBs, i.e., $\rho_{jk} = 1$, for $j = 1, ..., L$, $k \neq j$.

This means that

$$\sigma^2_X = Var(X) = Var(\sum_{j=1}^{L} X_j) = \left\{ \sum_{j=1}^{L} \sigma_j \right\}^2 \tag{A.27}$$

where σ^2_j is the variance of LOB j.

Combining Equations A.26 and A.27 we get

$$\sigma^2_X = \left\{ \sum_{j=1}^{L} (\sigma^2_{Vol} + \sigma^2_{trend})^{1/2} \right\}^2$$

and as the trend risk can be written in terms of the volatility risk (see Equation A.26b), we have

$$\sigma^2_X = \left\{ \sum_{j=1}^{L} (1 + b_j)^{1/2} \sigma_{Vol,j} \right\}^2 \tag{A.28}$$

Simplifying Equation A.28 and at the same time once more being prudent, we approximate

$$(1 + b_j)^{1/2} \doteq 1 + \frac{1}{2} b_j + o(b_j)$$

giving

$$\sigma^2_X = \left\{ \sum_{j=1}^{L} (1 + 0.5 b_j) \sigma_{Vol,j} \right\}^2$$

which gives Equation A.21.

A.1.2.3 Estimation of the Volatility Risks in Mortality, $\sigma^2_{Vol,j}$

As shown by the German Insurance Federation (GDV) in its description of a supervisory model for German life insurance undertakings, the volatility risks for mortality can be estimated by simple assumptions.

In its model the number of LOBs that gives the risks are $j = 1$, whole life risk; $j = 2$, endowment risk; $j = 3$, business interruption risk; and $j = 4$, other risks. The risk classes $j = 3$ and 4 are treated like the whole life risk.

For the *whole life risk*, we define C_R as the *capital at risk* and E as the *actual expenses*. For the *endowment risk*, we define DT as the *insurance provision minus insured whole life benefits* and I_e as the *actual inheritance earnings*.

If we approximate Y and q_x by

	$Y \doteq$	$q_x \doteq$
Whole life risk	$\dfrac{C_r}{n}$	$\dfrac{E}{C_r}$
Endowment risk	$\dfrac{DT}{n}$	$\dfrac{I_e}{DT}$

we get the following approximations of the volatility risks.

$$\sigma^2_{whole\_life} \approx \frac{(C_R - E)E}{n}$$

and

$$\sigma^2_{endowment} \approx \frac{(DT - I_e)I_e}{n}$$

A.1.3 Surrender and Lapse Risk

Risks related to surrender and lapse are mainly defined for life insurance. Since increases in interest rates lower the market value of bonds, and probably also increase surrender rates, it is assumed that this risk is highly correlated to the asset–liability risk in Section A.2.2.

A.1.3.1 The Capital Requirement

The part from the total capital requirement from the surrender and lapse risk is defined as a summation over line of business ($j = 1, ..., L$):

$$C_{slr} = k\sigma_{slr} = k\left\{\sum_{j=1}^{L}\sigma_{VOL,j} + 0.5\sum_{j=1}^{L}l_j\sigma_{VOL,j}\right\} \qquad (A.29)$$

where $\sigma_{VOL,j}$ is the volatility in the misspecification of the lapse ratio. It can be estimated by

$$\sigma_{VOL,j} \approx n_j U_j^+ \tfrac{1}{\sqrt{n_j}}\sqrt{p_j(1-p_j)} = \hat{X}_j^+ \tfrac{1}{\sqrt{n_j}}\sqrt{p_j(1-p_j)}$$

where $\hat{X}_j^+ = n_j U_j^+$ is an estimate of the total guaranteed surrender value *minus* the total best-estimate liability. The plus sign indicates the nonnegative part of the variable, i.e.,

$$\hat{X}_j^+ = \begin{cases} \hat{X} \text{ if } \hat{X} > 0 \\ 0 \text{ otherwise} \end{cases}$$

The trend parameters (the l_j values) and the lapse ratios (p_j) are defined on the EU level, and the volatilities $\sigma_{VOL,j}$ can be estimated from company data.

A.1.3.2 Background Theory

For policy i, $i = 1, ..., n$, we define

$$Y_i - S_i - V_i/(1 + r)$$

where S_i is the surrender value payable on lapsing the policy and V_i is the reserve, discounted by a factor r. The Y_i values are assumed independent and identically distributed (i.i.d.) with $E(Y) = \mu_Y$ and $Var(Y) = \sigma_Y^2$. Note that $nE(y) = $ the total guaranteed surrender value – the total best-estimate liability.

We now, in a way as similar to that for the biometric risk in Section A.1.2, define the true lapse ratio as p and the used lapse ratio as p'. The used lapse ratio could be defined as $p' = (1+l)p = p + lp$ in terms of the correct lapse ratio p, and l is a parameter defining the misspecification of the lapse ratio. Now we define the loss or profit as

$$X_{laps} = \sum_{i=1}^{n}Y_i = \sum_{i=1}^{n}S_i - \tfrac{1}{(1+r)}\sum_{i=1}^{n}V_i = S_{laps} - V_{laps}^*$$

The variance of the X_{laps}, $\sigma_X^2 = Var(X_{laps})$ is

$$\sigma_X^2 = Var(X_{laps}) = Var(\sum_{i=1}^n Y_i) = \sum_{i=1}^n Var(Y_i) \doteq nVar(Y_1) = nE(Y_1^2) - nE(Y_1)^2$$

(A.30)

Make the following approximations of $E(Y_1^2)$ and $E(Y_1)$:

$$E(Y_1^2) = U^2 p' + 0^2(1-q') = U^2(p+lp) = U^2 p + lU^2 p$$ (A.31a)

and

$$E(Y_1) = Up' + 0(1-p') = U(p+lp) = Up + lUp$$ (A.31b)

Using Equations A.31a and A.31b in Equation A.30 we get

$$n[E(Y_1^2) - E(Y_1)^2] = n[U^2 p + lU^2 p - (Up + lUp)^2]$$
$$= n[U^2 p(1-p) + lU^2 p(1-p) - l(1+k)U^2 p^2]$$

(A.32)

The variance of X_{laps} can thus be written as

$$\sigma_X^2 \doteq \sigma_{Vol}^2 + l\sigma_{Vol}^2 - nl(1+l)U^2 p^2$$ (A.33)

where $\sigma_{Vol}^2 = nU^2 p(1-p)$ describes the volatility and the other part a trend risk. The approximation U can be estimated by the nonnegative part of the mean difference between the guaranteed surrender value and the best-estimate liability.

In a prudent standard environment we can neglect the last term in Equation A.33 and write

$$\sigma_X^2 \doteq \sigma_{Vol}^2 + \sigma_{trend}^2 = (1+l)\sigma_{Vol}^2$$ (A.34a)

where

$$\sigma_{trend}^2 = l\sigma_{Vol}^2$$ (A.34b)

For each line of business (LOB), $j = 1, ..., L$, let $X_{j,laps}$ be the loss or profit resulting from lapses and $X = \sum_{j=1}^{L} X_{j,laps}$. For simplicity, we assume that there is a perfect correlation between the LOBs, i.e.,

$$\rho_{jk} = 1, \text{ for } j = 1, ..., L, \ k \neq j$$

This means that

$$\sigma_X^2 = Var(X) = Var\left(\sum_{j=1}^{L} X_{j,laps}\right) = \left\{\sum_{j=1}^{L} \sigma_{Xj}\right\}^2 \tag{A.35}$$

where σ_{Xj}^2 is the variance of LOB j as in Equation A.34a. Combining Equations A.34 and A.35 we get

$$\sigma_X^2 = \left\{\sum_{j=1}^{L} (\sigma_{Vol}^2 + \sigma_{trend}^2)^{1/2}\right\}^2$$

and as the trend risk can be written in terms of the volatility risk (see Equation A.34b)

$$\sigma_X^2 = \left\{\sum_{j=1}^{L} (1 + l_j)^{1/2} \sigma_{Vol,j}\right\}^2 \tag{A.36}$$

Simplifying Equation A.36, and at the same time once more being prudent, we approximate $(1 + l_j)^{1/2} \doteq 1 + \frac{1}{2} l_j + o(l_j)$

giving

$$\sigma_X^2 = \left\{\sum_{j=1}^{L} (1 + 0.5 l_j) \sigma_{Vol,j}\right\}^2$$

which gives Equation A.29.

A.1.4   Expense Risk (or Cost Risk)

The expense risk consists of, e.g., acquisition, administrative, and claim settlement costs. Some of the costs cannot be allocated to any particular line of business. They are called overhead expenses. The remaining costs are attributed to the specific line of business (LOB).

A.1.4.1   The Capital Requirement, C_{er}

The part of the total capital requirement from the expense risk is defined as

$$C_{er} = k\sigma_{er} = k\left\{\sum_{i=0}^{L} c_i^2(1+p_i)^2 E^2(X_i)\right\}^{1/2} \tag{A.37}$$

where the premium loading factors p_i and cost ratios c_i are defined on the EU level.
   X_0 is the total claim, $X_0 = \sum_{i=1}^{L} X_i$.

A.1.4.2   Background Theory

Let C be the total cost or expenses during the period under study and define

$$C = C_0 + \sum_{i=1}^{L} C_i = \sum_{i=0}^{L} C_i$$

where the costs C_i, $i = 1, ..., L$, are independent and C_0 is the overhead expenses. For simplicity, we ignore the dependence between C_0 and C_i, $i = 1, ..., L$. We assume that the costs C are defined as a ratio of the premium income, i.e.,

$$C_0 = c_0 P_{tot} \text{ and } C_i = c_i P_i$$

where $P_{tot} = \sum_{i=1}^{L} P_i$ are the sum of premiums over the different LOBs.

   Assume that the variance of P is mainly due to the underwriting cycle and that we can define $Var(P_{tot}) = \phi_0 E^2(X_0)$ and $Var(P_i) = \phi_i E^2(X_i)$, $i = 1, ..., L$, and with factors ϕ defined below. Hence,

$$Var(C) = \sum_{i=0}^{L} Var(C_i) = \sum_{i=0}^{L} c_i^2 \phi_i E^2(X_i) \tag{A.38}$$

where

$$\phi_i = Var(P_i) / E^2(X_i) = Var([1 + p_i]E(X_i)) / E^2(X_i) = Var([1 + p_i]) \approx (1 + p_i)^2$$

$$(A.39)$$

and the p values are loading factors. This gives us Equation A.37.

A.2 Market Risk

The discussion in Section A.2.2 is based mainly on a suggestion made by Rantala (2004a) to CEA's solvency working group.[1] In that paper Rantala suggested *a match between bonds and interest-sensitive fair-valued*[2] *liabilities*. In the model a mismatch of the form $d_B B - d_V V$ was assumed, where d_B, d_V are suitably defined durations of bond portfolio and the technical provisions, respectively. B is the market value of the bonds and V is the fair value of the technical provisions, based on the current market interest rate curve. Hence, if the amount and duration of technical provisions and the bond portfolio covering these technical provisions were equal, the resulting risk charge would be zero. Another working paper (FFSA, 2003) extended the proposed model to also include a duration analysis for equities and property. We will use these approaches as a start for our modeling. Here we will consider the duration of equities and property as the *holding period*, i.e., the *mean time* a company is holding its equities and property. The duration d will also be, in accordance with IAA (2004), split up into duration bands.

Before we consider the asset–liability mismatch risk (A/L risk) in Section A.2.2, we need a short introduction to the valuation of assets and duration analysis.

The risk we will study here is the market risk C_{MR} in terms of the A/L risk (Section A.2.2) and the currency risk (Section A.2.3).

A.2.1 Valuation of Assets and Duration Analysis

A.2.1.1 *General Discounted Cash Flow Model*

The general discounted cash flow model (DCF) can be defined as

$$W_0 = \sum_{t=1}^{n} \frac{CF_t}{\left(1 + r_t\right)^t} \qquad (A.40)$$

---

[1] The originating working paper is from March 2003.
[2] In the paper by Rantala (2004a) it is assumed to be *best estimate* instead of *fair value*.

where W_0 is the current, or present, value of the asset, CF_t is the expected cash flow at period t, r_t is the required rate of return for each period's cash flow (the discount rate), and n is the time to maturity of the asset.

The cash flow is usually defined per year. If it is paid out during the year, both the cash flow and the rate of return are divided by the number of periods per year; for example, a half-year payment with two periods per year (Equation A.40) becomes

$$W_0 = \sum_{t=1}^{n} \frac{CF_t/2}{\left(1+r_t/2\right)^t}$$

The *duration* is defined as the weighted average life of the asset:

$$D_n = \sum_{t=1}^{n} t w_t \tag{A.41}$$

where

$$w_t = \frac{1}{W_0}\frac{CF_t}{\left(1+r_t\right)^t}$$

Let us now assume $r_t = r, \forall t$, i.e., we assume a flat or level interest rate curve. The *modified duration* is defined as $D_{n,m} = Dn/(1+r)$ and $D_{n,m}W_0$ the modified dollar duration. The price *sensitivity* of the asset is defined as

$$\frac{dW_0}{dr} = -\sum_{t=1}^{n} \frac{tCF_t}{\left(1+r\right)^{t+1}} = -\frac{1}{(1+r)}D_n W_0 = -D_{n,m}W_0 \tag{A.42}$$

Using the approximation $dW_0/dr \approx \Delta W_0/\Delta r$ we get

$$\Delta W_0 \approx -D_{n,m}W_0\Delta r \tag{A.43}$$

or the relative change as $\Delta W_0/W_0 \approx -D_{n,m}\Delta r$. Paying attention to the units, if $D_{n,m}$ is measured in years, then Δr reters to change per year.

Note that $D_{n,m}$ measures the relative slope of the price–interest rate curve at a given point. This gives a straight-line (first-order) approximation to the price–interest rate curve and is a useful measure as a means of assessing risk and as a procedure of controlling it.

A.2.1.2 Bond Valuation

Assume that $n = T$ is the time to maturity for the bond. The market value of the bond is defined by

$$B_0 = \sum_{t=1}^{T} \frac{c_t^*}{(1+r)^t} + \frac{M}{(1+r)^T} = B_{0,cp} + B_{0,M} \qquad \text{(A.44a)}$$

Here we have used the following notation:

B_0 = Bond price, i.e., the total discounted value of the bond accounted for by the future payments

c_t^* = (Annual) coupon payment, i.e., the cash flow at time t

r = Yield

T = Time to maturity (in years)

M = Face value, i.e., the amount to be repaid as a lump sum at maturity

$B_{0,cp}$ = Bond price due to the coupon payment

$B_{0,M}$ = Bond price due to the payment at maturity

If we define $c_T = c_T^* + M$ and $c_t = c_t^*, t = 1,2,...T-1$, we get

$$B_0 = \sum_{t=1}^{T} \frac{c_t}{(1+r)^t} \qquad \text{(A.44b)}$$

A summary statistic of the effective maturity of the bond is the *Macaulay duration*. It was independently proposed by different authors, but first by Macaulay in 1938. From Equation A.41 we get

$$D_B = \frac{1}{B_0} \sum_{t=1}^{T} \frac{tc_t}{(1+r)^t} \qquad \text{(A.45)}$$

The sensitivity of the bond price can be expressed in terms of the Macaulay duration as (cf. Equation A.42)

$$\frac{dB_0}{dr} = -\sum_{t=1}^{T} \frac{tc_t}{(1+r)^{t+1}} = -\frac{1}{(1+r)} D_B B_0$$

It can also be written as

$$-\frac{1}{(1+r)}\left[D_{B,cp}+T\frac{M}{B_0(1+r)^T}\right]$$

where $D_{B,cp}$ is the Macaulay duration for the coupon payment period.

The first-order approximation (from the Taylor expansion) gives us (cf. Equation A.43)

$$\Delta B_0 \approx D_{B,m}B_0\Delta r \tag{A.46}$$

where $D_{B,m}$ is a modified Macaulay duration $(= -D_B/(1 + r))$.

The approximation in Equation A.46 is only good for small changes in the yield. This is the usual argument for improving Equation A.46 by introducing a second term, the so-called *convexity*. This is the suggestion made by IAA (2004, Appendix D, paragraph 7.15).

Let

$$B_0(r) = \sum c_t (1+r)^{-t}$$

$$B_0'(r) = -\sum tc_t (1+r)^{-t-1}$$

$$B_0''(r) = \sum t(t+1)c_t (1+r)^{-t-2}$$

Then from Taylor expansion we get

$$B_0(r+\Delta r) - B_0(r) \approx D_{B,m}B_0\Delta r + \frac{1}{2}C(\Delta r)^2 \tag{A.47}$$

where the first term is due to duration and the second to convexity.

$$C = \frac{1}{B_0(1+r)^2}\sum_{t=1}^{T}t(t+1)\frac{c_t}{(1+r)^t}$$

Note that Δr is the change in yield per year (or per period), depending on the units of $D_{B,m}$ and C. The second term approximates the percent price change due to convexity, i.e., the part of the price change due to the curvature of the price–yield relationship.

In the discussion above it was assumed that we had a single bond. In reality, we have a bond portfolio consisting of several different bonds with different yields and durations. The duration of the bond portfolio is equal to the weighted mean of the different durations, with weights equal to the relative bond price:

$$D = \sum_{i=1}^{m} \omega_i D_{Bi}$$

where $\omega_i = B_{0i} / B_0$ and $B_0 = \sum_{i=1}^{m} B_{0i}$ (for example, see Luenberger, 1998, Section 3.5, p. 62).

Later when we split the liabilities and the corresponding assets into duration bands, we could look at each duration band as if it constituted its own portfolio.

A.2.1.3 Valuation of Equities and Property

The cash flow of equities (or stocks) is denoted *dividends* (d_t). For property (or real estate) the term is *net operating income* (NOI_t). The discussion that follows applies to equities (or stock), but the same approach will be used for property, as this investment class resembles equities in many ways.

A.2.1.3.1 Dividend Discount Model

John Burr Williams introduced the dividend discount model (DDM) in a thesis in 1937. The idea was to forecast the future dividends a company would pay to its stockholders and then discount them using a suitable interest rate.

From Equation A.40 we get

$$S_0 = \sum_{t=1}^{n} \frac{d_t}{\left(1+r\right)^t} \tag{A.48}$$

where S_0 is the present value of the stock (the intrinsic or theoretical value), d_t is the dividend paid at t, and r is the required return (discount rate) for each year t.

Note that the maturity is perpetual (that means perhaps 20 years or so in reality). If a company buys equity after time H (the holding period for the seller), he assumes to get a profit from it in the future. S_0 can be written as

$\dfrac{d_1}{(1+r)} + S_1$, where S_1 is the sales price at the end of $t = 1$ equals Equation A.48, with $t = 2, 3, \ldots.$

Williams's basic model (Equation A.48) has been modified many times. Gordon and Shapiro introduced one of the most common modifications in 1956. In this model it is assumed that the dividends grow with a constant rate (g).

A.2.1.3.2 The Constant-Growth DDM (Gordon–Shapiro Model or Simply the Gordon Model)

Assume that $d_t = d_0 (1+g)^t$ where g is the constant-growth rate and d_0 is the most recent per-share dividend (and a value that is known for certainty). It is also assumed that $r > g$ and n is perpetual. From Equation A.48 we now get

$$S_0 = \frac{d_0(1+g)}{(r-g)} = \frac{d_1}{r-g} \tag{A.49}$$

A.2.1.3.3 Zero-Growth DDM

This is the simplest form of Equations A.48 and A.49 assuming constant dividends, i.e., zero growth. Assuming $g = 0$ gives us[3]

$$S_0 = \frac{d_0}{r} \tag{A.50}$$

Before looking at other modifications of the DDM we will shortly look at the duration and sensitivity. The duration can be defined as

$$D_{S,n} = \sum_{t=1}^{n} tw_t = \frac{1}{S_0} \sum_{t=1}^{n} \frac{td_t}{(1+r)^t}$$

We can now define the sensitivity of equities in terms of this general perpetual duration as

$$\frac{1}{S_0}\frac{dS_0}{dr} = -(1+r)^{-1} D_{S,n}$$

---

[3] The simplest way is to use Equation A.49. Using Equation A.48 we get $S_0 = d_0 \sum_{t=1}^{\infty} \frac{1}{(1+r)^t}$.

Multiplying both sides with $(1+r)$ gives us $(1+r)S_0 = d_0 + d_0 \sum_{t=1}^{\infty} \frac{1}{(1+r)^t}$ and subtracting both sides with S_0 gives $rS_0 - d_0$, and hence Equation A.50.

A.2.1.3.4 A Two-Part Dividend Stream

Assume that a company holds equities for a period of H years (holding period) and then sells them. For the holding period ($t = 1, 2, ..., H$) we can assume the DDM approach (Equation A.48), and for the future period ($t = H + 1, H + 2, ...$) we assume the constant-growth model (the Gordon–Shapiro model). This is also a common model for new technology equities, which are expected to grow rapidly for a few years and then to slow down as they approach maturity. For this model we have

$$S_0 = \sum_{t=1}^{H} \frac{d_t}{(1+r)^t} + \frac{d_H(1+g)}{(r-g)}\left(\frac{1}{1+r}\right)^H \tag{A.51a}$$

This valuation model is similar to the bond valuation model (Equation A.44a). Assume $d_H^* = d_H + \dfrac{d_H(1+g)}{r-g}$ and $d_t^* = d_t$ for $t = 1, 2, ..., H-1$. Then we have

$$S_{0,H} = \sum_{t=1}^{H} \frac{d_t^*}{(1+r)^t} \tag{A.51b}$$

We can now define the duration by

$$D_{S,H} = \sum_{t=1}^{H} tw_t = \frac{1}{S_{0,H}} \sum_{t=1}^{H} \frac{td_t}{(1+r)^t} \tag{A.52}$$

A first-order approximation (Taylor expansion) gives us the change in equity value as measured in terms of change in interest rate r (cf. Equation A.46):

$$\Delta S_{0,H} \approx D_{S,m} S_{0,H} \Delta r \tag{A.53}$$

where

$$D_{S,m} = -\frac{1}{1+r} D_{S,H}$$

The change in equity value is thus approximated by modified Macaulay duration of the equity's duration *times* the intrinsic equity value (present value) *times* the change in rate of return during the year. The estimation in Equation A.53 can be improved by using the second-order term as in

Equation A.47. Note that for the Gordon–Shapiro model (Equation A.49) we get the duration as $(1 + r)/(r - g)$, and therefore we have

$$\Delta S_{0,H} \approx -\Delta r d_0 \left(1+g\right)/\left(r-g\right)^2$$

A.2.1.3.5 Some Other Models

Estep and Hanson extended the Gordon–Shapiro model in 1980 to incorporate the effect of inflation by defining $(1 + r) = (1 + rr)(1 + ei)$ and $(1 + g) = (1 + rg)(1 + ftc)$, where rr = real rate of return, ei = expected rate of inflation, rg = real growth, and ftc = a flow-through coefficient. Other extensions include, e.g., property market cycles.

For a portfolio of equities and property we may use the same approach as for bonds (see Section A.2.1.2).

A.2.2 Asset–Liability Risk

The asset–liability risk results from the volatility inherent in market value of future cash flows from the insurer's assets and liabilities. We will only consider the retained business, as the risks related to reinsurance are taken into account by the credit risk. Nor will we consider surrender and lapse, as they also will be taken account of by the insurance risk.

In a standard approach (SA), the number of asset categories should not be large. It should be possible to enlarge the number of categories in an internal model. In the SA, the easiest way of classifying the categories would probably be into bonds, equities, property, cash, and others.

A.2.2.1 The Capital Requirement, C_{MR}

For each asset category $j = 1, ..., c$, let A_j be the asset amount invested in asset j. Let $\sigma(j)$ be the standard deviation of the return in asset[4] category j and $\rho(j,k)$ be the correlation coefficient between returns of categories j and k.

The part of the total capital requirement from the asset–liability risk is defined as

$$C_{MR} = k\sigma_{MR} = k\left\{ \sum_{k=1}^{8} A_{01}^2(d_k)\sigma^2(d_k) + \sum_{j=2}^{c} A_{0j}^2\sigma^2(j) + \sum_{j\ne}^{c}\sum_{l}^{k} A_{0j}A_{0l}\rho(j,l)\sigma(j)\sigma(l) \right\}^{1/2}$$

(A.54)

where the first term is the sum $(k = 1, ..., 8)$ and is a mismatch part depending on eight duration bands defined in Section 10.2 (see also Section A.2.2.7).

---

[4] For a portfolio of assets, for example, bonds, we assume the standard deviation to be calculated for the portfolio. The same holds for the correlation coefficients.

The durations of bonds are split up into eight duration bands, and for equities and property, we define bands of mean holding periods based on historical data. The second term is the total variance of the portfolio and the last term reflects the correlation between the assets in the portfolio.

The variances $\sigma^2(j)$ and the correlations $\rho(j,k)$ are, in the standard approach, defined on the EU level. The companies have only to list their assets in the c categories, defined by EU.

A.2.2.2 Background Theory

Define V_0 as the best estimate of liabilities, i.e., the technical provisions (TP), and A_0 as the assets covering V_0 + the solvency capital requirement at 0, SCR_0.

Hence, $SCR_0 = A_0 - V_0$.

The corresponding liabilities and assets after 1 year are V_1 and A_1, respectively.

Let the SCR at time 1 be

$$SCR_1 = A_1 - V_1 = (A_0 + I_1 + P_1 - C_1) - (V_0 + \Delta_1 V_0 + \Delta_2 V_0 + \Delta_3 V_0) \qquad \text{(A.55a)}$$

where I_1 is the return on assets during year 1, i.e., the sum of the investment income (i.e., the cash income of the assets) plus changes (+/−) in the value of the assets. This is also called the investment return during year 1. P_1 is the premium during year 1, and C_1 is the claims and expenses during year 1.

We also assume that $\Delta_1 V_0$ is the change in liabilities V_0 due to changes in interest rates, $\Delta_2 V_0$ is the change in liabilities V_0 due to interest rate guarantees and more or less binding bonus declarations (= BG, bonus guarantees), and $\Delta_3 V_0$ is the change in liabilities V_0 due to other changes, mainly the increase in V_0 due to net premiums (= $P_1 - C_1$).

With these notations we can write Equation A.55a as

$$SCR_1 = A'_1 - V'_1 = (A_0 + I_1) - (V_0 + \Delta_1 V_0 + BG) = (I_1 - \Delta_1 V_0) - (V_0 + BG - A_0) \qquad \text{(A.55b)}$$

In Equation A.55b only the first term $(I_1 - \Delta_1 V_0)$ is stochastic. The last term is deterministic.

Hence,

$$E[SCR_1] = E[A'_1 - V'_1] = E[I_1 - \Delta_1 V_0] - (V_0 + BG - A_0) \qquad \text{(A.56)}$$

The expectation $E[I_1 - \Delta_1 V_0]$ is the *expected mismatch* between return on assets and the change in liabilities due to the changes in interest rates. We denote it $E(MM)$ and call it the expected mismatch of changes. Rewriting Equation A.56 and letting $A_0 = V_0 + SCR_0$, we get

$$E[SCR_1] = SCR_0 + \{E(MM) - BG\} \qquad \text{(A.57)}$$

The expected solvency capital requirement from the asset–liability risk during year 1 can be written as the sum of the SCR during year 0 plus the difference between the expected mismatch of changes and bonus guarantee. Taking the variance of Equation A.55b gives

$$\text{Var}[SCR_1] = \text{Var}[I_1 - \Delta_1 V_0] = \text{Var}[I_1] + \text{Var}[\Delta_1 V_0] - 2\text{Cov}[I_1, \Delta_1 V_0] \quad (A.58)$$

If we look at this from a risk management perspective, we can consider the quantile at level α $Pr(SCR_1 < 0) = \alpha$. Using normal approximation we get

$$E[SCR_1] - k_\alpha \sigma \sqrt{(SCR_1)} \quad = 0$$

Denote by $\sigma(SCR_1)$ the standard deviation of SCR_1. This gives us

$$k_\alpha \sigma \sqrt{(SCR_1)} = E[SCR_1] = SCR_0 + \{E(MM) - BG\} \quad (A.59)$$

This expression gives us the part of the total solvency capital requirement that comes from the asset–liability risk, C_{MR}.

A.2.2.2.1 The Assets

We write the total asset as the sum of different assets categories $A_0^* = \sum_{j=1}^{c} A_{0j}$, and the corresponding random return rates are denoted by $i(j)$, $j = 1, ..., c$. Hence, the return on assets can be written as[5]

$$I_1 = \sum_{j=1}^{c} i(j) A_{0j}$$

We will now consider the return on the assets' bonds, equities, and property, i.e., look closer at $i(bonds) \times A_{bonds}$, $i(equities) \times A_{equities}$, and $i(property) \times A_{property}$. Later on we will assume a number of duration bands (see Section 10.2) and define a return rate curve with rates associated with each duration band.

A.2.2.3 Bonds

The change in the bond price due to a small change in the yield is, according to Equation A.46 in Section A.2.1.2,

---

[5] We assume that the random return rate for a portfolio of similar assets, for example, bonds, is the same irrespective of the bond type.

$$\Delta B_0 \approx D_{B,m} B_0 \Delta r \tag{A.60}$$

where $D_{B,m}$ is the modified Macaulay duration of the bond, Δr the change in

the yield, and $B_0 = \sum_{t=1}^{T} \dfrac{c_t}{(1+r)^t}$ is the market value of the bond as defined by

Equation A.44b.

We assume, as a simplification, that the volatility of the return on bonds is a result of a parallel shift in the prevailing interest rate curve, i.e.,

$$i_B = i_0 + \Delta i$$

where i_B is the new interest rate curve, i_0 is the present interest rate curve, and Δi is a random variable with $E(\Delta i) = 0$ and standard deviation $\sigma(\Delta i)$.

Later we will divide the interest rate curve into duration bands and also consider the bonds divided according to their duration (see Section 10.2).

The return on investments on bonds can be seen as the coupon payments during the year, and the change in value can thus be approximated by $[i(bonds) \times A_{bonds} =]$

$$I_B + \Delta i D_{B,m} B_0 \tag{A.61}$$

where I_B is the coupon rate income (e.g., in euros) on bonds.

The last term in Equation A.61 is the bond price (i.e., its present value) *times* its duration *times* the change in interest rate level. Equation A.61 includes some approximations:

- A simple model for the interest rate change (the change in interest rate occurs at the beginning of the year and coupon rate earnings after that).
- The change in bond value is computed by using the present value at the beginning of the year and omitting the cash flow during the year (such as coupon payments and their reinvestments, new premiums and their investment, etc.).
- The first-order approximation is used (cf. Section A.2.1.2).

These approximations can be improved when internal models are used, e.g., by using second-order approximation.

The change in the bond price is not fully explained by changes in yield, since the credit ratings also have effect.

A.2.2.4   Equities

Equities cannot be treated exactly as bonds as their maturity is defined to be perpetual, in general. But for a specific insurance company we could define a holding period H and write the present value of dividends over H years[6] (using a *dividend discount model* (DDM)) plus the ultimate sale price S_H (using the *Gordon–Shapiro model*), which could be incorporated in the ultimate dividend d_H. This is explained in Section A.2.1.3 and especially as the *two-part dividend stream*.

From Equation A.53 we get the change in equity price with respect to the discount rate r:

$$\Delta S_{0,H} \approx D_{S,m}S_{0,H}\Delta r \tag{A.62}$$

We can split the discount rate r into two terms: (1) the interest rate, $i_0 + \Delta i$, and (2) an economic sort of term or risk premium w. This means that a change in the interest rate curve leads to a change in the equities value at a constant risk premium.

The income and change in value of equities can be approximated by using the approximation of Equation A.61 [$i(equities) \times A_{equities} =$]

$$I_S + \Delta i D_{S,m}S_{0,H} + W_S \tag{A.63}$$

where I_s is the expected income on equities with the present interest rate levels, $W_s = \varepsilon_S S_{0,H}$ represents the variation of the equities' expected return that is not explained by Δi, and ε_S is a random variable, independent of Δi, and with $E(\varepsilon_S) = 0$ and $Var(\varepsilon_S) = \sigma^2(S)$.

A.2.2.5   Property

Property can be treated similar to equities. Instead of future dividends we have future *net operating income, NOI_t*. The discounted market value of property can be described using the two-part dividend stream defined by Equation A.51b in terms of property as[7]

$$P_{0,H} = \sum_{t=1}^{H} \frac{NOI_t}{(1+r)^t} \tag{A.64}$$

where $NOI^*_H = NOI_H + P_H$, where P_H is the sale price at $t = H$.

---

[6] In Chapter 10 we have assumed a mean holding period band (duration band) between 5 and 8 years, i.e., with a median duration of 6.5 years.
[7] In Chapter 10 we have assumed a mean holding period band (duration band) between 8 and 12 years, i.e., with a median duration of 10.0 years.

The change in property price with respect to the discounting factor can be approximated by (cf. Equation A.63)

$$I_P + \Delta i D_{P,m} P_{0,H} + W_P \qquad (A.65)$$

where $D_{P,m}$ is the modified Macaulay duration for property. As for equities, we split the discount rate r into $i_0 + \Delta i + w$, where w is an economic risk factor. Hence, $[i(property) \times A_{property} =]$

$$I_P + \Delta i D_{P,m} P_{0,H} + W_P \qquad (A.66)$$

where I_P is the expected net operating income from property with the present interest rate levels, $W_P = \varepsilon_P P_{0,H}$ represents the variation of the equities' expected return that is not explained by Δi, and ε_P is a random variable, independent of Δi, and with $E(\varepsilon_P) = 0$ and $Var(\varepsilon_P) = \sigma^2(P)$.

A.2.2.6   The Liabilities
We have to divide the liabilities, technical provisions, into duration bands according to the insurance contracts. The duration bands are, of course, the same as those for the bonds:

d_1 : [0–1] year (median duration: $md_1 = 0.5$)
d_2 : [1–2] years (median duration: $md_2 = 1.5$)
d_3 : [2–5] years (median duration: $md_3 = 3.5$)
d_4 : [5–8] years (median duration: $md_4 = 6.5$)
d_5 : [8–12] years (median duration: $md_5 = 10.0$)
d_6 : [12–16] years (median duration: $md_6 = 14.0$)
d_7 : [16–24] years (median duration: $md_7 = 20.0$)
d_8 : [24–] years (median duration, say: $md_8 = 28$)

By similar reasoning as for bonds, equities, and property we may write the change in liabilities as

$$\Delta_1 V_0 = \Delta i D_V V_0$$

where D_V is a mean duration measure of the liabilities V_0. This total change of liabilities is split up into

$$\Delta i(d_k) md_k V_0(d_k), k = 1, ..., 8$$

where $\Delta i(d_k)$ is a random variable with expectation zero and variance $\sigma^2(d_k)$, md_k is the median duration as defined above, and $V_0(d_k)$ the total technical provisions within the duration band k.

A.2.2.7 The Mismatch $I_1 - \Delta_1 V_0$

From the earlier discussion we have the mismatch

$$i(bonds) \times A_{bonds} + i(equities) \times A_{equities} + i(property) \times A_{property} +$$

$$\sum i(j)A_{0j} - \Delta_1 V_0 \text{ or in other terms}$$

$$I_1 - \Delta_1 V_0 = (I_B + \Delta i D_{B,m} B_0 + I_S + \Delta i D_{S,m} S_0 + W_s + I_P + \Delta i D_{P,m} P_0 + W_P + \sum_{j=4}^{c} i(j)A_{0j})$$

$$- \Delta i D_V V_0 = \sum_{j=4}^{c} i(j)A_{0j} + (I_B + I_S + I_P) + (W_s + W_P) + \Delta i \{D_{B,m} B_0 + D_{S,m} S_0 + D_{P,m} P_0$$

$$- D_V V_0\} \tag{A.67}$$

We may treat $\{D_{B,m} B_0 + D_{S,m} S_0 + D_{P,m} P_0 - D_V V_0\}$ as a *mismatch asset* category. The duration bands introduced in Section 10.2 may split the last term into a sum of a stochastic interest rate change (due to the duration band) *times* the mismatch of that duration[8] band.

Let $D_{B,m}$ be split into median durations md_k, $k = 1, ..., 8$, $D_{S,m} = md_4$, and $D_{P,m} = md_5$, as proposed in Section 10.2. The duration of the liability will also be divided according to their duration; i.e., D_V is split into median durations md_k, $k = 1, ..., 8$. The present value (at the end of the last year) of the bonds, equity, property, and liability will also be split into their duration bands. We denote these as $B_0(d_k)$, $S_0(d_4)P_0(d_5)$, and $V_0(d_k)$, $k = 1, ..., 8$. The change in mean interest rate Δi will be thought of as a parallel shift in the interest rate curve, and that curve will be divided into disjoint intervals according to the duration. For each of the eight intervals we will consider the stochastic variables $\Delta i(d_k)$ with expectation $E(\Delta i(d_k)) = 0$ and variance $Var(\Delta i(d_k)) = \sigma^2(d_k)$. The $\Delta i(d_k)$ values are assumed fully correlated.

Let

$$A_{01}(d_k) = md_k\{B_0(d_k) - V_0(d_k)\}, \ k = 1, 2, 3, 6, 7, 8$$

$$A_{01}(d_4) = md_4\{B_0(d_4) + S_0(d_4) - V_0(d_4)\}, \ k = 4$$

---

[8] In practice, it is usually better to interpret duration in terms of cash-flow maturity, i.e., projected asset cash flow and liability cash flow.

$$A_{01}(d_5) = md_5\{B_0(d_5) + P_0(d_5) - V_0(d_5)\} \ , \ k = 5$$

This implies that we can write Equation A.67 as

$$I_1 - \Delta_1 V_0 = \sum_{k=1}^{8} \Delta i(d_k) A_{01}(d_k) + \sum_{j=4}^{c} i(j) A_{0j} + (I_B + I_S + I_P) + \varepsilon_S S_{0,H} + \varepsilon_P P_{0,H} \quad (A.68)$$

where I_B, I_S, and I_P are fixed (nonstochastic) and ε_S and ε_P are independent of each other and of the other stochastic variables ($\Delta i(d_k)$ and $i(j)$).
This gives us[9]

$$Var[I_1 - \Delta_1 V_0] = Var(\sum_{k=1}^{8} \Delta i(d_k) A_{01}(d_k) + \sum_{j=4}^{c} i(j) A_{0j} + \varepsilon_S S_{0,H} + \varepsilon_P P_{0,H})$$

$$\sum_{k=1}^{8} A_{01}^2(d_k)\sigma^2(d_k) + \sum_{j=4}^{c} A_{0j}^2\sigma^2(j) + \sum_{j\neq}^{c}\sum_{l}^{c} A_{0j}A_{0l}\rho(j,l)\sigma(j)\sigma(l) + S_{0,H}^2\sigma^2(S) + P_{0,H}^2\sigma^2(P)$$

$$(A.69)$$

as A_{0j} is constants and we use the notation $Var[i(j)] = \sigma^2(j)$ and the correlation $\rho[i(j), i(k)] = \rho(j,k)$. We have taken the absolute value of $A_{01}(d_k)$, as the mismatch should be the distance between the asset part and the liability part. The first part in Equation A.69 depends on the assumption of full correlations. In the third term we also include the overall mismatch part from Equation A.67; i.e., the first sum in Equation A.69 is regarded as $\Delta i A_{01}$. The two last terms of Equation A.69 can also be written as $i(2)A_{02}$ and $i(3)A_{03}$, respectively, and hence the second term in Equation A.69 can be summed from 2 to c:

$$\sum_{k=1}^{8} A_{01}^2(d_k)\sigma^2(d_k) + \sum_{j=2}^{c} A_{0j}^2\sigma^2(j) + \sum_{j\neq}^{c}\sum_{l}^{c} A_{0j}A_{0l}\rho(j,l)\sigma(j)\sigma(l) \quad (A.70)$$

which gives us Equation A.54.

In Equation A.69 we have assumed that $\sigma(d_k)$ is the same for equities and bonds, with a median duration of 6.5 years (duration band 4), and is the same for property and bonds, with a median duration of 10 years (duration band 5). The volatility is usually highest for property and equities. One simple trick would be to weight the two volatilities with weights proportional to the present value.

---

[9] In the calculation we always use assets backing the liabilities and not assets backing the solvency margin, as this would increase the capital charge: the more capital backing the margin, the more charge.

Duration band 4, equities and bonds:

$$\sigma(d_4) = w_S\sigma(d_4, stocks) + (1 - w_S)\sigma(d_4, bonds) \text{ with } w_S = \frac{S_0(d_4)}{S_0(d_4) + B_0(d_4)}$$

Duration band 5, property and bonds:

$$\sigma(d_5) = w_P\sigma(d_5, property) + (1 - w_P)\sigma(d_5, bonds) \text{ with } w_P = \frac{P_0(d_5)}{P_0(d_5) + B_0(d_5)}$$

A.2.3 Currency Risk

Currency risk is to a large extent taken into account in the A/L risk above if the calculations are made by currencies. If there is a need for a specific term for the currency mismatch risk, it could simply be defined as a *coefficient times the mismatch*. A discussion of this risk is made in IAA (2004, pp. 139–140). If bonds are dominated in a foreign currency, the volatility of the exchange rate has to be accounted for. As stated in IAA (2004, p. 140), a typical standard deviation for returns of a floating foreign exchange (*FX*) rate is around 10%, i.e.,

$$\frac{1}{k_{1-\alpha}}C_{FX} \approx 0.1\, B_0$$

where B is the value of the bonds.

We will not include the currency risk in this model example.

A.3 Credit Risk

As suggested in Section 10.3 the credit risk will consist of the default credit risk (C_{dcr}; Section A.3.3), the concentration risk (C_{cor}; Section A.3.1), and the reinsurance risk (C_{rr}; Section A.3.2). As proposed in Section 10.6, we will use the following dependence structure:

$$C_{CR} = \sqrt{C_{dcr}^2 + C_{cor}^2 + C_{rr}^2} \tag{A.71}$$

A.3.1 Concentration Risk

The concentration risk deals with the risks of increased exposure to losses of both investment (concentration of assets) and catastrophic events (concentration of liabilities). For example, investment in a high proportion of specific equities can be considered as concentration. Concentration on one or two reinsurers is also a part of the concentration risk.

We consider the risk net of reinsurance.

A.3.1.1 The Capital Requirement, C_{cor}

The part of the total capital requirement from the concentration risk is defined as the sum of an asset concentration part and a liability concentration part:

$$C_{cor} = k \left\{ A^{*2} \{ max(a_i^*) \overline{a}^* - \overline{a}^{*2} \} + \{ M_X \overline{X}^* - \overline{X}^{*2} \} \right\}^{1/2}$$

(A.72)

where A^* is the total assets covering the liabilities. $a_i^* = A_i^* / A^*$, $0 \le a_i^* \le 1$, are the relative asset concentration categories, and $\overline{a}^*$ is the mean asset concentration. M_X is the estimated maximum loss and $\overline{X}^*$ is the estimated mean loss of catastrophic events.

A.3.1.2 Background Theory

We will consider the asset concentration risk and the liability concentration risk as independent of each other, i.e., Var(concentration) = Var(asset concentration) + Var(liability concentration).

A.3.1.2.1 Asset Concentration Risk

Assume that the assets are split up into different individual investments; i.e., all owned shares in one company are considered as one individual investment, A_i^*, $i = 1, \ldots, n$. We look at the relative part of each individual investment $a_i^* = A_i^* / A^*$, where $A^* = \sum_{j=1}^{n} A_j^*$, and assume that they are i.i.d.

We will also make a simplifying procedure by conditioning on A^* in the second term in Equation A.73.

The volatility in the assets can thus be described by

$$Var(A^*) = \sum_{i=1}^{n} Var(a_i^* A^*) \approx A^{*2}\{E(a_i^{*2}) - E^2(a_i^*)\} = A^{*2}\{E(a_i^{*2}) - \overline{a}^{*2}\} \leq A^{*2}\{max(a_i^*)\overline{a}^* - \overline{a}^{*2}\}$$

(A.73)

This gives us the first part of Equation A.72.

A.3.1.2.2 Liability Concentration Risk

Assume that the liabilities are split up into different liability categories depending on the risks; i.e., the risks or groups of risks that could be exposed to different catastrophic events are considered as liability category X_i^*, $i = 1$, ..., n. We look at the relative part of each liability category $x_i^* = X_i^* / X^*$, where $X^* = \sum_{j=1}^{n} X_j^*$, and assume that they are i.i.d. We also make the simplifying assumption that X^* is constant.

The volatility in the liability can thus be described by

$$Var(X^*) = \sum_{i=1}^{n} Var(x_i^* X^*) = \{E(X_i^{*2}) - E^2(X_i^*)\} = \{E(X_i^{*2}) - \overline{X}^{*2}\} \leq \{M_X \overline{X}^* - \overline{X}^{*2}\}$$

(A.74)

where M_X is the estimated maximum loss, net of reinsurance. This gives us the second part of Equation A.72.

A.3.1.2.3 An Alternative Approach

There are specific measures on concentration (see, e.g., Nygård and Sandström, 1981). One such measure is the Gini coefficient. It takes on the value zero when all entities ($n > 1$) are equal and the value 1 when the measuring is concentrated in one entity, all others having zero. We look at the asset concentration. Assume a minimum number of assets that it is possible to have. These categories can be the four discussed in Section 8.2: bonds, equities, property, and cash. This means that we will have $n \geq 4$.

The Gini coefficient can be written in many ways, but we use the following one and notation from A.3.1.2.1. Let $a_{(i)}^*$ be the ith rank-ordered relative part of each individual investment, i.e., $a_{(1)}^* < a_{(2)}^* ... < a_{(n)}^*$.

Ties are ignored and can be randomly given a rank:

$$R = 2\frac{1}{n-1}\sum_{i=1}^{n} ia_{(i)}^* - \frac{n+1}{n-1}, \quad R \in [0,1] \tag{A.75}$$

Assume that a company only has one bond category; i.e., its relative part of each individual investment will be 1, but we need at least three other asset categories whose relative parts will be zero. The Gini coefficient will then be $R = 2(1 \times 0 + 2 \times 0 + 3 \times 0 + 4 \times 1)/3 - 5/3 = 8/3 - 5/3 = 1$.

If we have five different assets that are equal, we get

$$R = 2(1 \times \tfrac{1}{5} + 2 \times \tfrac{1}{5} + 3 \times \tfrac{1}{5} + 4 \times \tfrac{1}{5} + 5 \times \tfrac{1}{5})/4 - 6/4 = 0$$

One way to use R is to multiply R with the total asset value A^* and then define its volatility by

$$A^* R \in [0, A^*] \tag{A.76}$$

and apply this instead of Equation A.73.

A.3.2 Reinsurance Counterparty Risk

The underwriting risk was considered the net of reinsurance. Reinsurance is discussed below. An alternative method is the Norwegian approach (see Section 6.11).

A.3.2.1 The Capital Requirement

For each reinsurance company, $i = 1, ..., r$, the regulator (EU level) gives a score, ω_i, indicating its rating. The regulator can, e.g., use company ratings from Standard & Poor's or Moody's. The scores are usually defined in the interval $0 < \omega_i < 1$.

The part of the total capital requirement from the reinsurance risk is defined as

$$C_{rr} = k\sigma_{rr} = k\left\{ \sum_{i=1}^{r}(1-\omega_i)\sigma_i^2 + \sum_{i=1}^{r}(1-\omega_i)P_{i,rr}^2 \right\}^{1/2} \tag{A.77}$$

where σ_i^2 shows the volatility in the receivables from reinsurer i and P_i^2 is the squared ceded premiums from reinsurer i. The last term addresses the counterparty risk inherent in reinsurance cover.

As the score ω_1 is less than 1, we see that there will always be capital requirements from the reinsurance cover. The higher the score, the less the risk factor, and hence the contribution to the solvency margin.

A.3.2.2 Background Theory

Let R be the balance of dealings with the reinsurers:

$$R = -P_{rr} + X_{rr}$$

where P_{rr} is the ceded reinsurance premiums during the year and X_{rr} is the receivables paid by the reinsurers. For simplicity, we assume that the ceded reinsurance premiums are fixed (and paid at the first day of the year) so that only the receivables are stochastic.

Assume that the company has business with r different reinsurers. Then we can write R as

$$R = -\sum_{i=1}^{r} P_{i,rr} + \sum_{i=1}^{r} X_{i,rr} \tag{A.78}$$

The total loss during the year is denoted X_{tot}, and hence the net loss can be written as $X = X_{tot} - B_{rr}$ or $B_{rr} = X_{tot} - X$, where B_{rr} is the recoveries that should be paid by the reinsurers according to contracts. The total loss can thus be written as $X_{tot} = X + B_{rr} = X + X_{rr} + (B_{rr} - X_{rr})$, where the last term reflects what the company may have to pay due to insolvency in one or more of the reinsurance companies.

The variance of Equation A.78, with the assumption that premiums are fixed and that there is independence between the reinsurers, is

$$Var(R) = \sum_{i=1}^{r} Var(X_{i,rr})$$

Assume that there is uncertainty in getting the recoveries from reinsurer i, $i = 1, \ldots, r$.

Let B_i be the recoveries from reinsurer i with $E(B_{i,rr}) = \mu_i$ and $Var(B_{i,rr}) = \sigma_i^2$. Introduce an indicator variable I_i, such that

$$I_i = \begin{cases} 1 & if \quad X_i = B_{i,rr} \\ 0 & if \quad X_i < B_{i,rr} \end{cases}$$

Now we have that

$$E(I_i^2) = E(I_i) = q_i$$

and

$$Var(I_i) = E(I_i^2) - E(I_i)^2 = q_i(1-q_i)$$

Define $X_{i,rr} = I_i B_{i,rr}$.
Now, using the conditional identity as in Section A.1.1.2 we have

$$Var(X_{i,rr}) = E[Var(I_i B_{i,rr} \mid I_i)] + Var[E(I_i B_{i,rr} \mid I_i)] = q_i \sigma_i^2 + q_i(1-q_i)\mu_i^2$$

This gives us

$$Var(R) = \sum_{i=1}^{r} q_i \sigma_i^2 + \sum_{i=1}^{r} q_i \mu_i^2 - \sum_{i=1}^{r} q_i^2 \mu_i^2 \qquad (A.79a)$$

A reinsurance company with a high risk factor will induce a large capital requirement for the cedant. We can therefore, in a prudent environment, drop the last term in Equation A.79a, giving

$$Var(R) = \sum_{i=1}^{r} q_i \sigma i + \sum_{i=1}^{r} q_i \mu_i^2 \qquad (A.79b)$$

The first term in Equations A.79a and A.79b shows the volatility in the receivables from the reinsurers. The μ_i values are the expected receivables from reinsurer i, and this should in the long run be equal to the premiums ceded. Hence, we can estimate μ_i with P_{rr}.

The risk factor should be set at the EU level for each reinsurer. If ratings of reinsurers are used, we can define $q_i = 1 - \omega_i$, where ω_i is the rating (or score) factor. This gives us Equation A.77.

A.3.3 Default Credit Risk

For the default credit risk we will use the standard approach used for banks according to Basel II (see Section 5.1). The main reason for doing so is that we do not want any possibility for arbitrage from the banking to the insurance sector, and vice versa.

The basis for the default credit risk will be Equation 5.5, i.e.,

$$C_{dcr}^2 = 0.08 \sum_{j,c} r_{jc} A_{jc} = \sum_j w_{jc} A_{jc}$$

where r_{jc} is the risk weight according to Table 5.2 and j the asset category and c the rating of the exposure and the risk factors $w_{jc} = 0.08 r_{jc}$. A_{jc} is the corresponding exposed asset.

A.4 Operational Risk

Operational risk is often seen as a residual risk. There is no absolute definition of the risk, but it can be viewed as including many hard-to-quantify risk categories, such as failures in management and control, failures in IT processes, human error, fraud (both internal and external), and legal and jurisdictional risks. It can therefore be seen as an isolated risk component. Sometimes it is regarded as the guarantee system in insurance.

In the draft New Basel Capital Accord from April 2003 (BIS, 2003), the measurement methodologies for banks are discussed. Three different methods of calculating the operational risk are presented (Chapter V) in an increasingly sophisticated and risk-sensitive way:

1. The basic indicator approach
2. The standardized approach
3. The advanced measurement approaches (AMA)

The last approach is indicated for the internal models. We will adopt the second approach for insurance companies.

The models presented in the draft from 2003 were changed in the final report (BIS, 2004; see also Section 5.1). However, we will use the simpler form presented in BIS (2003).

A.4.1 The Capital Requirement, C_{OR}

The capital requirement for the operational risk is defined as the sum over different lines of business ($i = 1, \ldots, L$)

$$C_{OR} = \sum_{i=1}^{L} \bar{B}_{i,3} \beta_i \tag{A.80}$$

where $\bar{B}_{i,3} = \dfrac{1}{3}\sum\limits_{j=1}^{3} B_{i,j}$ is the mean of the *gross* premium income during the three last years and β_i is a factor, $0 < \beta_i < 1$, set on the EU level.

Note that the basic indicator approach is just $\bar{B}_3\beta$, i.e., the mean of the total gross premium over the last three years multiplied by a EU factor β.

Appendix B

Insurance Classes

Non-Life Classes

The *insurance risk classes* were first defined in the annex to the *first non-life directive* (EEC, 1973). The system of risk classes was later extended by the *tourist assistance* directive (EEC, 1984) and the *credit and suretyship insurance* directive (EEC, 1987).

A.   Classification of Risks according to Classes of Insurance

1. *Accident* (including industrial injury and occupational diseases)
   - Fixed pecuniary benefits
   - Benefits in the nature of indemnity
   - Combinations of the two
   - Injury to passengers
2. *Sickness/health*
   - Fixed pecuniary benefits
   - Benefits in the nature of indemnity
   - Combinations of the two
3. *Land vehicles* (other than railway rolling stock)
   All damage to or loss of
   - Land motor vehicles
   - Land vehicles other than motor vehicles
4. *Railway rolling stock*
   All damage to or loss of railway rolling stock
5. *Aircraft*
   All damage to or loss of aircraft
6. *Ships* (sea, lake, and river and canal vessels)

All damage to or loss of

- River and canal vessels
- Lake vessels
- Sea vessels

7. *Goods in transit* (including merchandise, baggage, and all other goods)

   All damage to or loss of goods in transit or baggage, irrespective of the form of transport

8. *Fire and natural forces*

   All damage to or loss of property (other than property included in classes 3 to 7) due to

   - Fire
   - Explosion
   - Storm
   - Natural forces other than storm
   - Nuclear energy
   - Land subsidence

9. *Other damage to property*

   All damage to or loss of property (other than property included in classes 3, 4, 5, 6, and 7) due to hail or frost, and any event such as theft, other than those mentioned under 8

10. *Motor vehicle liability* (third-party liability, land-based vehicles)

    All liability arising out of the use of motor vehicles operating on the land (including carrier's liability)

11. *Aircraft liability* (third-party liability, aircraft)

    All liability arising out of the use of aircraft (including carrier's liability)

12. *Liability for ships* (sea, lake, and river and canal vessels) (third-party liability, seagoing, inland waterway, and river vessels)

    All liability arising out of the use of ships, vessels, or boats on the sea, lakes, rivers, or canals (including carrier's liability)

13. *General liability* (third-party liability, general)

    All liability other than those forms mentioned under numbers 10 to 12

14. *Credit*

    - Insolvency (general)
    - Export credit
    - Installment credit
    - Mortgages

- Agricultural credit
15. *Suretyship* (guarantees)
    - Suretyship (direct)
    - Suretyship (indirect)
16. *Miscellaneous financial loss*
    - Employment risks
    - Insufficiency of income (general)
    - Bad weather
    - Loss of benefits
    - Continuing general expenses
    - Unforeseen trading expenses
    - Loss of market value
    - Loss of rent or revenue
    - Indirect trading losses other than those mentioned above
    - Other financial loss (nontrading)
    - Other forms of financial loss
17. *Legal expenses*

    Legal expenses and costs of litigation
18. *Assistance to tourists*

    Assistance to individuals with difficulties on holiday or during an absence from their place of permanent or habitual residence

The risks included in a class may not be included in any other class except in the cases referred to in point C.

B.   Description of Authorizations Granted for More Than One Class of Insurance Where the Authorization Simultaneously Covers:

(a) Classes 1 and 2; it shall be named "Accident and Health Insurance"
(b) Classes 1 (fourth indent), 3, 7, and 10; it shall be named "Motor Insurance"
(c) Classes 1 (fourth indent), 4, 6, 7, and 12; it shall be named "Marine and Transport Insurance"
(d) Classes 1 (fourth indent), 5, 7, and 11; it shall be named "Aviation Insurance"
(e) Classes 8 and 9; it shall be named "Insurance against Fire and Other Damage to Property"
(f) Classes 10 to 13; it shall be named "Liability Insurance"

(g) Classes 14 and 15; it shall be named "Credit and Suretyship Insurance"

(h) All classes; it shall be named at the choice of the member state in question, which shall notifiy the other member states and the commission of its choice

C.   Ancillary Risks

An undertaking obtaining an authorization for a principal risk belonging to one class or a group of classes may also insure risks included in another class without an authorization being necessary for them if they:

- Are connected with the principal risk
- Concern the object that is covered against the principal risk
- Are covered by the contract insuring the principal risk

However, the risks included in classes 14 and 15 in point A may not be regarded as risks ancillary to other classes.

Life Classes

The *insurance risk classes*, as defined in the annex to the *first life directive* (EEC, 1979), are given below. The annex refers to Article 1 of the first life directive. We start with this article.

Article 1

This directive concerns the taking up and pursuit of the self-employed activity of direct insurance carried on by undertakings which are established in a member state or wish to become established there in the form of the activities defined below:

1. The following kinds of insurance where they are on a contractual basis:
    (a) life assurance, that is to say, the class of insurance which comprises, in particular, assurance on survival to a stipulated age only, assurance on death only, assurance on survival to a stipulated age or on earlier death, life assurance with return of premiums, marriage assurance, birth assurance;
    (b) annuities;

(c) supplementary insurance carried on by life assurance undertakings, that is to say, in particular, insurance against personal injury including incapacity for employment, insurance against death resulting from an accident and insurance against disability resulting from an accident or sickness, where these various kinds of insurance are underwritten in addition to life assurance;

(d) the type of insurance existing in Ireland and the United Kingdom known as permanent health insurance not subject to cancellation.

2. The following operations, where they are on a contractual basis, in so far as they are subject to supervision by the administrative authorities responsible for the supervision of private insurance and are authorized in the country concerned:

(a) tontines whereby associations of subscribers are set up with a view to jointly capitalizing their contributions and subsequently distributing the assets thus accumulated among the survivors or among the beneficiaries of the deceased;

(b) capital redemption operations based on actuarial calculation whereby, in return for single or periodic payments agreed in advance, commitments of specified duration and amount are undertaken;

(c) management of group pension funds, i.e. operations consisting, for the undertaking concerned, in managing the investments, and in particular the assets representing the reserves of bodies that effect payments on death or survival or in the event of discontinuance or curtailment of activity;

(d) the operations referred to in (c) where they are accompanied by insurance covering either conservation of capital or payment of a minimum interest;

(e) the operations carried out by insurance companies such as those referred to in chapter 1, title 4 of book iv of the French "code des assurances."

3. Operations relating to the length of human life, which are prescribed by or provided for in social insurance legislation, when they are affected or managed at their own risk by assurance undertakings in accordance with the laws of a member state.

Classes of Insurance

i. The insurance referred to in Article 1(1)(a), (b), and (c), excluding those referred to in ii and iii

ii. Marriage insurance, birth insurance

iii. The insurance referred to in Article 1(1)(a) and (b), which are linked to investment funds

iv. Permanent health insurance, referred to in Article 1(1)(d)

v. Tontines, referred to in Article 1(2)(a)

vi. Capital redemption operations, referred to in Article 1(2)(b)

vii. Management of group pension funds, referred to in Article 1(2)(c) and (d)

viii. The operations referred to in Article 1(2)(e)

ix. The operations referred to in Article 1(3)

Appendix C

From the Non-Life Directives

Solvency 0

Articles from the First Non-Life Directive (EEC, 1973)

Article 16

1. Each Member State shall require every undertaking whose head office is situated in its territory to establish an adequate solvency margin in respect of its entire business.

   The solvency margin shall correspond to the assets of the undertaking, free of all foreseeable liabilities, less any intangible items. In particular the following shall be considered:

   - the paid up share capital or, in the case of a mutual concern, the effective initial fund,

   - one-half of the share capital or the initial fund which is not yet paid up, once the paid-up part reaches 25% of this capital or fund,

   - reserves (statutory reserves and free reserves) not corresponding to underwriting liabilities,

   - any carry-forward of profits,

   - in the case of a mutual or mutual-type association with variable contributions, any claim which it has against its members by way of a call for supplementary contribution, within the financial year, up to one-half of the difference between the maximum contributions and the contributions actually called in, and subject to an over-riding limit of 50% of the margin,

   - at the request of, and upon proof being shown by the undertaking, and with the agreement of the supervisory authorities of each other Member State where it carries on its business, any hidden reserves resulting from under-estimation of assets or over-estimation of liabilities in the balance sheet, in so far as such hidden reserves are not of an exceptional nature.

Over-estimation of technical reserves shall be determined in relation to their amount calculated by the undertaking in conformity to national regulations; however, pending further coordination of technical reserves, an amount equivalent to 75% of the difference between the amount of the reserve for outstanding risks calculated at a flat rate by the undertaking by application of a minimum percentage in relation to premiums and the amount that would have been obtained by calculating the reserve contract by contract where the national law gives an option between the two methods, can be taken into account in the solvency margin up to 20%.

2. The solvency margin shall be determined on the basis either of the annual amount of premiums or contributions, or of the average burden of claims for the past three financial years. In the case, however, of undertakings, which essentially underwrite only one or more of the risks of storm, hail, frost, the last seven years shall be taken as the period of reference for the average burden of claims.

3. Subject to the provisions of Article 17, the amount of the solvency margin shall be equal to the higher of the following two results:

First result (premium basis):

• the premiums or contributions (inclusive of charges ancillary to premiums or contributions) due in respect of all direct business in the last financial year for all financial years, shall be aggregated,

• to this aggregate there shall be added the amount of premiums accepted for all reinsurance in the last financial year,

• from this sum there shall then be deducted the total amount of premiums or contributions cancelled in the last financial year, as well as the total amount of taxes and levies pertaining to the premiums or contributions entering into the aggregate.

The amount so obtained shall be divided into two portions, the first portion extending up to 10 million units of account, the second comprising the excess; 18% and 16% of these portions respectively shall be calculated and added together.

The first result shall be obtained by multiplying the sum so calculated by the ratio existing in respect of the last financial year between the amount of claims remaining to be borne by the undertaking after deduction of transfers for reinsurance and the gross amount of claims; this ratio may in no case be less than 50%.

Second result (claims basis):

• the amounts of claims paid in respect of direct business (without any deduction of claims borne by reinsurers and retrocessionaires) in the periods specified in (2) shall be aggregated,

- to this aggregate there shall be added the amount of claims paid in respect of reinsurances or retrocessions accepted during the same periods,
- to this sum there shall be added the amount of provisions or reserves for outstanding claims established at the end of the last financial year both for direct business and for reinsurance acceptances,
- from this sum there shall be deducted the amount of claims paid during the periods specified in (2),
- from the sum then remaining, there shall be deducted the amount of provisions or reserves for outstanding claims established at the commencement of the second financial year preceding the last financial year for which there are accounts, both for direct business and for reinsurance acceptances.

One-third, or one-seventh, of the amount so obtained, according to the period of reference established in (2), shall be divided into two portions, the first extending up to seven million units of account and the second comprising the excess; 26% and 23% of these portions respectively shall be calculated and added together.

The second result shall be obtained by multiplying the sum so obtained by the ratio existing in respect of the last financial year between the amount of claims remaining to be borne by the business after transfers for reinsurance and the gross amount of claims; this ratio may in no case be less than 50%.

4. The fractions applicable to the portions referred to in (3) shall each be reduced to a third in the case of health insurance practised on a similar technical basis to that of life assurance, if

- the premiums paid are calculated on the basis of sickness tables according to the mathematical method applied in insurance,
- a reserve is set up for increasing age,
- an additional premium is collected in order to set up a safety margin of an appropriate amount,
- the insurer may only cancel the contract before the end of the third year of insurance at the latest,
- the contract provides for the possibility of increasing premiums or reducing payments even for current contracts.

Article 17

1. One-third of the solvency margin shall constitute the guarantee fund.
2. (a) The guarantee fund may not, however, be less than:

- 400 000 units of account in the case where all or some of the risks included in one of the classes listed in point A of the Annex under Nos 10, 11, 12, 13, 14 and 15 are covered,
- 300 000 units of account in the case where all or some of the risks included in one of the classes listed in point A of the Annex under Nos 1, 2, 3, 4, 5, 6, 7, 8, and 16 are covered,
- 200 000 units of account in the case where all or some of the risks included in one of the classes listed in point A of the Annex under Nos 9 and 17 are covered;

(b) If the business carried on by the undertaking covers several classes or several risks, only that class or risk for which the highest amount is required shall be taken into account;

(c) Any Member State may provide for a one-fourth reduction of the minimum guarantee fund in the case of mutual associations and mutual-type associations.

Articles from the Third Non-Life Directive (EEC, 1992a)

Article 24

Article 16(1) of Directive 73/239/EEC shall be replaced by the following:

1. The home Member State shall require every insurance undertaking to establish an adequate solvency margin in respect of its entire business.

   The solvency margin shall correspond to the assets of the undertaking free of any foreseeable liabilities less any intangible items. In particular the following shall be included:

   - the paid-up share capital or, in the case of a mutual insurance undertaking, the effective initial fund plus any members' accounts which meet all the following criteria:

     (a) the memorandum and articles of association must stipulate that payments may be made from these accounts to members only insofar as this does not cause the solvency margin to fall below the required level, or, after the dissolution of the undertaking, if all the undertaking's other debts have been settled;

     (b) the memorandum and articles of association must stipulate, with respect to any such payments for reasons other than the individual termination of membership, that the competent authorities must be notified at least one month in advance and can prohibit the payment within that period and

     (c) the relevant provisions of the memorandum and articles of association may be amended only after the competent author-

ities have declared that they have no objection to the amendment, without prejudice to the criteria stated in (a) and (b);

- one-half of the unpaid share capital or initial fund, once the paid-up part amounts to 25% of that share capital or fund,
- reserves (statutory reserves and free reserves) not corresponding to underwriting liabilities,
- any profits brought forward,
- in the case of mutual or mutal-type association with variable contributions, any claim which it has against its members by way of a call for supplementary contribution, within the financial year, up to one-half of the difference between the maximum contributions and the contributions actually called in, and subject to a limit of 50% of the margin,
- at the request of and on the production of proof by the insurance undertaking, any hidden reserves arising out of the undervaluation of assets, insofar as those hidden reserves are not of an exceptional nature,
- cumulative preferential share capital and subordinated loan capital may be included but, if so, only up to 50% of the margin, no more than 25% of which shall consist of subordinated loans with a fixed maturity, or fixed-term cumulative preferential share capital, if the following minimum criteria are met:

  (a) in the event of the bankruptcy or liquidation of the insurance undertaking, binding agreements must exist under which the subordinated loan capital or preferential share capital ranks after the claims of all other creditors and is not to be repaid until all other debts outstanding at the time have been settled.

Subordinated loan capital must fulfil the following additional conditions:

(b) only fully paid-up funds may be taken into account;

(c) for loans with a fixed maturity, the original maturity must be at least five years. No later than one year before the repayment date the insurance undertaking must submit to the competent authorities for their approval a plan showing how the solvency margin will be kept at or brought to the required level at maturity, unless the extent to which the loan may rank as a component of the solvency margin is gradually reduced during at least the last five years before the repayment date. The competent authorities may authorize the early repayment of such loans provided application is made by the issuing insurance undertaking and its solvency margin will not fall below the required level;

(d) loans the maturity of which is not fixed must be repayable only subject to five years' notice unless the loans are no longer considered a component of the solvency margin or unless the prior consent of the competent authorities is specifically required for early repayment. In the latter event the insurance undertaking must notify the competent authorities at least six months before the date of the proposed repayment, specifying the actual and required solvency margins both before and after that repayment. The competent authorities shall authorize repayment only if the insurance undertaking's solvency margin will not fall below the required level;

(e) the loan agreement must not include any clause providing that in specified circumstances, other than the winding-up of the insurance undertaking, the debt will become repayable before the agreed repayment dates;

(f) the loan agreement may be amended only after the competent authorities have declared that they have no objection to the amendment;

- securities with no specified maturity date and other instruments that fulfil the following conditions, including cumulative preferential shares other than those mentioned in the preceding indent, up to 50% of the margin for the total of such securities and the subordinated loan capital referred to in the preceding indent:

(a) they may not be repaid on the initiative of the bearer or without the prior consent of the competent authority;

(b) the contract of issue must enable the insurance undertaking to defer the payment of interest on the loan;

(c) the lender's claims on the insurance undertaking must rank entirely after those of all non-subordinated creditors;

(d) the documents governing the issue of the securities must provide for the loss-absorption capacity of the debt and unpaid interest, while enabling the insurance undertaking to continue its business;

(e) only fully paid-up amounts may be taken into account.

Solvency I

New Article 16

Article 16 shall be replaced by the following:

1. Each Member State shall require of every insurance undertaking whose head office is situated in its territory an adequate available solvency margin in respect of its entire business at all times, which is at least equal to the requirements in this Directive.

2. The available solvency margin shall consist of the assets of the insurance undertaking free of any foreseeable liabilities, less any intangible items, including:

   (a) the paid-up share capital or, in the case of a mutual insurance undertaking, the effective initial fund plus any members' accounts which meet all the following criteria:

       (i)   the memorandum and articles of association must stipulate that payments may be made from these accounts to members only in so far as this does not cause the available solvency margin to fall below the required level, or, after the dissolution of the undertaking, if all the undertaking's other debts have been settled;

       (ii) the memorandum and articles of association must stipulate, with respect to any payments referred to in point (i) for reasons other than the individual termination of membership, that the competent authorities must be notified at least one month in advance and can prohibit the payment within that period;

       (iii) the relevant provisions of the memorandum and articles of association may be amended only after the competent authorities have declared that they have no objection to the amendment, without prejudice to the criteria stated in points (i) and (ii);

   (b) reserves (statutory and free) not corresponding to underwriting liabilities;

   (c) the profit or loss brought forward after deduction of dividends to be paid.

   The available solvency margin shall be reduced by the amount of own shares directly held by the insurance undertaking.

   For those insurance undertakings which discount or reduce their technical provisions for claims outstanding to take account of investment income as permitted by Article 60(1)(g) of Council Directive 91/674/EEC of 19 December 1991 on the annual accounts and consolidated accounts of insurance undertakings (7), the available solvency margin shall be reduced by the difference between the undiscounted technical provisions or technical provisions before deductions as disclosed in the notes on the accounts, and the discounted or technical provisions after deductions. This adjustment shall be made for all risks listed in point A of the Annex, except for risks listed under classes 1 and 2. For classes other than 1 and 2, no adjustment need be made in respect of the discounting of annuities included in technical provisions.

3. The available solvency margin may also consist of:

(a) cumulative preferential share capital and subordinated loan capital up to 50% of the lesser of the available solvency margin and the required solvency margin, no more than 25% of which shall consist of subordinated loans with a fixed maturity, or fixed-term cumulative preferential share capital, provided in the event of the bankruptcy or liquidation of the insurance undertaking, binding agreements exist under which the subordinated loan capital or preferential share capital ranks after the claims of all other creditors and is not to be repaid until all other debts outstanding at the time have been settled.

Subordinated loan capital must also fulfil the following conditions:

(i) only fully paid-up funds may be taken into account;

(ii) for loans with a fixed maturity, the original maturity must be at least five years. No later than one year before the repayment date the insurance undertaking must submit to the competent authorities for their approval a plan showing how the available solvency margin will be kept at or brought to the required level at maturity, unless the extent to which the loan may rank as a component of the available solvency margin is gradually reduced during at least the last five years before the repayment date. The competent authorities may authorise the early repayment of such loans provided application is made by the issuing insurance undertaking and its available solvency margin will not fall below the required level;

(iii) loans the maturity of which is not fixed must be repayable only subject to five years' notice unless the loans are no longer considered as a component of the available solvency margin or unless the prior consent of the competent authorities is specifically required for early repayment. In the latter event the insurance undertaking must notify the competent authorities at least six months before the date of the proposed repayment, specifying the available solvency margin and the required solvency margin both before and after that repayment. The competent authorities shall authorise repayment only if the insurance undertaking's available solvency margin will not fall below the required level;

(iv) the loan agreement must not include any clause providing that in specified circumstances, other than the winding-up of the insurance undertaking, the debt will become repayable before the agreed repayment dates;

(v) the loan agreement may be amended only after the competent authorities have declared that they have no objection to the amendment;

(b) securities with no specified maturity date and other instruments, including cumulative preferential shares other than those mentioned in point (a), up to 50% of the lesser of the available solvency margin and the required solvency margin for the total of such securities and the subordinated loan capital referred to in point (a) provided they fulfil the following:

(i)   they may not be repaid on the initiative of the bearer or without the prior consent of the competent authority;

(ii) the contract of issue must enable the insurance undertaking to defer the payment of interest on the loan;

(iii) the lender's claims on the insurance undertaking must rank entirely after those of all non-subordinated creditors;

(iv) the documents governing the issue of the securities must provide for the loss-absorption capacity of the debt and unpaid interest, while enabling the insurance undertaking to continue its business;

(v) only fully paid-up amounts may be taken into account.

4. Upon application, with supporting evidence, by the undertaking to the competent authority of the home Member State and with the agreement of that competent authority, the available solvency margin may also consist of:

(a) one-half of the unpaid share capital or initial fund, once the paid-up part amounts to 25% of that share capital or fund, up to 50% of the lesser of the available solvency margin and the required solvency margin;

(b) in the case of mutual or mutual-type association with variable contributions, any claim which it has against its members by way of a call for supplementary contribution, within the financial year, up to one-half of the difference between the maximum contributions and the contributions actually called in, and subject to a limit of 50% of the lesser of the available solvency margin and the required solvency margin. The competent national authorities shall establish guidelines laying down the conditions under which supplementary contributions may be accepted;

(c) any hidden net reserves arising out of the valuation of assets, in so far as such hidden net reserves are not of an exceptional nature.

5. Amendments to paragraphs 2, 3 and 4 to take into account developments that justify a technical adjustment of the elements eligible for

the available solvency margin, shall be adopted in accordance with the procedure laid down in Article 2 of Council Directive 91/675/EEC(8).

New Article 16a

The following article shall be inserted:

1. The required solvency margin shall be determined on the basis either of the annual amount of premiums or contributions, or of the average burden of claims for the past three financial years.

   In the case, however, of insurance undertakings, which essentially underwrite only one or more of the risks of credit, storm, hail or frost, the last seven financial years shall be taken as the reference period for the average burden of claims.

2. Subject to Article 17, the amount of the required solvency margin shall be equal to the higher of the two results as set out in paragraphs 3 and 4.

3. The premium basis shall be calculated using the higher of gross written premiums or contributions as calculated below, and gross earned premiums or contributions.

   Premiums or contributions in respect of the classes 11, 12 and 13 listed in point A of the Annex shall be increased by 50%.

   The premiums or contributions (inclusive of charges ancillary to premiums or contributions) due in respect of direct business in the last financial year shall be aggregated.

   To this sum there shall be added the amount of premiums accepted for all reinsurance in the last financial year.

   From this sum there shall then be deducted the total amount of premiums or contributions cancelled in the last financial year, as well as the total amount of taxes and levies pertaining to the premiums or contributions entering into the aggregate.

   The amount so obtained shall be divided into two portions, the first portion extending up to EUR 50 million, the second comprising the excess; 18% and 16% of these portions respectively shall be calculated and added together.

   The sum so obtained shall be multiplied by the ratio existing in respect of the sum of the last three financial years between the amount of claims remaining to be borne by the undertaking after deduction of amounts recoverable under reinsurance and the gross amount of claims; this ratio may in no case be less than 50%.

   With the approval of the competent authorities, statistical methods may be used to allocate the premiums or contributions in respect of the classes 11, 12 and 13.

4. The claims basis shall be calculated, as follows, using in respect of the classes 11, 12 and 13 listed in point A of the Annex, claims, provisions and recoveries increased by 50%.

   The amounts of claims paid in respect of direct business (without any deduction of claims borne by reinsurers and retrocessionaires) in the periods specified in paragraph 1 shall be aggregated.

   To this sum there shall be added the amount of claims paid in respect of reinsurances or retrocessions accepted during the same periods and the amount of provisions for claims outstanding established at the end of the last financial year both for direct business and for reinsurance acceptances.

   From this sum there shall be deducted the amount of recoveries effected during the periods specified in paragraph 1.

   From the sum then remaining, there shall be deducted the amount of provisions for claims outstanding established at the commencement of the second financial year preceding the last financial year for which there are accounts, both for direct business and for reinsurance acceptances. If the period of reference established in paragraph 1 equals seven years, the amount of provisions for claims outstanding established at the commencement of the sixth financial year preceding the last financial year for which there are accounts shall be deducted.

   One-third, or one-seventh, of the amount so obtained, according to the period of reference established in paragraph 1, shall be divided into two portions, the first extending up to EUR 35 million and the second comprising the excess; 26% and 23% of these portions respectively shall be calculated and added together.

   The sum so obtained shall be multiplied by the ratio existing in respect of the sum of the last three financial years between the amount of claims remaining to be borne by the undertaking after deduction of amounts recoverable under reinsurance and the gross amount of claims; this ratio may in no case be less than 50%.

   With the approval of the competent authorities, statistical methods may be used to allocate the claims, provisions and recoveries in respect of the classes 11, 12 and 13. In the case of the risks listed under class 18 in point A of the Annex, the amount of claims paid used to calculate the claims basis shall be the costs borne by the insurance undertaking in respect of assistance given. Such costs shall be calculated in accordance with the national provisions of the home Member State.

5. If the required solvency margin as calculated in paragraphs 2, 3 and 4 is lower than the required solvency margin of the year before, the required solvency margin shall be at least equal to the required solvency margin of the year before multiplied by the ratio of the

amount of the technical provisions for claims outstanding at the end of the last financial year and the amount of the technical provisions for claims outstanding at the beginning of the last financial year. In these calculations technical provisions shall be calculated net of reinsurance but the ratio may in no case be higher than 1.

6. The fractions applicable to the portions referred to in the sixth sub-paragraph of paragraph 3 and the sixth subparagraph of paragraph 4 shall each be reduced to a third in the case of health insurance practised on a similar technical basis to that of life assurance, if

   (a) the premiums paid are calculated on the basis of sickness tables according to the mathematical method applied in insurance;

   (b) a provision is set up for increasing age;

   (c) an additional premium is collected in order to set up a safety margin of an appropriate amount;

   (d) the insurance undertaking may cancel the contract before the end of the third year of insurance at the latest;

   (e) the contract provides for the possibility of increasing premiums or reducing payments even for current contracts.

New Article 17

Article 17 shall be replaced by the following:

1. One-third of the required solvency margin as specified in Article 16a shall constitute the guarantee fund. This fund shall consist of the items listed in Article 16(2), (3) and, with the agreement of the competent authority of the home Member State, (4)(c).

2. The guarantee fund may not be less than EUR 2 million. Where, however, all or some of the risks included in one of the classes 10 to 15 listed in point A of the Annex are covered, it shall be EUR 3 million.

   Any Member State may provide for a one-fourth reduction of the minimum guarantee fund in the case of mutual associations and mutual-type associations.

New Article 17a

The following article shall be inserted:

1. The amounts in euro as laid down in Article 16a(3) and (4) and Article 17(2) shall be reviewed annually starting 20 September 2003 in order to take account of changes in the European index of consumer prices comprising all Member States as published by Eurostat.

The amounts shall be adapted automatically by increasing the base amount in euro by the percentage change in that index over the period between the entry into force of this Directive and the review date and rounded up to a multiple of EUR 100 000.

If the percentage of change since the last adaptation is less than 5%, no adaptation shall take place.

2. The Commission shall inform annually the European Parliament and the Council of the review and the adapted amounts referred to in paragraph 1.

Appendix D

From the Life Directives

Solvency 0

Articles from the First Life Directive (EEC, 1979)

Article 18

Each Member State shall require of every undertaking whose head office is situated in its territory an adequate solvency margin in respect of its entire business.

The solvency margin shall consist of:

1. the assets of the undertaking, free of all foreseeable liabilities, less any intangible items; in particular the following shall be included:
   - the paid-up share capital or, in the case of a mutual concern, the paid-up amount of its fund,
   - one-half of the unpaid-up share capital or fund once 25% of such capital or fund are paid up,
   - statutory reserves and free reserves not corresponding to under-writing liabilities,
   - any carry-forward of profits;
2. in so far as authorized under national law, profit reserves appearing in the balance sheet where they may be used to cover any losses which may arise and where they have not been made available for distribution to policy-holders;
3. upon application, with supporting evidence, by the undertaking to the supervisory authority of the Member State in the territory of which its head office is situated and with the agreement of that authority:
   (a) an amount equal to 50% of the undertaking's future profits; the amount of the future profits shall be obtained by multiplying the estimated annual profit by a factor which represents the average period left to run on policies; the factor used may not exceed 10;

the estimated annual profit shall be the arithmetical average of the profits made over the last five years in the activities listed in Article 1.

The bases for calculating the factor by which the estimated annual profit is to be multiplied and the items comprising the profits made shall be defined by common agreement by the competent authorities of the Member States in collaboration with the commission. Pending such agreement, those items shall be determined in accordance with the laws of the Member State in the territory of which the undertaking (head office, agency or branch) carries on its activities.

When the competent authorities have defined the concept of profits made, the commission shall submit proposals for the harmonization of this concept by means of a directive on the harmonization of the annual accounts of insurance undertakings and providing for the coordination set out in Article 1(2) of directive 78/660/eec (7);

(b) where Zillmerizing is not practised or where, if practised, it is less than the loading for acquisition costs included in the premium, the difference between a non-Zillmerized or partially Zillmerized mathematical reserve and a mathematical reserve Zillmerized at a rate equal to the loading for acquisition costs included in the premium; this figure may not, however, exceed 3,5% of the sum of the differences between the relevant capital sums of life assurance activities and the mathematical reserves for all policies for which Zillmerizing is possible; the difference shall be reduced by the amount of any undepreciated acquisition costs entered as an asset;

(c) where approval is given by the supervisory authorities of the Member States concerned in which the undertaking is carrying on its activities any hidden reserves resulting from the underestimation of assets and over-estimation of liabilities other than mathematical reserves in so far as such hidden reserves are not of an exceptional nature.

Article 19

Subject to Article 20, the minimum solvency margin shall be determined as shown below according to the classes of insurance underwritten:

(a) For the kinds of insurance referred to in Article 1(1)(a) and (b) other than assurances linked to investment funds and for the operations referred to in Article 1(3), it must be equal to the sum of the following two results:

- *first result*:
  - a 4% fraction of the mathematical reserves, relating to direct business gross of re-insurance cessions and to re-insurance acceptances shall be multiplied by the ratio, for the last financial year, of the total mathematical reserves net of re-insurance cessions to the gross total mathematical reserves as specified above; that ratio may in no case be less than 85%;
- *second result*:
  - for policies on which the capital at risk is not a negative figure, a 0,3% fraction of such capital underwritten by the undertaking shall be multiplied by the ratio, for the last financial year, of the total capital at risk retained as the undertaking's liability after re-insurance cessions and retrocessions to the total capital at risk gross of re-insurance; that ratio may in no case be less than 50%.
  - For temporary assurance on death of a maximum term of three years the above fraction shall be 0,1%; for such assurance of a term of more than three years but not more than five years the above fraction shall be 0,15%.

(b) For the supplementary insurance referred to in Article 1(1)(c), it shall be equal to the result of the following calculation:
  - the premiums or contributions (inclusive of charges ancillary to premiums or contributions) due in respect of direct business in the last financial year in respect of all financial years shall be aggregated;
  - to this aggregate there shall be added the amount of premiums accepted for all reinsurance in the last financial year;
  - from this sum shall then be deducted the total amount of premiums or contributions cancelled in the last financial year as well as the total amount of taxes and levies pertaining to the premiums or contributions entering into the aggregate.

The amount so obtained shall be divided into two portions, the first extending up to 10 million units of account and the second comprising the excess; 18% and 16% of these portions respectively shall be calculated and added together.

The result shall be obtained by multiplying the sum so calculated by the ratio existing in respect of the last financial year between the amount of claims remaining to be borne by the undertaking after deduction of transfers for reinsurance and the gross amount of claims; this ratio may in no case be less than 50%.

In the case of the association of underwriters known as Lloyd's, the calculation of the solvency margin shall be made on the basis of net premiums, which shall be multiplied by flat-rate percentage fixed annually by the supervisory authority of the head-office Member State. This flat-rate percentage must be calculated on the basis of the most recent statistical data on commissions paid. The details together with the relevant calculations shall be sent to the supervisory authorities of the countries in whose territory Lloyd's is established.

(c) For permanent health insurance not subject to cancellation referred to in Article 1(1)(d), and for capital redemption operations referred to in Article 1(2)(b), it shall be equal to a 4% fraction of the mathematical reserves calculated in compliance with the conditions set out in the first result in (a) of this article.

(d) For tontines, referred to in Article 1(2)(a), it shall be equal to 1% of their assets.

(e) For assurances covered by Article 1(1)(a) and (b) linked to investment funds and for the operations referred to in Article 1(2)(c), (d) and (e) it shall be equal to:

- a 4% fraction of the mathematical reserves, calculated in compliance with the conditions set out in the first result in (a) of this article in so far as the undertaking bears an investment risk, and a 1% fraction of the reserves calculated in the fashion, in so far as the undertaking bears no investment risk provided that the term of the contract exceeds five years and the allocation to cover management expenses set out in the contract is fixed for a period exceeding five years, plus

- a 0,3% fraction of the capital at risk calculated in compliance with the conditions set out in the first subparagraph of the second result of (a) of this article in so far as the undertaking covers a death risk.

Article 20

1. One-third of the minimum solvency margin as specified in Article 19 shall constitute the guarantee fund. Subject to paragraph 2, at least 50% of this fund shall consist of the items listed in Article 18(1) and (2).

2. (a) The guarantee fund may not, however, be less than a minimum of 800 000 units of account.

   (b) Any Member State may provide for the minimum of the guarantee fund to be reduced to 600 000 units of account in the case of mutual associations and mutual-type associations and tontines.

   (c) For mutual associations referred to in the second sentence of the second indent of Article 3(2), as soon as they come within the scope of this directive, and for tontines, any Member State may permit the establishment of a minimum of the guarantee fund of 100 000 units of account to be increased progressively to the amount fixed in (b) by successive tranches of 100 000 units of account whenever the contributions increase by 500 000 units of account.

(d) The minimum of the guarantee fund referred to in (a), (b) and (c) must consist of the items listed in Article 18(1) and (2).

3. Mutual associations wishing to extend their business within the meaning of Article 8(2) or Article 10 may not do so unless they comply immediately with the requirements of paragraph 2(a) and (b) of this article.

Articles from the Third Life Directive (EEC, 1992b)

Article 25

Article 18, second subparagraph, point 1 of Directive 79/267/EEC shall be replaced by the following:

1. the assets of the undertaking free of any foreseeable liabilities, less any intangible items. In particular the following shall be included:

   • the paid-up share capital or, in the case of a mutual assurance undertaking, the effective initial fund plus any members' accounts which meet all the following criteria:

     (a) the memorandum and articles of association must stipulate that payments may be made from these accounts to members only in so far as this does not cause the solvency margin to fall below the required level, or, after the dissolution of the undertaking, if all the undertaking's other debts have been settled;

     (b) the memorandum and articles of association must stipulate, with respect to any such payments for reasons other than the individual termination of membership, that the competent authorities must be notified at least one month in advance and can prohibit the payment within that period;

     (c) the relevant provisions of the memorandum and articles of association may be amended only after the competent authorities have declared that they have no objection to the amendment, without prejudice to the criteria stated in (a) and (b),

   • one-half of the unpaid share capital or initial fund, once the paid-up part amounts to 25% of that share capital or fund,

   • reserves (statutory reserves and free reserves) not corresponding to underwriting liabilities,

- any profits brought forward,
- cumulative preferential share capital and subordinated loan capital may be included but, if so, only up to 50% of the margin, no more than 25% of which shall consist of subordinated loans with a fixed maturity, or fixed-term cumulative preferential share capital, if the following minimum criteria are met:
  - (a) in the event of the bankruptcy or liquidation of the assurance undertaking, binding agreements must exist under which the subordinated loan capital or preferential share capital ranks after the claims of all other creditors and is not to be repaid until all other debts outstanding at the time have been settled.
- Subordinated loan capital must also fulfil the following conditions:
  - (b) only fully paid-up funds may be taken into account;
  - (c) for loans with a fixed maturity, the original maturity must be at least five years. No later than one year before the repayment date the assurance undertaking must submit to the competent authorities for their approval a plan showing how the solvency margin will be kept at or brought to the required level at maturity, unless the extent to which the loan may rank as a component of the solvency margin is gradually reduced during at least the last five years before the repayment date. The competent authorities may authorize the early repayment of such loans provided application is made by the issuing assurance undertaking and its solvency margin will not fall below the required level;
  - (d) loans the maturity of which is not fixed must be repayable only subject to five years' notice unless the loans are no longer considered as a component of the solvency margin or unless the prior consent of the competent authorities is specifically required for early repayment. In the latter event the assurance undertaking must notify the competent authorities at least six months before the date of the proposed repayment, specifying the actual and required solvency margin both before and after that repayment. The competent authorities shall authorize repayment only if the assurance undertaking's solvency margin will not fall below the required level;
  - (e) the loan agreement must not include any clause providing that in specified circumstances, other than the winding-up of the assurance undertaking, the debt will become repayable before the agreed repayment dates;
  - (f) the loan agreement may be amended only after the competent authorities have declared that they have no objection to the amendment,

- securities with no specified maturity date and other instruments that fulfil the following conditions, including cumulative preferential shares other than those mentioned in the preceding indent, up to 50% of the margin for the total of such securities and the subordinated loan capital referred to in the preceding indent:

  (a) they may not be repaid on the initiative of the bearer or without the prior consent of the competent authority;

  (b) the contract of issue must enable the assurance undertaking to defer the payment of interest on the loan;

  (c) the lender's claims on the assurance undertaking must rank entirely after those of all non-subordinated creditors;

  (d) the documents governing the issue of the securities must provide for the loss-absorption capacity of the debt and unpaid interest, while enabling the assurance undertaking to continue its business;

  (e) only fully paid-up amounts may be taken into account.

Solvency I

Structure

Article 27: Available Solvency Margin

1. Each Member State shall require of every assurance undertaking whose head office is situated in its territory an adequate available solvency margin in respect of its entire business at all times which is at least equal to the requirements in this Directive.

2. The available solvency margin shall consist of the assets of the assurance undertaking free of any foreseeable liabilities, less any intangible items, including:

   (a) the paid-up share capital or, in the case of a mutual assurance undertaking, the effective initial fund plus any members' accounts which meet all the following criteria:

      (i) the memorandum and articles of association must stipulate that payments may be made from these accounts to members only in so far as this does not cause the available solvency margin to fall below the required level, or, after the dissolution of the undertaking, if all the undertaking's other debts have been settled;

(ii) the memorandum and articles of association must stipulate, with respect to any payments referred to in point (i) for reasons other than the individual termination of membership, that the competent authorities must be notified at least one month in advance and can prohibit the payment within that period;

(iii) the relevant provisions of the memorandum and articles of association may be amended only after the competent authorities have declared that they have no objection to the amendment, without prejudice to the criteria stated in points (i) and (ii);

(b) reserves (statutory and free) not corresponding to underwriting liabilities;

(c) the profit or loss brought forward after deduction of dividends to be paid;

(d) in so far as authorised under national law, profit reserves appearing in the balance sheet where they may be used to cover any losses which may arise and where they have not been made available for distribution to policy holders.

The available solvency margin shall be reduced by the amount of own shares directly held by the assurance undertaking.

3. The available solvency margin may also consist of:

(a) cumulative preferential share capital and subordinated loan capital up to 50% of the lesser of the available solvency margin and the required solvency margin, no more than 25% of which shall consist of subordinated loans with a fixed maturity, or fixed-term cumulative preferential share capital, provided that binding agreements exist under which, in the event of the bankruptcy or liquidation of the assurance undertaking, the subordinated loan capital or preferential share capital ranks after the claims of all other creditors and is not to be repaid until all other debts outstanding at the time have been settled.

Subordinated loan capital must also fulfil the following conditions:

(i) only fully paid-up funds may be taken into account;

(ii) for loans with a fixed maturity, the original maturity must be at least five years. No later than one year before the repayment date, the assurance undertaking must submit to the competent authorities for their approval a plan showing how the available solvency margin will be kept at or brought to the required level at maturity, unless the extent to which the loan may rank as a component of the available solvency margin is gradually reduced during at least the last five years before the repayment date. The competent authorities may authorise the early repayment of such loans provided application is made by the issuing assurance undertaking and its available solvency margin will not fall below the required level;

(iii) loans the maturity of which is not fixed must be repayable only subject to five years' notice unless the loans are no longer considered as a component of the available solvency margin or unless the prior consent of the competent authorities is specifically required for early repayment. In the latter event the assurance undertaking must notify the competent authorities at least six months before the date of the proposed repayment, specifying the available solvency margin and the required solvency margin both before and after that repayment. The competent authorities shall authorise repayment only if the assurance undertaking's available solvency margin will not fall below the required level;

(iv) the loan agreement must not include any clause providing that in specified circumstances, other than the winding-up of the assurance undertaking, the debt will become repayable before the agreed repayment dates;

(v) the loan agreement may be amended only after the competent authorities have declared that they have no objection to the amendment;

(b) securities with no specified maturity date and other instruments, including cumulative preferential shares other than those mentioned in point (a), up to 50% of the lesser of the available solvency margin and the required solvency margin for the total of such securities and the subordinated loan capital referred to in point (a) provided they fulfil the following:

(i) they may not be repaid on the initiative of the bearer or without the prior consent of the competent authority;

(ii) the contract of issue must enable the assurance undertaking to defer the payment of interest on the loan;

(iii) the lender's claims on the assurance undertaking must rank entirely after those of all non-subordinated creditors;

(iv) the documents governing the issue of the securities must provide for the loss-absorption capacity of the debt and unpaid interest, while enabling the assurance undertaking to continue its business;

(v) only fully paid-up amounts may be taken into account.

4. Upon application, with supporting evidence, by the undertaking to the competent authority of the home Member State and with the

agreement of that competent authority, the available solvency margin may also consist of:

(a) until 31 December 2009 an amount equal to 50% of the undertaking's future profits, but not exceeding 25% of the lesser of the available solvency margin and the required solvency margin. The amount of the future profits shall be obtained by multiplying the estimated annual profit by a factor which represents the average period left to run on policies. The factor used may not exceed six. The estimated annual profit shall not exceed the arithmetical average of the profits made over the last five financial years in the activities listed in Article 2(1).

Competent authorities may only agree to include such an amount for the available solvency margin:

   (i) when an actuarial report is submitted to the competent authorities substantiating the likelihood of emergence of these profits in the future; and

   (ii) in so far as that part of future profits emerging from hidden net reserves referred to in point (c) has not already been taken into account;

(b) where Zillmerising is not practised or where, if practised, it is less than the loading for acquisition costs included in the premium, the difference between a non-Zillmerised or partially Zillmerised mathematical provision and a mathematical provision Zillmerised at a rate equal to the loading for acquisition costs included in the premium. This figure may not, however, exceed 3,5% of the sum of the differences between the relevant capital sums of life assurance activities and the mathematical provisions for all policies for which Zillmerising is possible. The difference shall be reduced by the amount of any undepreciated acquisition costs entered as an asset;

(c) any hidden net reserves arising out of the valuation of assets, in so far as such hidden net reserves are not of an exceptional nature;

(d) one-half of the unpaid share capital or initial fund, once the paid-up part amounts to 25% of that share capital or fund, up to 50% of the lesser of the available and required solvency margin.

5. Amendments to paragraphs 2, 3 and 4 to take into account developments that justify a technical adjustment of the elements eligible for the available solvency margin shall be adopted in accordance with the procedure laid down in Article 65(2).

Article 28: Required Solvency Margin

1. Subject to Article 29, the required solvency margin shall be determined as laid down in paragraphs 2 to 7 according to the classes of assurance underwritten.

2. For the kinds of assurance referred to in Article 2(1)(a) and (b) other than assurances linked to investment funds and for the operations referred to in Article 2(3), the required solvency margin shall be equal to the sum of the following two results:

   (a) first result:

   a 4% fraction of the mathematical provisions relating to direct business and reinsurance acceptances gross of reinsurance cessions shall be multiplied by the ratio, for the last financial year, of the total mathematical provisions net of reinsurance cessions to the gross total mathematical provisions. That ratio may in no case be less than 85%;

   (b) second result:

   for policies on which the capital at risk is not a negative figure, a 0,3% fraction of such capital underwritten by the assurance undertaking shall be multiplied by the ratio, for the last financial year, of the total capital at risk retained as the undertaking's liability after reinsurance cessions and retrocessions to the total capital at risk gross of reinsurance; that ratio may in no case be less than 50%.

   For temporary assurance on death of a maximum term of three years the fraction shall be 0,1%. For such assurance of a term of more than three years but not more than five years the above fraction shall be 0,15%.

3. For the supplementary insurance referred to in Article 2(1)(c) the required solvency margin shall be equal to the required solvency margin for insurance undertakings as laid down in Article 16a of Directive 73/239/EEC, excluding the provisions of Article 17 of that Directive.

4. For permanent health insurance not subject to cancellation referred to in Article 2(1)(d), the required solvency margin shall be equal to:

   (a) a 4% fraction of the mathematical provisions, calculated in compliance with paragraph 2(a) of this article; plus

   (b) the required solvency margin for insurance undertakings as laid down in Article 16a of Directive 73/239/EEC, excluding the provisions of Article 17 of that Directive. However, the condition contained in Article 16a(6)(b) of that Directive that a provision be set up for increasing age may be replaced by a requirement that the business be conducted on a group basis.

5. For capital redemption operations referred to in Article 2(2)(b), the required solvency margin shall be equal to a 4% fraction of the mathematical provisions calculated in compliance with paragraph 2(a) of this article.

6. For tontines, referred to in Article 2(2)(a), the required solvency margin shall be equal to 1% of their assets.

7. For assurances covered by Article 2(1)(a) and (b) linked to investment funds and for the operations referred to in Article 2(2)(c), (d) and (e), the required solvency margin shall be equal to the sum of the following:

   (a) in so far as the assurance undertaking bears an investment risk, a 4% fraction of the technical provisions, calculated in compliance with paragraph 2(a) of this article;

   (b) in so far as the undertaking bears no investment risk but the allocation to cover management expenses is fixed for a period exceeding five years, a 1% fraction of the technical provisions, calculated in compliance with paragraph 2(a) of this article;

   (c) in so far as the undertaking bears no investment risk and the allocation to cover management expenses is not fixed for a period exceeding five years, an amount equivalent to 25% of the last financial year's net administrative expenses pertaining to such business;

   (d) in so far as the assurance undertaking covers a death risk, a 0,3% fraction of the capital at risk calculated in compliance with paragraph 2(b) of this article.

Article 29: Guarantee Fund

1. One-third of the required solvency margin as specified in Article 28 shall constitute the guarantee fund. This fund shall consist of the items listed in Article 27(2), (3) and, with the agreement of the competent authority of the home Member State, (4)(c).

2. The guarantee fund may not be less than a minimum of EUR 3 million.

   Any Member State may provide for a one-fourth reduction of the minimum guarantee fund in the case of mutual associations and mutual-type associations and tontines.

Article 30: Review of the Amount of the Guarantee Fund

1. The amount in euro as laid down in Article 29(2) shall be reviewed annually starting on 20 September 2003, in order to take account of

changes in the European index of consumer prices comprising all Member States as published by Eurostat.

The amount shall be adapted automatically, by increasing the base amount in euro by the percentage change in that index over the period between 20 March 2002 and the review date and rounded up to a multiple of EUR 100 000.

If the percentage change since the last adaptation is less than 5%, no adaptation shall take place.

2. The Commission shall inform annually the European Parliament and the Council of the review and the adapted amount referred to in paragraph 1.

Appendix E

IAIS: Insurance Principles, Standards, and Guidelines

This is a list of the principles, standards, and guidances that have been published by IAIS,[1] the International Association of Insurance Supervisors.

Principles

The IAIS sets out principles that are fundamental to effective insurance supervision. The principles identify areas in which the insurance supervisor should have authority or control. These form the basis from which standards are developed.

1. Insurance core principles and methodology

    (First approved September 1997; revised and approved October 2000; revised and approved October 2003)

2. Principles applicable to the supervision of international insurers and insurance groups and their cross-border business operations

    (First approved September 1997; revised and approved December 1999)

3. Principles for the conduct of insurance business

    (Approved December 1999)

4. Principles on the supervision of insurance activities on the Internet

    (Approved October 2000)

5. Principles on capital adequacy and solvency

    (Approved January 2002)

6. Principles on minimum requirements for supervision of reinsurers

    (Approved October 2002)

---

[1] The published principles and guidelines can be found at www.iaisweb.org.

Standards

Standards focus on particular issues and describe best or most prudent practices. In some cases, standards will set out best practices for a supervisory authority; in others, the papers describe the practices a well managed insurance company would be expected to follow, and thereby assist supervisors in assessing the practices that companies in their jurisdictions have in place.

1. Supervisory standard on licensing
   (Approved October 1998)
2. Supervisory standard on on-site inspections
   (Approved October 1998)
3. Supervisory standard on derivatives
   (Approved October 1998)
4. Supervisory standard on asset management by insurance companies
   (Approved December 1999)
5. Supervisory standard on group coordination
   (Approved October 2000)
6. Supervisory standard on the exchange of information
   (Approved January 2002)
7. Supervisory standard on the evaluation of the reinsurance cover
   (Approved January 2002)
8. Supervisory standard on supervision of reinsurers
   (Approved October 2003)

Guidances

Guidance papers are an adjunct to principles and standards. They are designed to assist supervisors, although sometimes they are addressed at insurance companies.

1. Guidance on insurance regulation and supervision for emerging market economies
   (Approved September 1997)
2. A model memorandum of understanding (to facilitate the exchange of information between financial supervisors)
   (Approved September 1997)

3. Guidance paper for fit and proper principles and their application
   (Approved October 2000)
4. Guidance paper on public disclosure by insurers
   (Approved January 2002)
5. Anti-money laundering guidance notes for insurance supervisors and insurance entities
   (Approved January 2002)
6. Solvency control levels guidance paper
   (Approved October 2003)
7. The use of actuaries as part of a supervisory model guidance paper
   (Approved October 2003)
8. Stress testing by insurers guidance paper
   (Approved October 2003)
9. Investment risk management
   (Approved October 2004)

Appendix F

From the Proposed Reinsurance Directive

The proposed directive for reinsurance (see COM, 2004) includes the following articles that are of relevance for the solvency assessment. Article 38, about the requirements for life reinsurance activities, have after consultation with the industry been changed.[1]

Chapter 3: Rules Relating to the Solvency Margin and to the Guarantee Fund

Section 1: Available Solvency Margin

Article 35: General Rule

Each Member State shall require of every reinsurance undertaking whose head office is situated in its territory an adequate available solvency margin in respect of its entire business at all times, which is at least equal to the requirements in this Directive.

Article 36: Eligible Items

1. The available solvency margin shall consist of the assets of the reinsurance undertaking free of any foreseeable liabilities, less any intangible items, including:

   (a) the paid-up share capital or, in the case of a mutual reinsurance undertaking, the effective initial fund plus any members' accounts which meet all the following criteria:

      (i) the memorandum and articles of association must stipulate that payments may be made from these accounts to members only in so far as this does not cause the available solvency

---

[1] Council secretariat: Proposal for a directive of the European Parliament and of the council on reinsurance and amending Council Directives 73/239/EEC and 92/49/EEC and Directives 98/78/EC and 2002/83/EC, September 29, 2004.

margin to fall below the required level, or, after the disso-
lution of the undertaking, if all the undertaking's other debts
have been settled;

(ii) the memorandum and articles of association must stipulate,
with respect to any payments referred to in point (i) for
reasons other than the individual termination of membership, that the competent authorities must be notified at least
one month in advance and can prohibit the payment within
that period;

(iii) the relevant provisions of the memorandum and articles of
association may be amended only after the competent authorities have declared that they have no objection to the
amendment, without prejudice to the criteria stated in points
(i) and (ii);

(b) statutory and free reserves not corresponding to underwriting
liabilities;

(c) the profit or loss brought forward after deduction of dividends
to be paid.

2. The available solvency margin shall be reduced by the amount of
own shares directly held by the reinsurance undertaking.

For those reinsurance undertakings which discount or reduce their
non-life technical provisions for claims outstanding to take account
of investment income as permitted by Article 60(1)(g) of Directive
91/674/EEC, the available solvency margin shall be reduced by the
difference between the undiscounted technical provisions or technical provisions before deductions as disclosed in the notes on the
accounts, and the discounted or technical provisions after deductions. This adjustment shall be made for all risks listed in point A
of the Annex to Directive 73/239/EEC, except for risks listed under
classes 1 and 2 of that Annex. For classes other than 1 and 2 of that
Annex, no adjustment need be made in respect of the discounting
of annuities included in technical provisions.

In addition to the deductions in subparagraphs 1 and 2, the available
solvency margin shall be reduced by the following items:

(a) participations which the reinsurance undertaking holds in the
following entities:

(i) insurance undertakings within the meaning of Article 6 of
Directive 73/239/EEC, Article 4 of Directive 2002/83/EC,
or Article 1(b) of Directive 98/78/EC,

(ii) reinsurance undertakings within the meaning of Article 3 of
this Directive or non-member-country reinsurance undertaking within the meaning of Article 1(l) of Directive 98/78/
EC,

(iii) insurance holding companies within the meaning of Article 1(i) of Directive 98/78/EC,

(iv) credit institutions and financial institutions within the meaning of Article 1(1) and (5) of Directive 2000/12/EC,

(v) investment firms and financial institutions within the meaning of Article 1(2) of Council Directive 93/22/EEC and of Article 2(4) and (7) of Council Directive 93/6/EEC;

(b) each of the following items which the reinsurance undertaking holds in respect of the entities defined in (a) in which it holds a participation:

(i) instruments referred to in paragraph 4,

(ii) instruments referred to in Article 27(3) of Directive 2002/83/EC,

(iii) subordinated claims and instruments referred to in Article 35 and Article 36(3) of Directive 2000/12/EC.

Where shares in another credit institution, investment firm, financial institution, insurance or reinsurance undertaking or insurance holding company are held temporarily for the purposes of a financial assistance operation designed to reorganise and save that entity, the competent authority may waive the provisions on deduction referred to under (a) and (b) of the fourth subparagraph.

As an alternative to the deduction of the items referred to in (a) and (b) of the fourth subparagraph which the reinsurance undertaking holds in credit institutions, investment firms and financial institutions, Member States may allow their reinsurance undertakings to apply mutatis mutandis methods 1, 2, or 3 of Annex I to Directive 2002/87/EC. Method 1 (Accounting consolidation) shall only be applied if the competent authority is confident about the level of integrated management and internal control regarding the entities which would be included in the scope of consolidation. The method chosen shall be applied in a consistent manner over time.

Member States may provide that, for the calculation of the solvency margin as provided forby this Directive, reinsurance undertakings subject to supplementary supervision in accordance with Directive 98/78/EC or to supplementary supervision in accordance with Directive 2002/87/EC, need not deduct the items referred to in (a) and (b) of the fourth subparagraph which are held in credit institutions, investment firms, financial institutions, insurance or reinsurance undertakings or insurance holding companies which are included in the supplementary supervision.

For the purposes of the deduction of participations referred to in this paragraph, participation shall mean a participation within the meaning of Article 1(f) of Directive 98/78/EC.

3. The available solvency margin may also consist of:

(a) cumulative preferential share capital and subordinated loan capital up to 50% of the available solvency margin or the required solvency margin, whichever is the smaller, no more than 25% of which shall consist of subordinated loans with a fixed maturity, or fixed-term cumulative preferential share capital, provided in the event of the bankruptcy or liquidation of the reinsurance undertaking, binding agreements exist under which the subordinated loan capital or preferential share capital ranks after the claims of all other creditors and is not to be repaid until all other debts outstanding at the time have been settled.

Subordinated loan capital must also fulfil the following conditions:

(i)   only fully paid-up funds may be taken into account;

(ii) for loans with a fixed maturity, the original maturity must be at least five years. No later than one year before the repayment date the reinsurance undertaking must submit to the competent authorities for their approval a plan showing how the available solvency margin will be kept at or brought to the required level at maturity, unless the extent to which the loan may rank as a component of the available solvency margin is gradually reduced during at least the last five years before the repayment date. The competent authorities may authorise the early repayment of such loans provided application is made by the issuing reinsurance undertaking and its available solvency margin will not fall below the required level;

(iii) loans the maturity of which is not fixed must be repayable only subject to five years' notice unless the loans are no longer considered as a component of the available solvency margin or unless the prior consent of the competent authorities is specifically required for early repayment. In the latter event the reinsurance undertaking must notify the competent authorities at least six months before the date of the proposed repayment, specifying the available solvency margin and the required solvency margin both before and after that repayment. The competent authorities shall authorise repayment only if the reinsurance undertaking's available solvency margin will not fall below the required level;

(iv) the loan agreement must not include any clause providing that in specified circumstances, other than the winding-up of the reinsurance undertaking, the debt will become repayable before the agreed repayment dates;

      (v) the loan agreement may be amended only after the competent authorities have declared that they have no objection to the amendment;

  (b) securities with no specified maturity date and other instruments, including cumulative preferential shares other than those mentioned in point (a), up to 50% of the available solvency margin or the required solvency margin, whichever is the smaller, for the total of such securities and the subordinated loan capital referred to in point (a) provided they fulfil the following:

      (i)   they may not be repaid on the initiative of the bearer or without the prior consent of the competent authority;

      (ii) the contract of issue must enable the reinsurance undertaking to defer the payment of interest on the loan;

      (iii) the lender's claims on the reinsurance undertaking must rank entirely after those of all non-subordinated creditors;

      (iv) the documents governing the issue of the securities must provide for the loss absorption capacity of the debt and unpaid interest, while enabling the reinsurance undertaking to continue its business;

      (v) only fully paid-up amounts may be taken into account.

4. Upon application, with supporting evidence, by the undertaking to the competent authority of the home Member State and with the agreement of that competent authority, the available solvency margin may also consist of:

  (a) one-half of the unpaid share capital or initial fund, once the paid-up part amounts to 25% of that share capital or fund, up to 50% of the available solvency margin or the required solvency margin, whichever is the smaller;

  (b) in the case of non-life mutual or mutual-type association with variable contributions, any claim which it has against its members by way of a call for supplementary contribution, within the financial year, up to one-half of the difference between the maximum contributions and the contributions actually called in, and subject to a limit of 50% of the available solvency margin or the required solvency margin, whichever is the smaller. The competent national authorities shall establish guidelines laying down the conditions under which supplementary contributions may be accepted;

  (c) any hidden net reserves arising out of the valuation of assets, in so far as such hidden net reserves are not of an exceptional nature.

5. In addition, with respect to life reinsurance activities, the available solvency margin may, upon application, with supporting evidence,

by the undertaking to the competent authority of the home Member State and with the agreement of that competent authority, consist of:

(a) until 31 December 2009 an amount equal to 50% of the undertaking's future profits, but not exceeding 25% of the available solvency margin or the required solvency margin, whichever is the smaller; the amount of the future profits shall be obtained by multiplying the estimated annual profit by a factor which represents the average period left to run on policies; the factor used may not exceed six; the estimated annual profit shall not exceed the arithmetical average of the profits made over the last five financial years in the activities listed in Article 2(1) of Directive 2002/83/EC.

Competent authorities may only agree to include such an amount for the available solvency margin:

(i) when an actuarial report is submitted to the competent authorities substantiating the likelihood of emergence of these profits in the future; and

(ii) in so far as that part of future profits emerging from hidden net reserves referred to in paragraph 5(c) has not already been taken into account;

(b) where Zillmerising is not practised or where, if practised, it is less than the loading for acquisition costs included in the premium, the difference between a non-Zillmerised or partially Zillmerised mathematical provision and a mathematical provision Zillmerised at a rate equal to the loading for acquisition costs included in the premium; this figure may not, however, exceed 3,5% of the sum of the differences between the relevant capital sums of life assurance activities and the mathematical provisions for all policies for which Zillmerising is possible; the difference shall be reduced by the amount of any undepreciated acquisition costs entered as an asset.

6. Amendments to paragraphs 1 to 5 to take into account developments that justify a technical adjustment of the elements eligible for the available solvency margin, shall be adopted in accordance with the procedure laid down in Article 55(2) of this Directive.

Section 2: Required Solvency Margin

Article 37: Required Solvency Margin for Non-Life Reinsurance Activities

1. The required solvency margin shall be determined on the basis either of the annual amount of premiums or contributions, or of the average burden of claims for the past three financial years.

In the case, however, of reinsurance undertakings which essentially underwrite only one or more of the risks of credit, storm, hail or frost, the last seven financial years shall be taken as the reference period for the average burden of claims.

2. Subject to Article 40, the amount of the required solvency margin shall be equal to the higher of the two results as set out in paragraphs 3 and 4.

3. The premium basis shall be calculated using the higher of gross written premiums or contributions as calculated below, and gross earned premiums or contributions.

    Premiums or contributions in respect of the classes 11, 12 and 13 listed in point A of the Annex to Directive 73/239/EEC shall be increased by 50%.

    Premiums or contributions in respect of classes other than 11, 12 and 13 listed in point A of the Annex to Directive 73/239/EEC, may be enhanced up to 50%, for specific reinsurance activities or contract types, in order to take account of the specificities of these activities or contracts, in accordance with the procedure referred to in Article 55(2) of this Directive. The premiums or contributions, inclusive of charges ancillary to premiums or contributions, due in respect of reinsurance business in the last financial year shall be aggregated.

    From this sum there shall then be deducted the total amount of premiums or contributions cancelled in the last financial year, as well as the total amount of taxes and levies pertaining to the premiums or contributions entering into the aggregate.

    The amount so obtained shall be divided into two portions, the first portion extending up to EUR 50 million, the second comprising the excess; 18% and 16% of these portions respectively shall be calculated and added together.

    The sum so obtained shall be multiplied by the ratio existing in respect of the sum of the last three financial years between the amount of claims remaining to be borne by the reinsurance undertaking after deduction of amounts recoverable under retrocession and the gross amount of claims; this ratio may in no case be less than 50%.

    With the approval of the competent authorities, statistical methods may be used to allocate the premiums or contributions.

4. The claims basis shall be calculated, as follows, using in respect of the classes 11, 12 and 13 listed in point A of the Annex to Directive 73/239/EEC, claims, provisions and recoveries increased by 50%.

    Claims provisions and recoveries in respect of classes other than 11, 12 and 13 listed in point A of the Annex to Directive 73/239/EEC, may be enhanced up to 50%, for specific reinsurance activities or contract types, in order to take account of the specificities of these activities or contracts, in accordance with the procedure referred to in Article 55(2) of this Directive.

The amounts of claims paid, without any deduction of claims borne by retrocessionaires, in the periods specified in paragraph 1 shall be aggregated.

To this sum there shall be added the amount of provisions for claims outstanding established at the end of the last financial year.

From this sum there shall be deducted the amount of recoveries effected during the periods specified in paragraph 1.

From the sum then remaining, there shall be deducted the amount of provisions for claims outstanding established at the commencement of the second financial year preceding the last financial year for which there are accounts. If the period of reference established in paragraph 1 equals seven years, the amount of provisions for claims outstanding established at the commencement of the sixth financial year preceding the last financial year for which there are accounts shall be deducted.

One-third, or one-seventh, of the amount so obtained, according to the period of reference established in paragraph 1, shall be divided into two portions, the first extending up to EUR 35 million and the second comprising the excess; 26% and 23% of these portions respectively shall be calculated and added together.

The sum so obtained shall be multiplied by the ratio existing in respect of the sum of the last three financial years between the amount of claims remaining to be borne by the undertaking after deduction of amounts recoverable under retrocession and the gross amount of claims; this ratio may in no case be less than 50%.

With the approval of the competent authorities, statistical methods may be used to allocate claims, provisions and recoveries.

5. If the required solvency margin as calculated in paragraphs 2, 3 and 4 is lower than the required solvency margin of the year before, the required solvency margin shall be at least equal to the required solvency margin of the year before multiplied by the ratio of the amount of the technical provisions for claims outstanding at the end of the last financial year and the amount of the technical provisions for claims outstanding at the beginning of the last financial year. In these calculations technical provisions shall be calculated net of retrocession but the ratio may in no case be higher than 1.

6. The fractions applicable to the portions referred to in the sixth subparagraph of paragraph 3 and the sixth subparagraph of paragraph 4 shall each be reduced to a third in the case of reinsurance of health

insurance practised on a similar technical basis to that of life assurance, if

(a) the premiums paid are calculated on the basis of sickness tables according to the mathematical method applied in insurance;

(b) a provision is set up for increasing age;

(c) an additional premium is collected in order to set up a safety margin of an appropriate amount;

(d) the insurance undertaking may cancel the contract before the end of the third year of insurance at the latest;

(e) the contract provides for the possibility of increasing premiums or reducing payments even for current contracts.

Article 38: Required Solvency Margin for Life Reassurance Activities

1. The required solvency margin for life reassurance activities shall be determined according to Article 37 of this Directive.

2. Notwithstanding paragraph 1, the home Member State may provide that for reinsurance classes of assurance business covered by Article 2(1)(a) of Directive 2002/83/EC linked to investment funds or participating contracts and for the operations referred to in Article 2(1)(b), 2(2)(c), (d) and (e) of Directive 2002/83/EC the required solvency margin shall be determined in accordance with Article 28 of Directive 2002/83/EC.

Article 39: Required Solvency Margin for a Reinsurance Undertaking Conducting Simultaneously Non-Life and Life Reinsurance

1. The home Member State shall require that every reinsurance undertaking conducting both non-life reinsurance and life reinsurance business shall have an available solvency margin to cover the total sum of required solvency margins in respect of both non-life and life reinsurance activities which shall be determined in accordance with Articles 37 and 38 respectively.

2. If the available solvency margin does not reach the level required in paragraph 1, the competent authorities shall apply the measures provided for in Articles 42 and 43.

Section 3: Guarantee Fund

Article 40: Amount of the Guarantee Fund

1. One-third of the required solvency margin as specified in Articles 37 to 39 shall constitute the guarantee fund. This fund shall consist of the items listed in Article 36(1) to (3) and, with the agreement of the competent authority of the home Member State, (4)(c).

2. The guarantee fund may not be less than a minimum of EUR 3 million.

   Any Member State may provide that as regards captive reinsurance undertakings, the minimum guarantee fund be not less than EUR 1 million.

Article 41: Review of the Amount of the Guarantee Fund

1. The amounts in euro as laid down in Article 40(2) shall be reviewed annually starting [date of implementation laid down in Article 61(1)] in order to take account of changes in the European index of consumer prices comprising all Member States as published by Eurostat.

   The amounts shall be adapted automatically by increasing the base amount in euro by the percentage change in that index over the period between the entry into force of this Directive and the review date and rounded up to a multiple of EUR 100 000.

   If the percentage of change since the last adaptation is less than 5%, no adaptation shall take place.

2. The Commission shall inform annually the European Parliament and the Council of the review and the adapted amounts referred to in paragraph 1.

Appendix G

Annex I and Annex II in the Insurance Group Directive

The Insurance Group Directive is given in COM (1998). In Annex I calculation methods for the adjusted solvency requirement are given, and in Annex II supplementary supervision for insurance undertakings that are subsidiaries of a holding company, a reinsurance undertaking, etc.

Annex I : Calculation of the Adjusted Solvency of Insurance Undertakings

1. Choice of Calculation Method and General Principles
   A. Member States shall provide that the calculation of the adjusted solvency of insurance undertakings referred to in Article 2(1) shall be carried out according to one of the methods described in point 3. A Member State may, however, provide for the competent authorities to authorise or impose the application of a method set out in point 3 other than that chosen by the Member State.
   B. **Proportionality**

      The calculation of the adjusted solvency of an insurance undertaking shall take account of the proportional share held by the participating undertaking in its related undertakings.

      'Proportional share' means either, where method 1 or method 2 described in point 3 is used, the proportion of the subscribed capital that is held, directly or indirectly, by the participating undertaking or, where method 3 described in point 3 is used, the percentages used for the establishment of the consolidated accounts.

However, whichever method is used, when the related undertaking is a subsidiary undertaking and has a solvency deficit, the total solvency deficit of the subsidiary has to be taken into account.

However, where, in the opinion of the competent authorities, the responsibility of the parent undertaking owning a share of the capital is limited strictly and unambiguously to that share of the capital, such competent authorities may give permission for the solvency deficit of the subsidiary undertaking to be taken into account on a proportional basis.

C. Elimination of Double Use of Solvency Margin Elements

C.1. *General Treatment of Solvency Margin Elements*

Regardless of the method used for the calculation of the adjusted solvency of an insurance undertaking, the double use of elements eligible for the solvency margin among the different insurance undertakings taken into account in that calculation must be eliminated.

For this purpose, when calculating the adjusted solvency of an insurance undertaking and where the methods described in point 3 do not provide for it, the following amounts shall be eliminated:

- the value of any asset of that insurance undertaking which represents the financing of elements eligible for the solvency margin of one of its related insurance undertakings,

- the value of any asset of a related insurance undertaking of that insurance undertaking which represents the financing of elements eligible for the solvency margin of that insurance undertaking,

- the value of any asset of a related insurance undertaking of that insurance undertaking which represents the financing of elements eligible for the solvency margin of any other related insurance undertaking of that insurance undertaking.

C.2. *Treatment of Certain Elements*

Without prejudice to the provisions of section C.1:

- profit reserves and future profits arising in a related life assurance undertaking of the insurance undertaking for which the adjusted solvency is calculated, and

- any subscribed but not paid-up capital of a related insurance undertaking of the insurance undertaking for which the adjusted solvency is calculated, may only be included in the calculation in so far as they are eligible for covering the solvency margin requirement of that related undertaking. However, any subscribed but not paid-up capital which represents a potential obligation on the part of the participating undertaking shall be entirely excluded from the calculation.

Any subscribed but not paid-up capital of the participating insurance undertaking which represents a potential obligation on the part of a related insurance undertaking shall also be excluded from the calculation.

Any subscribed but not paid-up capital of a related insurance undertaking which represents a potential obligation on the part of another related insurance undertaking of the same participating insurance undertaking shall be excluded from the calculation.

C.3. *Transferability*

If the competent authorities consider that certain elements eligible for the solvency margin of a related insurance undertaking other than those referred to in section C.2 cannot effectively be made available to cover the solvency margin requirement of the participating insurance undertaking for which the adjusted solvency is calculated, those elements may be included in the calculation only in so far as they are eligible for covering the solvency margin requirement of the related undertaking.

C.4. The sum of the elements referred to in sections C.2 and C.3 may not exceed the solvency margin requirement of the related insurance undertaking.

D. **Elimination of the Intra-Group Creation of Capital**

When calculating adjusted solvency, no account shall be taken of any element eligible for the solvency margin arising out of reciprocal financing between the insurance undertaking and:

- a related undertaking,
- a participating undertaking,
- another related undertaking of any of its participating undertakings.

Furthermore, no account shall be taken of any element eligible for the solvency margin of a related insurance undertaking of the insurance undertaking for which the adjusted solvency is calculated when the element in question arises out of reciprocal financing with any other related undertaking of that insurance undertaking.

In particular, reciprocal financing exists when an insurance undertaking, or any of its related undertakings, holds shares in, or makes loans to, another undertaking which, directly or indirectly, holds an element eligible for the solvency margin of the first undertaking.

E. The competent authorities shall ensure that the adjusted solvency is calculated with the same frequency as that laid down by Directives 73/239/EEC and 79/267/EEC for calculating the solvency margin of insurance undertakings. The value of the assets and liabilities shall be assessed according to the relevant provisions of Directives 73/239/EEC, 79/267/EEC and 91/674/EEC (1).

2. Application of the Calculation Methods

   2.1. **Related Insurance Undertakings**

   The adjusted solvency calculation shall be carried out in accordance with the general principles and methods set out in this Annex.

   In the case of all methods, where the insurance undertaking has more than one related insurance undertaking, the adjusted solvency calculation shall be carried out by integrating each of these related insurance undertakings.

   In cases of successive participations (for example, where an insurance undertaking is a participating undertaking in another insurance undertaking which is also a participating undertaking in an insurance undertaking), the adjusted solvency calculation shall be carried out at the level of each participating insurance undertaking which has at least one related insurance undertaking.

   Member States may waive calculation of the adjusted solvency of an insurance undertaking:

   - if the undertaking is a related undertaking of another insurance undertaking authorised in the same Member State, and that related undertaking is taken into account in the calculation of the adjusted solvency of the participating insurance undertaking, or

   - if the insurance undertaking is a related undertaking either of an insurance holding company or of a reinsurance undertaking which has its registered office in the same Member State as the insurance undertaking, and both the holding insurance company or the reinsurance undertaking and the related insurance undertaking are taken into account in the calculation carried out.

   Member States may also waive calculation of the adjusted solvency of an insurance undertaking if it is a related insurance undertaking of another insurance undertaking, a reinsurance undertaking or an insurance holding company which has its registered office in another Member State, and if the competent authorities of the Member States concerned have agreed to grant exercise of the supplementary supervision to the competent authority of the latter Member State.

In each case, the waiver may be granted only if the competent authorities are satisfied that the elements eligible for the solvency margins of the insurance undertakings included in the calculation are adequately distributed between those undertakings.

Member States may provide that where the related insurance undertaking has its registered office in a Member State other than that of the insurance undertaking for which the adjusted solvency calculation is carried out, the calculation shall take account, in respect of the related undertaking, of the solvency situation as assessed by the competent authorities of that other Member State.

2.2. Related Reinsurance Undertakings

When calculating the adjusted solvency of an insurance undertaking which is a participating undertaking in a reinsurance undertaking, this related reinsurance undertaking shall be treated, solely for the purposes of the calculation, by analogy with a related insurance undertaking, applying the general principles and methods described in this Annex.

To this end, a notional solvency requirement shall be established for each related reinsurance undertaking on the basis of the same rules as are laid down in Article 16(2) to (5) of Directive 73/239/EEC or Article 19 of Directive 79/267/EEC. However, in the event of significant difficulty in applying these rules, the competent authorities may permit the notional life solvency requirement to be calculated on the basis of the first result as set out in Article 16(3) of Directive 73/239/EEC. The same elements as are found in Article 16(1) of Directive 73/239/EEC and in Article 18 of Directive 79/267/EEC shall be recognised as eligible for the notional solvency margin. The value of the assets and liabilities shall be assessed according to the same rules as are laid down in those Directives and in Directive 91/674/EEC.

2.3. Intermediate Insurance Holding Companies

When calculating the adjusted solvency of an insurance undertaking which holds a participation in an insurance undertaking, a related reinsurance undertaking, or an insurance undertaking in a non-member country through an insurance holding company, the situation of the intermediate insurance holding company is taken into account. For the sole purpose of this calculation, to be undertaken in accordance with the general principles and methods described in this Annex, this insurance holding company shall be treated as if it were an insurance undertaking subject to a zero solvency requirement and were subject to the same conditions as are laid down in Article 16(1) of Directive 73/239/EEC or in Article 18 of Directive 79/267/EEC in respect of elements eligible for the solvency margin.

2.4. Related Insurance or Reinsurance Undertakings Having Their Registered Office in Non-Member Countries

A. *Related Non-Member-Country Insurance Undertakings*

When calculating the adjusted solvency of an insurance undertaking which is a participating undertaking in a non-member-country insurance undertaking, the latter shall be treated solely for the purposes of the calculation, by analogy with a related insurance undertaking, by applying the general principles and methods described in this Annex.

However, where the non-member country in which that undertaking has its registered office makes it subject to authorisation and imposes on it a solvency requirement at least comparable to that laid down in Directive 73/239/EEC or 79/267/EEC, taking into account the elements of cover of that requirement, Member States may provide that the calculation shall take into account, as regards that undertaking, the solvency requirement and the elements eligible to satisfy that requirement as laid down by the non-member country in question.

B. *Related Non-Member-Country Reinsurance Undertakings*

Notwithstanding section 2.2, when calculating the adjusted solvency of an insurance undertaking which is a participating undertaking in a reinsurance undertaking with its registered office in a non-member country, and subject to the same conditions as those set out in point A above, Member States may provide that the calculation shall take account, as regards the latter undertaking, of the own-funds requirement and the elements eligible to satisfy that requirement as laid down by the non-member country in question. Where only the insurance undertakings of that non-member country are subject to such provisions, the notional own-funds requirement on the related reinsurance undertaking and the elements eligible to satisfy that notional requirement may be calculated as if the undertaking in question were a related insurance undertaking of that non-member country.

2.5. Non-Availability of the Necessary Information

Where information necessary for calculating the adjusted solvency of an insurance undertaking, concerning a related undertaking with its registered office in a Member State or a non-member country, is not available to the competent authorities, for whatever reason, the book value of that undertaking in the participating insurance undertaking shall be deducted from the elements eligible for the adjusted solvency margin. In that case, the unrealised gains connected with such participation shall not be allowed as an element eligible for the adjusted solvency margin.

3. Calculation Methods

Method 1: Deduction and aggregation method

The adjusted solvency situation of the participating insurance undertaking is the difference between:

(i) the sum of:

    (a) the elements eligible for the solvency margin of the participating insurance undertaking, and

    (b) the proportional share of the participating insurance undertaking in the elements eligible for the solvency margin of the related insurance undertaking

and

(ii) the sum of:

    (a) the book value in the participating insurance undertaking of the related insurance undertaking, and

    (b) the solvency requirement of the participating insurance undertaking, and

    (c) the proportional share of the solvency requirement of the related insurance undertaking.

Where the participation in the related insurance undertaking consists, wholly or in part, of an indirect ownership, then item (ii)(a) shall incorporate the value of such indirect ownership, taking into account the relevant successive interests; and items (i)(b) and (ii)(c) shall include the corresponding proportional shares of the elements eligible for the solvency margin of the related insurance undertaking and of the solvency requirement of the related insurance undertaking, respectively.

Method 2: Requirement deduction method

The adjusted solvency of the participating insurance undertaking is the difference between:

(i) the sum of the elements eligible for the solvency margin of the participating insurance undertaking, and

(ii) the sum of:

    (a) the solvency requirement of the participating insurance undertaking, and

(b) the proportional share of the solvency requirement of the related insurance undertaking.

When valuing the elements eligible for the solvency margin, participations within the meaning of this Directive are valued by the equity method, in accordance with the option set out in Article 59(2)(b) of Directive 78/660/EEC.

Method 3: Accounting Consolidation-Based Method

The calculation of the adjusted solvency of the participating insurance undertaking shall be carried out on the basis of the consolidated accounts. The adjusted solvency of the participating insurance undertaking is the difference between:

the elements eligible for the solvency margin calculated on the basis of consolidated data, and

(a) either the sum of the solvency requirement of the participating insurance undertaking and of the proportional shares of the solvency requirements of the related insurance undertakings, based on the percentages used for the establishment of the consolidated accounts,

(b) or the solvency requirement calculated on the basis of consolidated data.

The provisions of Directives 73/239/EEC, 79/267/EEC and 91/674/EEC shall apply for the calculation of the elements eligible for the solvency margin and of the solvency requirement based on consolidated data.

Annex II: Supplementary Supervision for Insurance Undertakings That Are Subsidiaries of an Insurance Holding Company, a Reinsurance Undertaking, or a Non-Member-Country Insurance Undertaking

1. In the case of two or more insurance undertakings referred to in Article 2(2) which are the subsidiaries of an insurance holding company, a reinsurance undertaking or a non-member-country insurance undertaking and which are established in different Member States, the competent authorities shall ensure that the method described in this Annex is applied in a consistent manner.

The competent authorities shall exercise the supplementary supervision with the same frequency as that laid down by Directives 73/239/EEC and 79/267/EEC for calculating the solvency margin of insurance undertakings.

2. Member States may waive the calculation provided for in this Annex with regard to an insurance undertaking:

   - if that insurance undertaking is a related undertaking of another insurance undertaking and if it is taken into account in the calculation provided for in this Annex carried out for that other undertaking,

   - if that insurance undertaking and one or more other insurance undertakings authorised in the same Member State have as their parent undertaking the same insurance holding company, reinsurance undertaking or non-member-country insurance undertaking, and the insurance undertaking is taken into account in the calculation provided for in this Annex carried out for one of these other undertakings,

   - if that insurance undertaking and one or more other insurance undertakings authorised in other Member States have as their parent undertaking the same insurance holding company, reinsurance undertaking or non-member-country insurance undertaking, and an agreement granting exercise of the supplementary supervision covered by this Annex to the supervisory authority of another Member State has been concluded in accordance with Article 4(2).

   In the case of successive participations (for example: an insurance holding company or a reinsurance undertaking which is itself owned by another insurance holding company, a reinsurance undertaking or a non-member-country insurance undertaking), Member States may apply the calculations provided for in this Annex only at the level of the ultimate parent undertaking of the insurance undertaking which is an insurance holding company, a reinsurance undertaking or a non-member-country insurance undertaking.

3. The competent authorities shall ensure that calculations analogous to those described in Annex I are carried out at the level of the insurance holding company, reinsurance undertaking or non-member-country insurance undertaking.

   The analogy shall consist in applying the general principles and methods described in Annex I at the level of the insurance holding company, reinsurance undertaking or non-member-country insurance undertaking.

   For the sole purpose of this calculation, the parent undertaking shall be treated as if it were an insurance undertaking subject to:

   - a zero solvency requirement where it is an insurance holding company,

   - a notional solvency requirement as provided for in section 2.2 of Annex I where it is a reinsurance undertaking, or as provided for in section 2.4(B) of Annex I where it is a reinsurance undertaking with its registered office in a non-member country,

- a solvency requirement determined according to the principles of section 2.4(A) of Annex I, where it is a non-member-country insurance undertaking,

and is subject to the same conditions as laid down in Article 16(1) of Directive 73/239/EEC or in Article 18 of Directive 79/267/EEC as regards the elements eligible for the solvency margin.

4. Non-Availability of the Necessary Information

Where information necessary for the calculation provided for in this Annex, concerning a related undertaking with its registered office in a Member State or a non-member country, is not available to the competent authorities, for whatever reason, the book value of that undertaking in the participating undertaking shall be deducted from the elements eligible for the calculation provided for in this Annex. In that case, the unrealised gains connected with such participation shall not be allowed as an element eligible for the calculation.

Appendix H

From the Financial Conglomerates Directive

The Financial Conglomerates Directive (COM, 2002d) includes amendments to the first non-life and first life directives and also to the Insurance Group Directive.

Amendments to the Non-Life Directive (EEC, 1973)

Article 22 (part of)

2. the following subparagraphs shall be added to Article 16(2):

    The available solvency margin shall also be reduced by the following items:

    (a) participations which the insurance undertaking holds in

    - insurance undertakings within the meaning of Article 6 of this Directive, Article 6 of First Directive 79/267/EEC of 5 March 1979 on the coordination of laws, regulations and administrative provisions relating to the taking up and pursuit of the business of direct life assurance(17), or Article 1(b) of Directive 98/78/EC of the European Parliament and of the Council(18),
    - reinsurance undertakings within the meaning of Article 1(c) of Directive 98/78/EC,
    - insurance holding companies within the meaning of Article 1(i) of Directive 98/78/EC,
    - credit institutions and financial institutions within the meaning of Article 1(1) and (5) of Directive 2000/12/EC of the European Parliament and of the Council(19),
    - investment firms and financial institutions within the meaning of Article 1(2) of Directive 93/22/EEC(20) and of Article 2(4) and (7) of Directive 93/6/EEC(21);

(b) each of the following items which the insurance undertaking holds in respect of the entities defined in (a) in which it holds a participation:

- instruments referred to in paragraph 3,
- instruments referred to in Article 18(3) of Directive 79/267/EEC,
- subordinated claims and instruments referred to in Article 35 and Article 36(3) of Directive 2000/12/EC.

Where shares in another credit institution, investment firm, financial institution, insurance or reinsurance undertaking or insurance holding company are held temporarily for the purposes of a financial assistance operation designed to reorganise and save that entity, the competent authority may waive the provisions on deduction referred to under (a) and (b) of the fourth subparagraph.

As an alternative to the deduction of the items referred to in (a) and (b) of the fourth subparagraph which the insurance undertaking holds in credit institutions, investment firms and financial institutions, Member States may allow their insurance undertakings to apply mutatis mutandis methods 1, 2, or 3 of Annex I to Directive 2002/87/EC of the European Parliament and of the Council of 16 December 2002 on the supplementary supervision of credit institutions, insurance undertakings and investment firms in a financial conglomerate(22). Method 1 (Accounting consolidation) shall only be applied if the competent authority is confident about the level of integrated management and internal control regarding the entities which would be included in the scope of consolidation. The method chosen shall be applied in a consistent manner over time.

Member States may provide that, for the calculation of the solvency margin as provided for by this Directive, insurance undertakings subject to supplementary supervision in accordance with Directive 98/78/EC or to supplementary supervision in accordance with Directive 2002/87/EC, need not deduct the items referred to in (a) and (b) of the fourth subparagraph which are held in credit institutions, investment firms, financial institutions, insurance or reinsurance undertakings or insurance holding companies which are included in the supplementary supervision.

For the purposes of the deduction of participations referred to in this paragraph, participation shall mean a participation within the meaning of Article 1(f) of Directive 98/78/EC.

Amendments to the Life Directive (EEC, 1979)

Article 23 (part of)

2. the following subparagraphs shall be added to Article 18(2):

The available solvency margin shall also be reduced by the following items:

(a) participations which the assurance undertaking holds, in

- insurance undertakings within the meaning of Article 6 of this Directive, Article 6 of Directive 73/239/EEC(23), or Article 1(b) of Directive 98/78/EC of the European Parliament and of the Council(24),

- reinsurance undertakings within the meaning of Article 1(c) of Directive 98/78/EC,

- insurance holding companies within the meaning of Article 1(i) of Directive 98/78/EC,

- credit institutions and financial institutions within the meaning of Article 1(1) and (5) of Directive 2000/12/EC of the European Parliament and of the Council(25),

- investment firms and financial institutions within the meaning of Article 1(2) of Directive 93/22/EEC(26) and of Articles 2(4) and 2(7) of Directive 93/6/EEC(27);

(b) each of the following items which the assurance undertaking holds in respect of the entities defined in (a) in which it holds a participation:

- instruments referred to in paragraph 3,

- instruments referred to in Article 16(3) of Directive 73/239/EEC,

- subordinated claims and instruments referred to in Article 35 and Article 36(3) of Directive 2000/12/EC.

Where shares in another credit institution, investment firm, financial institution, insurance or reinsurance undertaking or insurance holding company are held temporarily for the purposes of a financial assistance operation designed to reorganise and save that entity, the competent authority may waive the provisions on deduction referred to under (a) and (b) of the third subparagraph.

As an alternative to the deduction of the items referred to in (a) and (b) of the third subparagraph which the insurance undertaking holds in credit institutions, investment firms and financial institutions, Member States may allow their insurance undertakings to apply mutatis mutandis methods 1, 2, or 3 of Annex I to Directive 2002/

87/EC of the European Parliament and of the Council of 16 December 2002 on the supplementary supervision of credit institutions, insurance undertakings and investment firms in a financial conglomerate(28). Method 1 (Accounting consolidation) shall only be applied if the competent authority is confident about the level of integrated management and internal control regarding the entities which would be included in the scope of consolidation. The method chosen shall be applied in a consistent manner over time.

Member States may provide that, for the calculation of the solvency margin as provided for by this Directive, insurance undertakings subject to supplementary supervision in accordance with Directive 98/78/EC or to supplementary supervision in accordance with Directive 2002/87/EC, need not deduct the items referred to in (a) and (b) of the third subparagraph which are held in credit institutions, investment firms, financial institutions, insurance or reinsurance undertakings or insurance holding companies which are included in the supplementary supervision.

For the purposes of the deduction of participations referred to in this paragraph, participation shall mean a participation within the meaning of Article 1(f) of Directive 98/78/EC.

Amendments to the Insurance Group Directive (COM, 1998)

5. in Annex I.1.B. the following paragraph shall be added:

   Where there are no capital ties between some of the undertakings in an insurance group, the competent authority shall determine which proportional share will have to be taken account of.

6. in Annex I.2. the following point shall be added:

   2.4a. Related credit institutions, investment firms and financial institutions

   When calculating the adjusted solvency of an insurance undertaking which is a participating undertaking in a credit institution, investment firm or financial institution, the rules laid down in Article 16(1) of Directive 73/239/EEC and in Article 18 of Directive 79/267/EEC on the deduction of such participations shall apply mutatis mutandis, as well as the provisions on the ability of Member States under certain conditions to allow alternative methods and to allow such participations not to be deducted.

The calculation of supplementary capital adequacy requirements is given in Annex I to COM (2002d).

___

Annex I: Capital Adequacy

The calculation of the supplementary capital adequacy requirements of the regulated entities in a financial conglomerate referred to in Article 6(1) shall be carried out in accordance with the technical principles and one of the methods described in this Annex.

Without prejudice to the provisions of the next paragraph, Member States shall allow their competent authorities, where they assume the role of coordinator with regard to a particular financial conglomerate, to decide, after consultation with the other relevant competent authorities and the conglomerate itself, which method shall be applied by that financial conglomerate.

Member States may require that the calculation be carried out according to one particular method among those described in this Annex if a financial conglomerate is headed by a regulated entity which has been authorised in that Member State. Where a financial conglomerate is not headed by a regulated entity within the meaning of Article 1, Member States shall authorise the application of any of the methods described in this Annex, except in situations where the relevant competent authorities are located in the same Member State, in which case that Member State may require the application of one of the methods.

I.   Technical Principles

1. *Extent and form of the supplementary capital adequacy requirements calculation*

   Whichever method is used, when the entity is a subsidiary undertaking and has a solvency deficit, or, in the case of a non-regulated financial sector entity, a notional solvency deficit, the total solvency deficit of the subsidiary has to be taken into account. Where in this case, in the opinion of the coordinator, the responsibility of the parent undertaking owning a share of the capital is limited strictly and unambiguously to that share of the capital, the coordinator may give permission for the solvency deficit of the subsidiary undertaking to be taken into account on a proportional basis.

   Where there are no capital ties between entities in a financial conglomerate, the coordinator, after consultation with the other relevant competent authorities, shall determine which proportional share will have to be taken into account, bearing in mind the liability to which the existing relationship gives rise.

2. *Other technical principles*

Regardless of the method used for the calculation of the supplementary capital adequacy requirements of regulated entities in a financial conglomerate as laid down in Section II of this Annex, the coordinator, and where necessary other competent authorities concerned, shall ensure that the following principles will apply:

(i) the multiple use of elements eligible for the calculation of own funds at the level of the financial conglomerate (multiple gearing) as well as any inappropriate intra-group creation of own funds must be eliminated; in order to ensure the elimination of multiple gearing and the intra-group creation of own funds, competent authorities shall apply by analogy the relevant principles laid down in the relevant sectoral rules;

(ii) pending further harmonisation of sectoral rules, the solvency requirements for each different financial sector represented in a financial conglomerate shall be covered by own funds elements in accordance with the corresponding sectoral rules; when there is a deficit of own funds at the financial conglomerate level, only own funds elements which are eligible according to each of the sectoral rules (cross-sector capital) shall qualify for verification of compliance with the additional solvency requirements;

where sectoral rules provide for limits on the eligibility of certain own funds instruments, which would qualify as cross-sector capital, these limits would apply mutatis mutandis when calculating own funds at the level of the financial conglomerate;

when calculating own funds at the level of the financial conglomerate, competent authorities shall also take into account the effectiveness of the transferability and availability of the own funds across the different legal entities in the group, given the objectives of the capital adequacy rules;

where, in the case of a non-regulated financial sector entity, a notional solvency requirement is calculated in accordance with section II of this Annex, notional solvency requirement means the capital requirement with which such an entity would have to comply under the relevant sectoral rules as if it were a regulated entity of that particular financial sector; in the case of asset management companies, solvency requirement means the capital requirement set out in Article 5a(1)(a) of Directive 85/611/EEC; the notional solvency requirement of a mixed financial holding company shall be calculated according to the sectoral rules of the most important financial sector in the financial conglomerate.

II. Technical Calculation Methods

Method 1: "Accounting consolidation" method

The calculation of the supplementary capital adequacy requirements of the regulated entities in a financial conglomerate shall be carried out on the basis of the consolidated accounts.

The supplementary capital adequacy requirements shall be calculated as the difference between:

(i) the own funds of the financial conglomerate calculated on the basis of the consolidated position of the group; the elements eligible are those that qualify in accordance with the relevant sectoral rules; and

(ii) the sum of the solvency requirements for each different financial sector represented in the group; the solvency requirements for each different financial sector are calculated in accordance with the corresponding sectoral rules.

The sectoral rules referred to are in particular Directives 2000/12/ EC, Title V, Chapter 3, as regards credit institutions, 98/78/EC as regards insurance undertakings, and 93/6/EEC as regards credit institutions and investment firms.

In the case of non-regulated financial sector entities which are not included in the aforementioned sectoral solvency requirement calculations, a notional solvency requirement shall be calculated.

The difference shall not be negative.

Method 2: "Deduction and aggregation" method

The calculation of the supplementary capital adequacy requirements of the regulated entities in a financial conglomerate shall be carried out on the basis of the accounts of each of the entities in the group.

The supplementary capital adequacy requirements shall be calculated as the difference between:

(i) the sum of the own funds of each regulated and non-regulated financial sector entity in the financial conglomerate; the elements eligible are those which qualify in accordance with the relevant sectoral rules; and

(ii) the sum of

- the solvency requirements for each regulated and non-regulated financial sector entity in the group; the solvency requirements shall be calculated in accordance with the relevant sectoral rules, and
- the book value of the participations in other entities of the group.

In the case of non-regulated financial sector entities, a notional solvency requirement shall be calculated. Own funds and solvency requirements shall be taken into account for their proportional share as provided for in Article 6(4) and in accordance with Section I of this Annex.

The difference shall not be negative.

Method 3: "Book value/requirement deduction" method

The calculation of the supplementary capital adequacy requirements of the regulated entities in a financial conglomerate shall be carried out on the basis of the accounts of each of the entities in the group.

The supplementary capital adequacy requirements shall be calculated as the difference between:

(i) the own funds of the parent undertaking or the entity at the head of the financial conglomerate; the elements eligible are those which qualify in accordance with the relevant sectoral rules, and

(ii) the sum of

- the solvency requirement of the parent undertaking or the head referred to in (i), and

- the higher of the book value of the former's participation in other entities in the group and these entities' solvency requirements; the solvency requirements of the latter shall be taken into account for their proportional share as provided for in Article 6(4) and in accordance with Section I of this Annex.

In the case of non-regulated financial sector entities, a notional solvency requirement shall be calculated. When valuing the elements eligible for the calculation of the supplementary capital adequacy requirements, participations may be valued by the equity method in accordance with the option set out in Article 59(2)(b) of Directive 78/660/EEC.

The difference shall not be negative.

Method 4: Combination of methods 1, 2 and 3

Competent authorities may allow a combination of methods 1, 2 and 3, or a combination of two of these methods.

Appendix I

Prudent Person Rule

Directive 2003/41/EC of the European Parliament and of the Council of June 3, 2003, on the activities and supervision of institutions for occupational retirement provision (*Official Journal*, L 235, 23/09/2003, pp. 0010–0021).

Article 18: Investment Rules

1. Member States shall require institutions located in their territories to invest in accordance with the "prudent person" rule and in particular in accordance with the following rules:

   (a) the assets shall be invested in the best interests of members and beneficiaries. In the case of a potential conflict of interest, the institution, or the entity, which manages its portfolio, shall ensure that the investment is made in the sole interest of members and beneficiaries;

   (b) the assets shall be invested in such a manner as to ensure the security, quality, liquidity and profitability of the portfolio as a whole.

   Assets held to cover the technical provisions shall also be invested in a manner appropriate to the nature and duration of the expected future retirement benefits;

   (c) the assets shall be predominantly invested on regulated markets. Investment in assets, which are not admitted to trading on a regulated financial market, must in any event be kept to prudent levels;

   (d) investment in derivative instruments shall be possible insofar as they contribute to a reduction of investment risks or facilitate efficient portfolio management. They must be valued on a prudent basis, taking into account the underlying asset, and included in the valuation of the institution's assets. The institution shall also avoid excessive risk exposure to a single counterparty and to other derivative operations;

(e) the assets shall be properly diversified in such a way as to avoid excessive reliance on any particular asset, issuer or group of undertakings and accumulations of risk in the portfolio as a whole.

Investments in assets issued by the same issuer or by issuers belonging to the same group shall not expose the institution to excessive risk concentration;

(f) investment in the sponsoring undertaking shall be no more than 5% of the portfolio as a whole and, when the sponsoring undertaking belongs to a group, investment in the undertakings belonging to the same group as the sponsoring undertaking shall not be more than 10% of the portfolio.

When the institution is sponsored by a number of undertakings, investment in these sponsoring undertakings shall be made prudently, taking into account the need for proper diversification.

Member States may decide not to apply the requirements referred to in points (e) and (f) to investment in government bonds.

2. The home Member State shall prohibit the institution from borrowing or acting as a guarantor on behalf of third parties. However, Member States may authorise institutions to carry out some borrowing only for liquidity purposes and on a temporary basis.

3. Member States shall not require institutions located in their territory to invest in particular categories of assets.

4. Without prejudice to Article 12, Member States shall not subject the investment decisions of an institution located in their territory or its investment manager to any kind of prior approval or systematic notification requirements.

5. In accordance with the provisions of paragraphs 1 to 4, Member States may, for the institutions located in their territories, lay down more detailed rules, including quantitative rules, provided they are prudentially justified, to reflect the total range of pension schemes operated by these institutions.

In particular, Member States may apply investment provisions similar to those of Directive 2002/83/EC.

However, Member States shall not prevent institutions from:

(a) investing up to 70% of the assets covering the technical provisions or of the whole portfolio for schemes in which the members bear the investment risks in shares, negotiable securities treated as shares and corporate bonds admitted to trading on regulated markets and deciding on the relative weight of these securities in their investment portfolio. Provided it is prudentially justified, Member States may, however, apply a lower limit to institutions which provide retirement products with a long-term interest rate guarantee, bear the investment risk and themselves provide for the guarantee;

(b) investing up to 30% of the assets covering technical provisions in assets denominated in currencies other than those in which the liabilities are expressed;

(c) investing in risk capital markets.

6. Paragraph 5 shall not preclude the right for Member States to require the application to institutions located in their territory of more stringent investment rules also on an individual basis provided they are prudentially justified, in particular in the light of the liabilities entered into by the institution.

7. In the event of cross-border activity as referred in Article 20, the competent authorities of each host Member State may require that the rules set out in the second subparagraph apply to the institution in the home Member State. In such case, these rules shall apply only to the part of the assets of the institution that corresponds to the activities carried out in the particular host Member State. Furthermore, they shall only be applied if the same or stricter rules also apply to institutions located in the host Member State.

The rules referred to in the first subparagraph are as follows:

(a) the institution shall not invest more than 30% of these assets in shares, other securities treated as shares and debt securities which are not admitted to trading on a regulated market, or the institution shall invest at least 70% of these assets in shares, other securities treated as shares, and debt securities which are admitted to trading on a regulated market;

(b) the institution shall invest no more than 5% of these assets in shares and other securities treated as shares, bonds, debt securities and other money and capital market instruments issued by the same undertaking and no more than 10% of these assets in shares and other securities treated as shares, bonds, debt securities and other money and capital market instruments issued by undertakings belonging to a single group;

(c) the institution shall not invest more than 30% of these assets in assets denominated in currencies other than those in which the liabilities are expressed.

To comply with these requirements, the home Member State may require ring-fencing of the assets.

References

AAA (2000): Recommendation of the American Academy of Actuaries' Life-Risk Based Capital Committee on Changes to the Covariance Treatment of Common Stock. Paper presented to NAIC, December. Available at www.actuary.org.

AAA (2002a): Fair Valuation of Insurance Liabilities: Principles and Methods, Public Policy Monograph. American Academy of Actuaries, September. Available at www.actuary.org.

AAA (2002b): Comparison of the NAIC Life, P&C and Health RBC Formulas. Report to NAIC from the Academy Joint RBC Task Force, February 12. Available at www.actuary.org.

Abbink, M. and M. Saker (2002): Getting to Grips with Fair Value. Presented to the Staple Inn Actuarial Society, March 5. Available at http://www.sias.org.uk/prog.html.

Ajne, M. (2004): Proposal for a modernised solvency system for insurance undertakings. In *Swedish Society of Actuaries Centennial Book, Stockholm, 2004.*

APRA (1999a): Study of the Prudential Supervisory Requirements for General Insurers in Australia. APRA, September. Report. Available at http://www.apra.gov.au.

APRA (1999b): A Statutory Liability Valuation Standard for General Insurers. APRA, September. Paper. Available at http://www.apra.gov.au.

APRA (1999c): A New Statutory Solvency Standard for General Insurers. APRA, September. Paper. Available at http://www.apra.gov.au.

APRA (1999d): Prudential Supervision of Conglomerates, Policy Discussion Paper. APRA, March. Available at http://www.apra.gov.au.

APRA (1999e): Prudential Supervision of Conglomerates, Policy Discussion Paper. APRA, November. Available at http://www.apra.gov.au.

APRA (2000a): A Proposed Reform to the Prudential Supervision of General Insurance Companies in Australia, Policy Discussion Paper. APRA, April. Available at http://www.apra.gov.au.

APRA (2000b): Harmonising Prudential Standards: A Principles-Based Approach, Policy Discussion Paper. APRA, December. Available at http://www.apra.gov.au.

APRA (2000c): Policy Framework for the Prudential Supervision of Conglomerate Groups Containing Authorised Deposit-Taking Institutions, Policy Information Paper. APRA, April. Available at http://www.apra.gov.au.

APRA (2001a): Prudential Supervision of General Insurance, Policy Discussion Paper. APRA, March. Available at http://www.apra.gov.au.

APRA (2001b): General Insurance Reform Act 2001. APRA, assented to September 19. Available at http://www.apra.gov.au.

APRA (2001c): Capital Adequacy and Exposure Limits for Conglomerate Groups including ADIs, Discussion Paper. APRA, October. Available at http://www.apra.gov.au.

APRA (2003): Prudential Supervision of General Insurance, Stage 2 Reforms, Discussion Paper. APRA, November. Available at http://www.apra.gov.au.

Artzner, P. (1999): Application of coherent risk measures to capital requirements in insurance. *North American Actuarial Journal*, (2)11–25.

Artzner, P., F. Delbaen, J.-M. Eber, and D. Heath (1999): Coherent measures of risk. *Mathematical Finance*, 9, 203–228.

Azzalini, A. (2003): azzalini.stat.unipd.it/SN/Intro/intro.html (Web site for skew normal distributions).

Bateup, R. and I. Reed (2001): Research and Data Analysis Relevant to the Development of Standards and Guidelines on Liability Valuation for General Insurance. The Institute of Actuaries of Australia, Tillinghast-Towers Perrin Report, November 20.

Beard, R.E., T. Pentikäinen, and E. Pesonen (1984): *Risk Theory, The Stochastic Basis of Insurance*, 3rd ed. Chapman & Hall, London.

Benjamin, B. (1977): *General Insurance*. Heinemann, London (published for the Institute of Actuaries and the Faculty of Actuaries).

BIS (1987): Proposals for international convergence of capital measurement and capital standards. Committee on Banking Regulations and Supervisory Practices, December.

BIS (1988): International Convergence of Capital Measurement and Capital Standards. Basel Committee on Banking Supervision, Regulation Paper, July.

BIS (1996): Amendment to the Capital Accord to Incorporate Market Risks. Basel Committee on Banking Supervision, January.

BIS (1999): A New Capital Adequacy Framework, Consultative Paper. Basel Committee on Banking Supervision, January.

BIS (2000) Stress Testing by Large Financial Institutions: Current Practice and Aggregation Issues. Committee on the Global Financial Systems, CGFS Publications No. 14, Basel Committee on Banking Supervision, April.

BIS (2001): The New Basel Capital Accord (a package of 11 consultative papers). Basel Committee on Banking Supervision, January.

BIS (2003): The New Basel Capital Accord, Consultative Document. Basel Committee on Banking Supervision, Bank for International Settlements, April.

BIS (2004): International Convergence of Capital Measurement and Capital Standards, A Revised Framework. Basel Committee on Banking Supervision, Bank for International Settlements, June.

Blalock, H.M. (1961): *Causal Inference in Nonexperimental Research*. The University of North Carolina Press, Chapel Hill.

Campagne, C. (1961): *Standard minimum de solvabilité applicable aux entreprises d'assurances*, Report of the OECE,[1] March 11. Reprinted in *Het Verzekerings-Archief deel XLVIII*, 1971–1974.

CAS (2004): *Fair Value of P&C Liabilities: Practical Implications*. Casualty Actuarial Society, Arlington, VA.

CEA (2005): Solvency Assessment Models Compared, essential groundwork for the Solvency II Project. Produced by the CEA and Mercer Oliver Wyman in cooperation with all European Insurance Markets, February.

---

[1] Organization Europeenne de Cooperation Economique.

Clark, P.K., P.H. Hinton, E.J. Nicholson, L. Storey, G.G. Wells, and M.G. White (2003): The implications of fair value accounting for general insurance companies. *British Actuarial Journal*, 9, 1007–1059.

Collings, S. and G. White (2001): APRA Risk Margin Analysis. Paper presented at the Institute of Actuaries of Australia XIIIth General Insurance Seminar, November 25–28.

COM (1998): Directive 98/78/EC of the European Parliament and of the Council of October 27 on the supplementary supervision of insurance undertakings in an insurance group. OJ L 330 5.12.1998.

COM (2000a): Proposal for a directive of the European Parliament and of the Council amending Council Directive 79/267/EEC as regards the solvency margin requirements for life insurance undertakings. COM(2000) 617. 2000/0249 (COD).

COM (2000b): Proposal for a directive of the European Parliament and of the Council amending Council Directive 73/239/EEC as regards the solvency margin requirements for non-life insurance undertakings. COM(2000) 634. 2000/0251 (COD).

COM (2002a): Directive 2002/12/EC of the European Parliament and of the Council of March 5 amending Council Directive 79/267/EEC as regards the solvency margin requirements for life insurance undertakings.

COM (2002b): Directive 2002/13/EC of the European Parliament and of the Council of March 5 amending Council Directive 73/239/EEC as regards the solvency margin requirements for non-life insurance undertakings.

COM (2002c): Directive 2002/83/EC of the European Parliament and of the Council of November 5 concerning life insurance. OJ L 345 19.12.2002, pp. 0001–0051.

COM (2002d): Directive 2002/87/EC of the European Parliament and of the Council of December 16 on the supplementary supervision of credit institutions, insurance undertakings and investment firms in a financial conglomerate and amending Council Directives 73/239/EEC, 79/267/EEC, 92/49/EEC, 92/96/EEC, 93/6/EEC, and 93/22/EEC, and Directives 98/78/EC and 2000/12/EC of the European Parliament and of the Council. OJ L 35 11.2.2003.

COM (2003a): Directive 2003/51/EC of the European Parliament and of the Council of June 18 amending Directives 78/660/EEC, 83/349/EEC, 86/635/EEC, and 91/674/EEC on the annual and consolidated accounts of certain types of companies, banks and other financial institutions and insurance undertakings. OJ L 178/16-22 17.7.2003, pp. 0016–0022.

COM (2003b): Directive 2003/41/EC of the European Parliament and of the Council of June 3 on the activities and supervision of institutions for occupational retirement provision. OJ L 235 23.09.2003, pp. 0010–0021.

COM (2004): Proposal for a Directive of the European Parliament and of the Council on reinsurance and amending Council Directives 73/239/EEC and 92/49/EEC and Directives 98/78/EC and 2002/83/EC. COM(2004) 273 final. 2004/0097 (COD).

CONSLEG (2003): Consolidated Text: Council Directive 91/674/EEC and Directive 2003/51/EC. CONSLEG 1991L0674 17.07.2003.

Cummins, J.D. and R.D. Phillips (2003): Estimating The Cost of Equity Capital for Property-Liability Insurers. Report. The Wharton Financial Institutions Center, 03-31, June 23. See also http://papers.ssrn.com. Forthcoming in *Journal of Risk and Insurance*.

Daykin, C.D. (1984): The Development of Concepts of Adequacy and Solvency in Non-Life Insurance in the EEC. Paper presented at the 22nd International Congress of Actuaries, Sydney, pp. 299–309.

Daykin, C.D., G.D. Bernstein, S.M. Coutts, E.R.F. Devitt, G.B. Hey, D.I.W. Reynolds, and P.D. Smith (1987): Assessing the solvency and financial strength of a general insurance company. *Journal of the Institute of Actuaries*, 114, 227–310.

Daykin, C.D., E.R. Devitt, M.R. Kahn, and J.P. McCaughan (1984): The solvency of general insurance companies. *Journal of the Institute of Actuaries*, 111, 279–336.

Daykin, C.D. and G.B. Hey (1990): Managing uncertainty in a general insurance company. *Journal of the Institute of Actuaries*, 117, 173–277.

Daykin, C.D., T. Pentikäinen, and M. Pesonen (1994): *Practical Risk Theory for Actuaries*. Chapman & Hall, London.

De Mori, Bruno (1965): Possibilite d'etablir des bases techniques acceptables pour le calcul d'une marge minimum de solvabilite des enterprises d'assurances contre les dommages. *ASTIN Bulletin*, III, 286–313.

De Wit, G.W. and W.M. Kastelijn (1980): The solvency margin in non-life insurance companies. *ASTIN Bulletin*, 11, 136–144.

Djehiche, B. and P. Hörfelt (2004): Standard Approaches to Asset and Liability Risk, Research Report. Fraunhofer Chalmers Research Centre, Gothenburg, FCC, November.

EC (1997): Report to the Insurance Committee on the Need for Further Harmonisation of the Solvency Margin (presented by the commission). COM(1997) 398. 24.07.1997.

EC (1999): Solvency Margin Review, Commission Services' Working Document. Annex to DIV 9049 (06/99). XV/2025/99/Rev.1.

EC (2000): ART Market Study, Final Report, Study Contract ETD/99/B5-3000/C/51, October 2. European Commission, Internal Market DG.

EEC (1973): First council directive of 24 July 1973 on the coordination of laws, regulations and administrative provisions relating to the taking-up and pursuit of the business of direct insurance other than life assurance. *Official Journal of the European Communities*, L 228/3 (73/239/EEC).

EEC (1978): Fourth Council Directive 78/660/EEC of 25 July 1978 based on Article 54(3)(g) of the treaty on the annual accounts of certain types of companies. *Official Journal of the European Communities*, L 222, 14/08/1978, pp. 0011–0031.

EEC (1979): First council directive of 5 March 1979 on the coordination of laws, regulations and administrative provisions relating to the taking-up and pursuit of the business of direct life insurance. *Official Journal of the European Communities*, L 63/1 (79/267/EEC).

EEC (1983): Seventh Council Directive 83/349/EEC of 13 June 1983 based on Article 54(3)(g) of the treaty on consolidated accounts. *Official Journal of the European Communities*, L 193, 18/07/1983, pp. 0001–0017.

EEC (1984): Council directive of December 10 1984 amending, particularly as regards tourist assistance, the first directive (73/239/EEC) on the coordination of laws, regulations and administrative provisions relating to the taking-up and pursuit of the business of direct insurance other than life insurance. *Official Journal of the European Communities*, L 339, 27/12/1984 (84/641/EEC).

EEC (1987): Council directive of 22 June 1987 amending, as regards credit insurance and suretyship insurance, first directive (73/239/EEC) on the coordination of laws, regulations and administrative provisions relating to the taking-up and pursuit of the business of direct insurance other than life insurance. *Official Journal of the European Communities*, L 185, 04/07/1987 (87/343/RRC).

EEC (1988): Second council directive of 22 June 1988 on the coordination of laws, regulations and administrative provisions relating to direct insurance other than life insurance and laying down provisions to facilitate the effective exercise of freedom to provide services and amending Directive 73/239/EEC. *Official Journal of the European Communities*, L 172/1 (88/357/EEC).

EEC (1990): Council directive of 8 November 1990 on the coordination of laws, regulations and administrative provisions relating to direct life assurance, laying down provisions to facilitate the effective exercise of freedom to provide services and amending Directive 79/267/EEC. *Official Journal of the European Communities*, L 330/50 (90/619/EEC).

EEC (1991): Council Directive 91/674/EEC of 19 December 1991 on the annual accounts and consolidated accounts of insurance undertaking. *Official Journal of the European Communities*, L 374, 31/12/91. pp. 0007–0031.

EEC (1992a): Council Directive 92/49/EEC of 18 June 1992 on the coordination of laws, regulations and administrative provisions relating to direct insurance other than life insurance and amending Directives 73/239/EEC and 88/357/EEC (third non-life insurance directive). Official Journal of the European Communities, No. L228, August.

EEC (1992b): Council Directive 92/96/EEC of 10 November 1992 on the coordination of laws, regulations and administrative provisions relating to direct life insurance and amending Directives 79/267/EEC and 90/619/EEC (third life insurance directive). Official Journal of the European Communities, No. L360, December.

FFSA (2003): Extension Proposal for Asset Liability Risk, Working Paper CEA (standard approach drafted by the chairman, Jukka Rantala). Féderation Francaise des Sociétés d'Assurance, 30/07/03.

Finanstilsynet (2003): Vejledning til indberetningsskemater til oplysning af kapitalforhold og risici I lvs-, skades- og genforsikringsselskaber, tvægående pensionskasser samt arbejdsskadeselskaber. *Finanstilsynet, Økonomi- og erhvervsministeriet*, December 8.

Finger, C.C. (2001): The one-factor creditmetrics model in the New Basel Capital Accord. *RiskMetrics Journal*, 2, 9–18.

FSA (2002a): Individual Capital Adequacy Standards, Consultation Paper CP136. Financial Services Authority, U.K., May. See also www.fsa.gov.uk.

FSA (2002b): Integrated Prudential Sourcebook: Feedback on Chapters of CP97 Applicable to Insurance Firms and Supplementary Consultation, Consultation Paper CP143. Financial Services Authority, U.K., July. See also www.fsa.gov.uk.

FSA (2003a): Enhanced Capital Requirements and Individual Capital Assessments for Non-Life Insurers, Consultation Paper CP190. Financial Services Authority, U.K., July. See also www.fsa.gov.uk.

FSA (2003b): Enhanced Capital Requirements and Individual Capital Assessments for Life Insurers, Consultation Paper CP195. Financial Services Authority, U.K., August. See also www.fsa.gov.uk.

FSA (2004): Integrated Prudential Sourcebok for Insurers, Policy Statement 04/16. Financial Services Authority, June.

FSAJ (2003): Presentation material for a Meeting of Solvency and Actuarial Issues Subcommittee of IAIS, September 3–5. Financial Services Agency, Japan.

Furrer, H. (2004): On the Calculation of the Risk Margin within the SST, Working Paper. Swiss Life, August.

GDV (2002a): Supervisory Model for German Insurance Undertakings (LIFE) (description of the model). Gesamtverband der Deutschen Versicherungswirtschaft e.V., GDV, July.

GDV (2002b): Supervisory Model for German Insurance Undertakings (Property/Casualty) (description of the model). Gesamtverband der Deutschen Versicherungswirtschaft e.V., GDV, July.

Girard, L.N. (2002): An approach to fair valuation of insurance liabilities using the firm's cost of capital. *North American Actuarial Journal*, 6, 18–46.

Gordy, M.B. (2003): A risk-factor model foundation for ratings-based bank capital rules. *Journal of Financial Intermediation*, 12, 199–232.

Gordy, M.B. (2004): Model foundations for the supervisory formula approach. In *Structured Credit Products: Pricing, Rating, Risk Management and Basel II*, W. Perraudin, Ed. Risk Books, London.

Gutterman, S. (1999): The Valuation of Future Cash Flows, Actuarial Issues Paper. American Academy of Actuaries (www.actuary.org). Also published in Vanderhoof, I.T. and E.I. Altman, Eds. (2000): *The Fair Value of Insurance Business*. Kluwer Academic Publishers, Netherlands.

Hägg, G. (1998): An Institutional Analysis of Insurance Regulation, The Case of Sweden. Thesis, Department of Economics, Lund Economic Studies 75, Lund, Sweden.

Hairs, C.J., D.J. Belsham, N.M. Bryson, C.M. George, D.J.P. Hare, D.A. Smith, and S. Thompson (2002): Fair valuation of liabilities. *British Actuarial Journal*, 8, 203–340.

Heckman, P.E. and G.G. Meyers (1983): The Calculation of aggregate loss distribution from claim severity and claim count distributions. In *Proceedings of Casualty Actuarial Society (PCAS)*, Vol. LXX, pp. 22–61 (includes a discussion by G. Venter).

Henty, J. (2003): *The Actuarial Profession in Europe: 25 years of the Groupe Consultatif*. Groupe Consultatif Actuariel Européen, Oxford.

Hesselager, O. and T. Witting (1988): A credibility model with random fluctuations in delay probabilities for the prediction of IBNR claims. *ASTIN Bulletin*, 18, 79–90.

Horsmeier, H. et al. (1998): Report on the oncoming revision of the EU solvency regime. *Transactions of the 26th ICA*, 3, 179–197.

IAA (2000): Comments to the IASC's insurance issues paper. These papers are available at http://www.actuaries.org/public/en/documents/submissions.cfm.

IAA (2004): *A Global Framework for Insurer Solvency Assessment*. IAA, Ontario.

IAIS (2002): Principles on Capital Adequacy and Solvency, Principles 5. IAIS, January.

IAIS (2003a): Glossary of Terms. IAIS, September. (A new version was published in February 2005.)

IAIS (2003b): Solvency Control Levels Guidance Paper, IAIS Guidance Paper 6. IAIS, October.

IAIS (2003c): Insurance Core Principles and Methodology, Principles 1. IAIS, October (first approved in 1997, third approval in 2003).

IAIS (2003d): Stress Testing by Insurers, Guidance paper No. 8, October.
IAIS (2004a): Enhancing Transparency and Disclosure in the Reinsurance Sector. Report. Task Force Re. IAIS, March.
IAIS (2004b): Guidance Paper on Investment Risk Management. IAIS, October.
IAIS (2004c): A New Framework for Insurance Supervision. Towards a Common Structure and Common Standards for the Assessment of Insurer Solvency. IAIS, October.
IAIS (2005): Towards a Common Structure and Common Standards for the Assessment of Insurer Solvency. Cornerstones for the Formulation of Regulatory Financial Requirements, Draft 11. IAIS, February.
IASB (2001): Draft Statement of Principles (DSOP) on Insurance Contracts. IASB, November 16.
IASB (2003): Exposure Draft ED 5 Insurance Contracts. IASB.
IASB (2004): IFRS 4 Insurance Contracts. IASB (including IFRS 4 Insurance Contracts; Basis for Conclusions; and Implementation Guidance).
IASC (1999): Issue papers on insurance accounting. Available at http://www.iasb.org/current/iasb.asp?showPageContent=no&xml=16_61_67_01012004.htm.
Jarvis, S., F. Southall, and E. Varnell (2001): Modern Valuation Techniques. Paper presented to the Staple Inn Actuarial Society, February 6. Available at http://www.sias.org.uk/prog.html.
Johnson, J. and S. Kotz (1970): *Distributions in Statistics, Continuous Univariate Distributions: 1*. John Wiley & Sons, New York.
Joint Forum (1999): Supervision of financial conglomerates. Papers prepared by the Joint Forum on Financial Conglomerates, February.
Joint Forum (2003a): Trends in Risk Integration and Aggregation. Joint Forum, August.
Joint Forum (2003b): Operational Risk Transfer across Financial Sectors. Joint Forum, August.
Jørgensen, P.L. (2004): On accounting standards and fair valuation of life insurance and pension liabilities. *Scandinavian Actuarial Journal*, 5, 372–394.
Kastelijn, W.M. and J.C.M. Remmerswaal (1986): *Solvency*, Surveys of Actuarial Studies 3. Nationale-Nederlanden N.V., Rotterdam, Netherlands.
KPMG (2002): Study into the Methodologies to Assess the Overall Financial Position of an Insurance Undertaking from the Perspective of Prudential Supervision, Contract ETD/2000/BS-3001/C/45. KPMG, May.
Kotz, S., N.L. Johnson, and C.B. Read (1985): *Encyclopedia of Statistical Sciences*. John Wiley & Sons, New York.
Kredittilsynet (2000a): Notes on Risk-Theoretic Methods Applied for Estimating Minimum Requirements for Technical Provisions in Non-Life Insurance, The Norwegian Case, DT/N/83/99/Rev.1. The Banking, Insurance and Securities Commission of Norway, April (presented to the Manghetti group).
Kredittilsynet (2000b): Insurance Solvency Supervison in Norway. The Banking, Insurance and Securities Commission of Norway, August (presented to OECD Insurance Committee).
Kristiansen, A. (1996): Minimum Requirement for Technical Provisions in Non-Life Insurance, The Norwegian Case. Lecture presented at the XXVII ASTIN Colloquium, Copenhagen, September 4 (unpublished).
Kroisandt, G. (2003): Insurance Companies: Premiums and Risk Measures, Research Report. Fraunhofer Chalmers Research Centre, Gothenburg, FCC, March 27.

Landsman, Z. and E.A. Valdez (2003): Tail conditional expectations for elliptical distributions. *North American Actuarial Journal*, 7 (4) 55–71.

Large, A. (2004): Why We Should Worry about Liquidity. *Financial Times*, November 11, p. 15.

Lommele, J.A. and M.G. McCarter (1998): Is the "Best Estimate" Best? Issues in Recording a Liability for Unpaid Claims, Unpaid Losses and Loss Adjustment Expenses. In CAS Forum Fall 1998, Philadelphia, pp. 211–227.

Luenberger, D.G. (1998): *Investment Science*. Oxford University Press, Oxford.

Madrid report (2003): Internal Control for Insurance Undertakings. CEIOPS, December. Available at www.ceiops.org.

Manghetti report (2001): Technical Provisions in Non-Life Insurance. Conference of Insurance Supervisory Authorities of the Member States of the European Union (now CEIOPS), DT/I/223/00.Rev 2 (the WG chaired by Giovanni Manghetti, Italy). Available at www.ceiops.org.

MARKT (1999): The Review of the Overall Financial Position of an Insurance Undertaking (Solvency II Review), MARKT/2095/99. EC Internal Market DG.

MARKT (2001a): Note to the Solvency Subcommittee of the Insurance Committee, Solvency II: Presentation of the Proposed Work, MARKT/2027/01. EC DG Internal Market.

MARKT (2001b): Note to the Solvency Subcommittee, Banking Rules: Relevance for the Insurance Sector? MARKT/2056/01. EC DG Internal Market.

MARKT (2001c): Note to the Solvency Subcommittee, Risk-Based Capital System, MARKT/2085/01. EC DG Internal Market.

MARKT (2002a): Note to the Solvency Subcommittee, Risk Models of Insurance Companies or Groups, MARKT/2515/02. EC DG Internal Market.

MARKT (2002b): Solvency II: Review of the Work, MARKT/2518/02. EC DG Internal Market.

MARKT (2002c): Discussion Note to the Members of the IC Solvency Subcommittee, Current and Future Solvency Work in the IAIS and within the Actuarial Profession from a Solvency II Point of View, MARKT/2520/02. EC DG Internal Market.

MARKT (2002d): Note to the Members of the IC Solvency Subcommittee, Considerations on the Links between the Solvency 2 Project and the Extension of the 'Lamfalussy' Approach to Insurance Regulation, MARKT/2519/02. EC DG Internal Market.

MARKT (2002e): Report of the Working Group on Life Assurance to the IC Solvency Subcommittee, MARKT/2528/02. EC DG Internal Market.

MARKT (2002f): Report of the Working Group on Non-Life Insurance to the IC Solvency Subcommittee, MARKT/2529/02. EC DG Internal Market.

MARKT (2002g): Solvency II: Review of Work (November 2002), MARKT/2536/02. EC DG Internal Market.

MARKT (2002h): Paper for the Solvency Subcommittee, Considerations on the Design of a Future Prudential Supervisory System, MARKT/2535/02. EC DG Internal Market.

MARKT (2003a): Design of a Future Prudential Supervisory System in the EU: Recommendations by the Commissions Services, MARKT/2509/03. EC DG Internal Market.

MARKT (2003b): Reflections on the General Outline of a Framework Directive and Mandates for Further Technical Work, MARKT/2539/03. EC DG Internal Market.

MARKT (2003c): Note to the Solvency Subcommittee, Questionnaire Concerning the Paper MARKT/2535/02 (January 9, 2003), MARKT/2500/03, EC DG Internal Market.

MARKT (2003d): Solvency II: Orientation Debate: Design of a Future Prudential Supervisory System in the EU (Recommendations by the Commission Services), MARKT/2503/03 (Insurance Committee).

MARKT (2003e): Draft Decision Points and Conclusions of the 32nd Meeting of the Insurance Committee, MARKT/2511/03, Brussels, April 9 (Insurance Committee).

MARKT (2003f): Solvency II: Update on Developments and Issues for Consideration by the Insurance Committee, MARKT/2530/03 (Insurance Committee).

MARKT (2004a): Organisation of Work, Discussion on Pillar 1 Work Areas and Suggestions of Further Work on Pillar 2 for CEIOPS: Issues Paper for the Meeting of the IC Solvency Subcommittee on 12 March 2004, MARKT/2543/03. EC DG Internal Market.

MARKT (2004b): Further Issues for Discussion and Suggestions for Preparatory Work for CEIOPS: Issues Paper for the Meeting of the IC Solvency Subcommittee on 22 April 2004, MARKT/2502/04. EC DG Internal Market.

MARKT (2004c): Solvency II: Road Map for the Development of Future Work: Proposed Framework for Consultation and Proposed First Wave of Specific Calls for Advice from CEIOPS, MARKT/2506/04. EC Insurance Committee.

MARKT (2004d): Note to the Members of the IC Solvency Subcommittee, The Draft Second Wave Calls for Advice from CEIOPS and Stakeholders Consultation on Solvency II, MARKT/2515/04, October 11, 2004.

MARKT (2004e): Solvency II: Proposed Second Wave of Specific Calls for Advice from CEIOPS, MARKT/2519/04, November 2004. (Annex to MARKT/2506/04, Specific Calls for Advice from CEIOPS (Second Wave), November 15, 2004.)

MARKT (2005): Solvency II: Consultative Document, Draft Specific Calls for Advice from CEIOPS (Third Wave), MARKT/2501/05, February 2005.

MAS (2003): Risk-Based Capital Framework for Insurance Business, Monetary Authority of Singapore, Consultation Paper 14-2003, November 2003.

MAS (2004a): Insurance Act (Chapter 142), Insurance (Valuation and Capital) Regulations 2004. Monetary Authority of Singapore.

MAS (2004b): Insurance Act (Chapter 142), Insurance (Accounts and Statements) Regulations 2004. Monetary Authority of Singapore.

Müller report (1997): Report of the Working Group "Solvency of Insurance Undertakings" Set Up by the Conference of the European Union Member States, DT/D/209/97. Available at www.ceiops.org.

Norberg, R. (1986): A contribution to modelling of IBNR claims. *Scandinavian Actuarial Journal*, 3/4, 155–203.

Norberg, R. (1993): A Solvency Study in Life Insurance. Paper presented at III AFIR Colloquium, Rome, March 30–April 3.

Norberg, R. and B. Sundt (1985): Draft of a system for solvency control in non-life insurance. *ASTIN Bulletin*, 15, 149–169.

Nygård, F. and A. Sandström (1981): *Measuring Income Inequality*, Stockholm Studies in Statistics 1. Almqvist & Wiksell International, Stockholm.

OSFI (1999): Supervisory Framework: 1999 and Beyond. Office of the Superintendent of Financial Institutions, Ottawa, Canada.

OSFI (2003): Minimum Capital Test (MCT) for Federally Regulated Property and Casualty Insurance Companies, Guideline A. Office of the Superintendent of Financial Institutions, Ottawa, Canada, July.

OSFI (2004): Minimum Continuing Capital and Surplus Requirements (MCCSR) for Life Insurance Companies, Guideline A. Office of the Superintendent of Financial Institutions, Ottawa, Canada, October.

Panjer, H. (2002): Measurement of Risk, Solvency Requirements and Allocation of Capital within Financial Conglomerates. Paper presented at AFIR/ICA, Cancun.

Pentikäinen, T. (1952): On the net retention and solvency of insurance companies. *Skandinavisk Aktuarietidskrift*, 35, 71–92.

Pentikäinen, T. (1967): On the solvency of insurance companies. *ASTIN Bulletin*, IV, 236–247.

Pentikäinen, T., Ed. (1982): *Solvency of Insurers and Equalization Reserves*, Vol. I, *General Aspects*. Insurance Publ. Co. Ltd., Helsinki.

Pentikäinen, T. (1984): Aspects on the solvency of insurers. In *22nd International Congress of Actuaries*, Sydney, T3, pp. 61–73.

Pentikäinen, T., H. Bonsdorff, M. Pesonen, T. Pukkila, A. Ranne, J. Rantala, M. Ruohonen, and S. Sarvamaa (1994): On the Asset Models as Part of All-Company Insurance Analysis (FIM-Group). Paper presented at IV AFIR Colloquium, Orlando, April 20–22.

Pentikäinen, T., H. Bonsdorff, M. Pesonen, J. Rantala, and M. Ruohonen (1989): *Insurance Solvency and Financial Strength*. Finnish Insurance Training and Publishing Company Ltd., Helsinki.

Pool, B. (1990): *The Creation of the Internal Market in Insurance*. Commission of the European Communities, Brussels.

PVK (2001): Principles for a Financial Assessment Framework. Pensioen- & Verzekeringskamer, Amsterdam, September. Available at www.pvk.nl.

PVK (2004): Financial Assessment Framework, Consultation Document, Apeldoorn, October 21. Available at www.dnb.nl.

PWC (2003): Basel: Hopes and Fears: A European Banking View of the Practical Application of Pillars 2. PricewaterhouseCoopers (05/03).

Ramlau-Hansen, H. (1982): An application of credibility theory to solvency margins: some comments on a paper by G.W. De Wit and W.M. Kastelijn. *ASTIN Bulletin*, 13, 37–45.

Ramlau-Hansen, H. (1988): A solvency study in non-life insurance. Part 1. Analyses of fire, windstorm, and glass claims (pp. 3–34). Part 2. solvency margin requirements (pp. 35–60). *Scandinavian Actuarial Journal*, 1/2.

Rantala, J., Ed. (1982): *Solvency of Insurers and Equalization Reserves*, Vol. II, *Risk Theoretical Model*. Insurance Publ. Comp. Ltd., Helsinki.

Rantala, J. (2003): A Standard Approach for the Underwriting Risk in Non-Life, CEA, Annex 2, 23.04.03 JR, Working Paper.

Rantala, J. (2004a): A Standard Approach for Asset/Liability Risk in Life Insurance, CEA, Annex 1, 12.01.04 JR, Working Paper (revised; first version presented to CEA during Spring 2003).

Rantala, J. (2004b): Some Remarks to the Treatment of the AL-Risk in the SA, 9.1.04, JR, Working Paper.

Rantala, J. (2004c): Illustrations of Magnitude of Some Parameters in the Standard Approach, 19.4.04, JR, Working Paper.

Rantala, J. and M. Vesterinen (1995): A Report on Assessing the Solvency of Insurance Companies, CEA Annex to CP 004 (03/95), 7.3.

Resti, A. (2002): The New Basel Capital Accord, Structure, Possible Changes and Micro- and Macroeconomic Effects, CEPS Research Report 30. September.

Schlude, H. (1979): The EEC solvency rules for non-life insurance companies. *Nordisk FörsäkringsTidskrift*, 1, 24–30.

Schmeiser, H. (2004): New risk-based capital standards in the European Union: a proposal based on empirical data. *Risk Management and Insurance Review*, 7, 41–52.

Sharma (2002): Prudential Supervision of Insurance Undertakings. Paper presented at Conference of Insurance Supervisory Services of the Member States of the European Union (now CEIOPS), December (the working group chaired by Paul Sharma, U.K.). Available at www.ceiops.org.

SOU (2003a): Principer för ett moderniserat solvenssystem för försäkringsbolag. Delbetänkande av Placeringsutredningen, Ministry of Finance, SOU 2003:13.

SOU (2003b): Förslag till ett moderniserat solvenssystem för försäkringsbolag. Slutbetänkande av Placeringsutredningen, Ministry of Finance. SOU 2003:84.

SST (2004a): White Paper on the Swiss Solvency Test. Swiss Federal Office for Private Insurance, November.

SST (2004b): The Risk Margin for the Swiss Solvency Test (Draft). Swiss Federal Office for Private Insurance, September 17.

Sutherland-Wong, C. and M. Sherris (2004): Risk-Based Regulatory Capital for Insurers: A Case Study. Paper presented at XIV AFIR Colloquium, Boston, November 8–10.

Tripp, M.H., H.L. Bradley, R. Devitt, G.C. Orron, G.L. Overton, L.M. Pryor, and R.A. Shaw (2004): Quantifying Operational Risk in General Insurance Companies, Developed by a GIRO Working Party. Presented to the Institute of Actuaries, March 22.

Tuomikoski, J. (2000): Financial Solidity of Pension Insurance Companies and Pension Funds within the Finnish Employment Pension System. Paper presented at X AFIR Colloquium, Tromsø, Norway, June 20–23.

Van Broekhoven, H. (2002): Market value of liabilities mortality risk: a practical model. *North American Actuarial Journal*, 6, 95–106.

Wang, S.S. (1998): Aggregation of correlated risk portfolios: models and algorithms. *CAS*, LXXXV, 848–939. See also the discussion by G. Meyers in *CAS*, LXXXV, 781–805, 1999.

Watson, W. (2003): Calibration of the General Insurance Risk Based Capital Model, Ref. 2001575. Prepared by Watson Wyatt LLP (Actuaries & Consultants), July 25. See also www.fsa.gov.uk.

Watson, W. (2004): Calibration of the Enhanced Capital Requirement for With-Profit Life Insurers, Ref. 2004-0129. Prepared by Watson Wyatt LLP (Actuaries & Consultants), June. See also www.fsa.gov.uk.

Wirch, J.L. (1997): Value-at-Risk for Risk Portfolios. Paper presented at Actuarial Research Conference, Calgary, Alberta, Canada, August.

Wirch, J.L. and M.R. Hardy (1999): A synthesis of risk measures for capital adequacy. *Insurance: Mathematics and Economics*, 25, 337–347.

Wolthuis, H. and M.J. Goovaerts, Eds. (1997): *Reserving and Solvency in Insurance in the EC*, 2nd ed. Institute of Actuarial Science and Econometrics (IAE), Universiteit van Amsterdam.

Index

9 780367 392147